D1605604

Lectures in Applied Mathematics

Proceedings of the Summer Seminar, Boulder, Colorado, 1960

VOLUME I — LECTURES IN STATISTICAL MECHANICS
G. E. Uhlenbeck and G. W. Ford with E. W. Montroll

VOLUME II — MATHEMATICAL PROBLEMS OF RELATIVISTIC PHYSICS
I. E. Segal with G. W. Mackey

VOLUME III — PERTURBATION OF SPECTRA IN HILBERT SPACE
K. O. Friedrichs

VOLUME IV — QUANTUM MECHANICS
R. Jost

Proceedings of the Summer Seminar, Ithaca, New York, 1963

VOLUME V — SPACE MATHEMATICS, PART 1
J. Barkley Rosser, Editor

VOLUME VI — SPACE MATHEMATICS, PART 2
J. Barkley Rosser, Editor

VOLUME VII — SPACE MATHEMATICS, PART 3
J. Barkley Rosser, Editor

Proceedings of the Summer Seminar, Ithaca, New York, 1965

VOLUME VIII — RELATIVITY THEORY AND ASTROPHYSICS
1. RELATIVITY AND COSMOLOGY
Jürgen Ehlers, Editor

VOLUME IX — RELATIVITY THEORY AND ASTROPHYSICS
2. GALACTIC STRUCTURE
Jürgen Ehlers, Editor

VOLUME X — RELATIVITY THEORY AND ASTROPHYSICS
3. STELLAR STRUCTURE
Jürgen Ehlers, Editor

Proceedings of the Summer Seminar, Stanford, California, 1967

VOLUME XI — MATHEMATICS OF THE DECISION SCIENCES, PART 1
George B. Dantzig and Arthur F. Veinott, Jr., Editors

VOLUME XII — MATHEMATICS OF THE DECISION SCIENCES, PART 2
George B. Dantzig and Arthur F. Veinott, Jr., Editors

Volume 12
Lectures in Applied Mathematics

MATHEMATICS OF THE DECISION SCIENCES

PART 2

George B. Dantzig and Arthur F. Veinott, Jr., EDITORS

1968
American Mathematical Society, Providence, Rhode Island

Prepared by the American Mathematical Society under contracts AT(30-1)-3164, Mod. No. 3, with U. S. Atomic Energy Commission, and PH-43-67-712 with the National Institute of Health and the Air Force Office of Scientific Research, and grants GZ-403 with the National Science Foundation and Nonr(G)-00003-67 task NR 047-065.

International Standard Book Number 0-8218-1112-6
Library of Congress Catalog Card Number 62-21481

Printed in the United States of America

Reprinted with corrections 1970

Contents

PART II

VI. CONTROL THEORY

VII. MATHEMATICAL ECONOMICS

VIII. DYNAMIC PROGRAMMING

PART I

I. LINEAR PROGRAMMING

II. PIVOT THEORY AND QUADRATIC PROGRAMS

III. Convex Polyhedra and Integer Programs

IV. Combinatorics

V. Nonlinear Programming

VI
CONTROL THEORY

Lucien W. Neustadt

A Survey of Certain Aspects of Control Theory[1]

1. **Problem statement.** In this article, we shall be concerned with a class of problems that have been given the name of optimal control problems. One of the first problems of this type that was extensively studied (see, for example [1, Chapter I, §2]) may be formulated as follows.

Consider the differential equation

$$dx(t)/dt = f(x(t), u(t), t), \qquad t_0 \leqq t \leqq t_1, \tag{1.1}$$

where $x \in R^n$ (i.e., x is an n-dimensional vector), $u(t)$ is a bounded measurable function from the interval $[t_0, t_1]$ into R^r, and f is a function from $R^n \times R^r \times [t_0, t_1]$ into R^n which is continuous in all of its arguments and possesses continuous first partial derivatives with respect to the components of x. Evidently, the solution of Equation (1.1), if it exists (by a solution, we here mean an absolutely continuous function $x(t)$ that satisfies (1.1) for almost all $t \in [t_0, t_1]$), depends on the choice of the function $u(t)$ and of the initial condition $x(t_0)$.

In the parlance of control systems theory, the vector x is the "state vector" and u is the "control vector", or, as is sometimes said, x is the "output" and u is the "input". In a typical, concrete system, the components of x describe the instantaneous state, or

[1] This work was supported by the United States Air Force Office of Scientific Research under Grant No. AF-AFOSR-1029-67.

condition, of the system, and the components of u describe the instantaneous state of the system controllers, or that part of the system which is subject to regulation. For example, if the "system" being controlled is a space vehicle, x may be a six-dimensional vector, three of whose components describe the position of the vehicle c.g., and the other three of which describe the c.g. velocity, u may be a three-dimensional vector whose components are those of the applied thrust vector, and Equation (1.1) states that the derivative of position is velocity and that mass times acceleration equals the sum of gravitational and propulsive (i.e., due to thrust) forces.

In the usual concrete control system, the following constraints are imposed on the above-described problem. A set $U \subset R^r$, together with vectors x_0 and x_1 in R^n, are given, and one is to find a control function $u(t)$ from $[t_0, t_1]$ into U such that the solution of Equation (1.1), with this function $u(t)$ and with the initial condition $x(t_0) = x_0$, exists and satisfies the boundary condition $x(t_1) = x_1$. In other words, the system is to be transferred from a given initial state (x_0) to a given final state (x_1) by means of a control function $u(t)$ whose values stay within prescribed limits (the set U). If U coincides with R^r, then the control function is clearly unconstrained. However, in most actual problems, U is a bounded (typically closed) set, corresponding to the fact that controllers ordinarily have certain maximum capabilities.

For mathematical convenience, we shall confine ourselves to control functions u which are measurable and bounded. A measurable, bounded control function $u(t)$ defined for $t_0 \leqq t \leqq t_1$ and taking on its values in U will be called an *admissible control*. If the solution of Equation (1.1) for a particular admissible control $u(t)$ and initial condition $x(t_0) = x_0$ exists and assumes the boundary value $x(t_1) = x_1$, then we shall say that u *transfers* our system from x_0 to x_1.

For given x_0 and x_1, it is to be expected that there will be many admissible controls that can transfer the system from x_0 to x_1, and one is therefore tempted into trying to find that control that is, in some sense, best. A typical "cost function"—which determines when one control is better that another—is of the form

$$\int_{t_0}^{t_1} f^0(x(t), u(t), t)\, dt, \tag{1.2}$$

where f^0 is a given scalar-valued function from $R^n \times R^r \times [t_0, t_1]$ with the same continuity and differentiability properties as f.

We can now state our optimal control problem: From among the admissible controls $u(t)$ that transfer (through Equation (1.1)) the system from the given state x_0 to the given state x_1, find one for which the integral (1.2) takes on its least value.

It is now convenient to pose the above problem in an equivalent, but slightly different, form. Namely, let X denote the $(n+1)$-vector (x^0, x) (where x is an n-vector and x^0 is a scalar), let $X_0 = (0, x_0) \in R^{n+1}$, and let F be the function from $R^n \times R^r \times [t_0, t_1]$ into R^{n+1} defined by $F(x,u,t) = (f^0(x,u,t), f(x,u,t))$. For notational convenience, we shall consider F to be a function from $R^{n+1} \times R^r \times [t_0, t_1]$, by considering that the argument x (an n-vector) consists of the last n components of the $(n+1)$-vector X.

Our original problem may now be reformulated as follows:

PROBLEM 1.1. Given the differential equation

$$\dot{X}(t) = F(X(t), u(t), t), \qquad t_0 \leqq t \leqq t_1, \tag{1.3}$$

the vector $X_0 \in R^{n+1}$ and the vector $x_1 \in R^n$, find an admissible control $u(t)$ such that the corresponding solution of (1.3), with initial value $X(t_0) = X_0$, exists, such that the last n coordinates of the vector $X(t_1)$ coincide with x_1, and such that the first coordinate of $X(t_1)$ takes on its least possible value.

Since, in the sequel, it will not be necessary for us to distinguish between x and X or between f and F, we shall henceforth revert to the x, f notation.

The following is now a natural generalization of the just-described optimal control problem:

PROBLEM 1.2. Given the differential Equation (1.1), functions γ_i, $i = 1, \cdots, m_1$, from R^n into R^1, and functions $\bar{\gamma}_i$, $i = 0, 1, \cdots, m_2$, from R^n into R^1, find an absolutely continuous function $x(t)$ from $[t_0, t_1]$ into R^n and an admissible control $u(t)$ such that (a) Equation (1.1) is satisfied for almost all t, (b) $\gamma_i(x(t_0)) = 0$ for $i = 1, \cdots, m_1$, (c) $\bar{\gamma}_i(x(t_1)) = 0$ for $i = 1, \cdots, m_2$, and (d) $\bar{\gamma}_0(x(t_1))$ is minimized. We shall suppose that the functions γ_i and $\bar{\gamma}_i$ have continuous first partial derivatives.

It is evident that Problem 1.1 is a special case of Problem 1.2 (if we make some obvious minor notational changes). The generalization exhibited by passing from the one problem to the other is not merely of academic interest, since a wide class of problems arising in applications exhibit the more general features of Problem 1.2. Physically speaking, the constraints of Problem 1.2 represent the fact that the initial and final system states need not be fully prescribed, but rather that only certain functions of these states need vanish.

In certain other control problems that arise in applications, the initial and/or final times (t_0 and t_1) may themselves be free (rather that preassigned), there may be constraints involving values of x at times intermediate between t_0 and t_1, and some of these constraints may involve inequalities as well as equalities. Thus we are led to the yet more general problem given below.

PROBLEM 1.3. Given the differential Equation (1.1) and real-valued functions γ_i, $i = 0, 1, \cdots, m_1, m_1 + 1, \cdots, m_2$, (which we assume to be continuously differentiable) defined on $R^{(n+1)k}$, find numbers $s_1, \cdots, s_k$ with $t_0 \leqq s_1 \leqq \cdots \leqq s_k \leqq t_1$, an absolutely continuous function $x(t)$ from $[s_1, s_k]$ into R^n and an admissible control $u(t)$ such that (a) Equation (1.1) is satisfied for almost all $t \in [s_1, s_k]$, (b) $\gamma_i(x(s_1), \cdots, x(s_k), s_1, \cdots, s_k) = 0$ for $i = 1, \cdots, m_1$, (c) $\gamma_i(x(s_1), \cdots, x(s_k), s_1, \cdots, s_k) \leqq 0$ for $i = m_1 + 1, \cdots, m_2$, and (d) $\gamma_0(x(s_1), \cdots, x(s_k), s_1, \cdots, s_k)$ is minimized.

It is evident that Problems 1.1 and 1.2 are special cases of Problem 1.3.

There are yet other concrete problems which require constraints that are similar to, but in a sense stronger than, the constraints (c) of Problem 1.3. Namely, suppose that there is given a twice continuously differentiable, real-valued function γ defined on R^n and that the following constraint is present:

$$\gamma(x(t)) \leqq 0 \quad \text{for all } t \text{ on which } x(t) \text{ is defined}$$

or equivalently,

$$\sup_t \gamma(x(t)) \leqq 0. \tag{1.4}$$

Constraints of this type are referred to as restrictions on the phase coordinates. Physically, they may represent, for example, that a space vehicle must never enter the atmosphere of a planet, pass too close to the sun, etc.

Having now described a general class of optimal control problems, let us discuss what questions arise in connection with them.

The first is that of existence: For a given problem, is there an "optimal" solution? For example, there may be a minimizing sequence of solutions, but no actual minimum. This problem has been extensively investigated. One of the most recent contributions is that of Cesari [**2**], which also contains a very complete bibliography of earlier work in this field.

The second question is that of finding necessary conditions, i.e. assuming a solution to our problem exists, what conditions must it satisfy? One wishes these conditions to be as strong as possible, so that they will yield a maximum amount of information about possible solutions. The remainder of this article will be devoted to this question.

The third question deals with the finding of computational algorithms for actually solving concrete optimal control problems. Since "closed form" solutions are impossible to obtain except in the simplest cases, most of the research in this field has dealt with successive approximation schemes for obtaining solutions. The interested reader is referred to references [**3**]—[**5**], each of which also contains an extensive bibliography.

The final question is that of sufficient conditions: What conditions can be determined which, when they are satisfied, will ensure that a solution of the problem has been found? To be useful, such conditions should be as weak as possible. The most desirable conditions are those which are both necessary and sufficient. Except for problems which are essentially linear and convex (see, e.g. [**6**]), there has been little success achieved in this area.

2. **Mathematical programming problems.** In this section we shall state a general mathematical programming problem, and shall state necessary conditions which every solution of such problems satisfies—provided certain hypotheses are fulfilled. In the next section we shall show the connection between the general problems described in this and the preceding sections.

A general mathematical programming problem may be formulated as follows.

PROBLEM 2.1. Given a set L in a real linear vector space S, and real-valued functions (i.e., functionals) ϕ_i ($i = -\mu, \cdots, 0, \cdots, m$, where μ and m are given nonnegative integers) defined on L, find an element $x_0 \in L$ which minimizes ϕ_0 on the set of all $x \in L$

that satisfy the constraints $\phi_i(x) \leqq 0$ for $i = -\mu, \cdots, -1$, and $\phi_i(x) = 0$ for $i = 1, \cdots, m$.

In other words, $x_0 \in L$ is a solution of our problem if $\phi_i(x_0) \leqq 0$ for $i < 0$, $\phi_i(x_0) = 0$ for $i > 0$, and if $\phi_0(x) \geqq \phi_0(x_0)$ for all $x \in L$ such that $\phi_i(x) \leqq 0$ for $i < 0$ and $\phi_i(x) = 0$ for $i > 0$.

If $L = S = R^n$ for some n (i.e., if S is finite-dimensional and L coincides with S), and if the ϕ_i are continuously differentiable on S, then it is not difficult to show (see e.g. [7, §2.2]) that a solution x_0 of our problem must satisfy the following "multiplier rule":

There exist real numbers λ_i, $i = -\mu, \cdots, m$, not all zero, such that

$$(2.1) \qquad \sum_{i=-\mu}^{m} \lambda_i \phi_i'(x_0) = 0, \qquad (1)$$

$$(2.2) \qquad \lambda_i \leqq 0 \quad \text{for } i \leqq 0, \qquad (2)$$

$$(2.3) \qquad \lambda_i = 0 \quad \text{if } i < 0 \text{ and } \phi_i(x_0) < 0. \qquad (3)$$

In (2.1), $\phi_i'(x_0)$ denotes the gradient of ϕ_i evaluated at x_0.

This rule is, in essence, a slight generalization of the classical Lagrange multiplier rule.

If L, instead of coinciding with all of R^n, is only a convex set in R^n, our multiplier rule remains in force in a modified form in that (2.1) must be replaced by the inequality

$$(2.4) \qquad \sum_{i=-\mu}^{m} \lambda_i \phi_i'(x_0) \cdot (x - x_0) \leqq 0 \quad \text{for all } x \in L.$$

In this form, the multiplier rule is essentially the same as one of the well-known Kuhn-Tucker necessary conditions.

In order to consider optimal control problems, it will be necessary for us to remove most of the restrictions we have imposed on our programming problem. Namely, we shall no longer be able to assume that S is finite-dimensional or that L is convex, and we must give a general definition of what we mean by a "gradient" of the functionals ϕ_i.

If we assume that L has a suitable convex "approximation" M, and that the ϕ_i are, in a certain weak sense, "differentiable" we shall, on the one hand, be able to obtain a generalized multiplier rule, and, on the other hand, be able to apply this rule to obtain necessary conditions for the optimal control problems described in §1.

More precisely we shall make the following assumptions regarding x_0, L and the ϕ_i:

ASSUMPTION 2.1. S is a linear topological vector space; i.e. S is a linear vector space and also a topological space such that the operations of addition and scalar multiplication are each continuous.

ASSUMPTION 2.2. There is a convex set $M \subset S$ with the following property: If $A = \{x_1, \cdots, x_r\}$ is any finite subset of M, η is an arbitrary positive number, and N is any neighborhood of 0 in S, then there exist a number ϵ, with $0 < \epsilon < \eta$, and a continuous map θ from the convex hull of A into L (ϵ and θ may depend on A, η and N) such that

$$((\theta(x) - x_0)/\epsilon) \in x + N \quad \text{for each } x \in \text{convex hull of } A.$$

ASSUMPTION 2.3. The functions ϕ_i for $i > 0$ are continuous on S, and there are linear functionals h_i defined on S, for $i = 1, \cdots, m$, such that, for each $i > 0$,

$$\frac{\phi_i(x_0 + \epsilon\bar{x}) - \phi_i(x_0)}{\epsilon} \xrightarrow[\bar{x} \to x;\, \epsilon \to 0]{} h_i(x) \quad \text{for every } x \in S.$$

ASSUMPTION 2.4. There are convex functionals h_i defined on S for $i = -\mu, \cdots, 0$, such that, for each $i \leqq 0$,

$$\frac{\phi_i(x_0 + \epsilon\bar{x}) - \phi_i(x_0)}{\epsilon} \xrightarrow[\bar{x} \to x;\, \epsilon \to 0+]{} h_i(x) \quad \text{for every } x \in S.$$

Assumption 2.2, roughly speaking, states that the convex set M is "tangent" to the set L at x_0. Assumptions 2.3 and 2.4 state that the h_i are certain types of differentials of the ϕ_i at x_0. Note that we require a much weaker differentiability condition for the ϕ_i with $i \leqq 0$ than for the ϕ_i with $i > 0$, in that in the latter (1) the limit need only exist as $\epsilon \to 0$ through positive values, and (2) the differentials h_i need only be convex, rather than linear as in the former.

If Assumptions 2.1—2.4 are satisfied, the following generalized (or abstract) multiplier rule is valid.

THEOREM 2.1. *Let x_0 be a solution of Problem 2.1, and suppose that Assumptions 2.1—2.4 are satisfied. Then there exist real numbers λ_i, $i = -\mu, \cdots, m$, not all zero, such that*

$$(1) \qquad \sum_{i=-\mu}^{m} \lambda_i h_i(x) \leqq 0 \quad \textit{for all } x \in M,$$

(2) $$\lambda_i \leqq 0 \quad \textit{for } i \leqq 0,$$

(3) $$\lambda_i = 0 \quad \textit{if } i < 0 \textit{ and } \phi_i(x_0) = 0.$$

In addition, there exist linear functions l_i for $i = -\mu, \cdots, 0$ such that

(4) $$\sum_{i=-\mu}^{0} \lambda_i l_i(x) + \sum_{i=1}^{m} \lambda_i h_i(x) \leqq 0 \quad \textit{for all } x \in M,$$

(5) $$l_i(x) \leqq h_i(x) \quad \textit{for all } x \in S \textit{ and each } i \leqq 0.$$

Theorem 2.1 is easily seen to be a special case of the maximum principle whose proof is presented in [**8**] (also see [**9**]).

In the next sections we shall apply Theorem 2.1 to obtain necessary conditions for the optimal control problem described in §1.

We note that Theorem 2.1 can be proved with Assumptions 2.1—2.4 somewhat relaxed, but for our purposes, the form of the theorem presented above is sufficient.

3. **Optimal control problems as mathematical programming problems.** In this section we shall show how Problem 1.2 can be put into the setting of Problem 2.1, and shall indicate how, under the assumptions that were made in §1, Assumptions 2.1—2.4 are satisfied.

Although we shall leave the details to the reader, it can be shown in a similar manner that Problem 1.3 is also a special case of Problem 2.1, and that, in addition, Assumptions 2.1—2.4 are also satisfied.

Thus, let S be the linear vector space of continuous functions from $[t_0, t_1]$ into R^n. We shall define a norm on s as follows:

(3.1) $$\| x \| = \max_{t_0 \leqq t \leqq t_1} | x(t) | .$$

In (3.1), the single vertical bars denote the Euclidean length of a vector in R^n. For notational convenience, we shall denote elements of S (which, we emphasize, are functions) by the letter x; the symbol $x(t)$ is to be understood as the value of the function x at the time t.

We define the functionals ϕ_i, $i = 0, 1, \cdots, m_1, m_1 + 1, \cdots, m_1 + m_2 = m$, on S as follows:

$$\phi_i(x) = \gamma_i(x(t_0)), \qquad \text{for } i = 1, \cdots, m_1,$$
$$\phi_{m_1+i}(x) = \bar{\gamma}_i(x(t_1)), \qquad \text{for } i = 1, \cdots, m_2,$$
$$\phi_0(x) = \bar{\gamma}_0(x(t_1)).$$

Finally, let L be the set of all absolutely continuous functions in S that satisfy Equation (1.1) for almost all t and some admissible control function u.

It is now evident that Problem 1.2 has been put in the form of Problem 2.1 where, in the latter, $\mu = 0$. Note that here S is not finite-dimensional, and that L is in general not convex.

In order to be able to apply Theorem 2.1 and obtain necessary conditions for the solution of our problem, we must verify that Assumptions 2.1—2.4 are satisfied.

Let us define the topology on S to be the norm topology. Then, Assumption 2.1 is evidently satisfied. Each of the functionals ϕ_i satisfies Assumption 2.3 (so that ϕ_0 certainly satisfies assumption 2.4), as is easily seen, with the h_i given by

$$\begin{aligned} h_i(x) &= \gamma_i'(x_0(t_0)) \cdot x(t_0) \qquad \text{for } i = 1, \cdots, m_1, \\ h_{m_1+i}(x) &= \bar{\gamma}_i'(x_0(t_1)) \cdot x(t_1) \qquad \text{for } i = 1, \cdots, m_2, \\ h_0(x) &= \bar{\gamma}_0'(x_0(t_1)) \cdot x(t_1). \end{aligned} \tag{3.2}$$

Finally, Assumption 2.2 is satisfied if, for M, we choose the convex hull of all functions x in S that are given by the formula

$$x(t) = \Phi(t) \left\{ \xi + \int_{t_0}^{t} \Phi^{-1}(\tau) \left[f(x_0(\tau), u(\tau), \tau) - f(x_0(\tau), u_0(\tau), \tau) \right] d\tau \right\} \qquad t_0 \leqq t \leqq t_1, \tag{3.3}$$

where ξ is an arbitrary vector in R^n; $u(\cdot)$ is an arbitrary admissible control; $u_0(\cdot)$ is the "optimal" control corresponding to the "optimal" trajectory $x_0(\cdot)$, i.e. u_0 is an admissible control such that

$$\dot{x}_0(t) = f(x_0(t), u_0(t), t) \quad \text{for almost all } t \in [t_0, t_1]; \tag{3.4}$$

and $\Phi(t)$ is the absolutely continuous $n \times n$ *matrix-valued function* defined on $[t_0, t_1]$ that satisfies the relations

$$\begin{aligned} \dot{\Phi}(t) &= f_x(x_0(t), u_0(t), t)\, \Phi(t) \quad \text{for almost all } t, \\ \Phi(t_0) &= \text{the identity matrix.} \end{aligned} \tag{3.5}$$

(f_x denotes the $n \times n$ Jacobian matrix of first partial derivatives of f.)

The proof that M thus defined satisfies Assumption 2.2 is given in [**10**, §3] using results and ideas first introduced in [**11**].

We point out that the same result can be obtained if we replace the class of admissible controls by any one of a large number of other classes (e.g., instead of requiring our controls to be such that $u(t) \in U$ for each t, we may instead require that $\int_{t_0}^{t_1} |u(t)|^2 \, dt \leqq 1$).

Finally, let us take note of what is to be done if Problem 1.2 is modified by the presence of an additional constraint, of the type of (1.4). If we then define the functional ϕ_{-1} on S by the relation

$$\phi_{-1}(x) = \sup_{t_0 \leqq t \leqq t_1} \gamma(x(t)),$$

it is clear that we again have a special case of Problem 2.1 (but now with $\mu = 1$). Further, the functional ϕ_{-1} satisfies Assumption 2.4 (although it does *not* satisfy the stronger differentiability condition of Assumption 2.3), with the convex functional h_{-1} given by the relation

$$h_{-1}(x) = \sup_{t \in I_e} \gamma'(x_0(t)) \cdot x(t), \tag{3.6}$$

where

$$I_e = \{t \colon t_0 \leqq t \leqq t_1, \gamma(x_0(t)) = \sup_{t_0 \leqq s \leqq t_1} \gamma(x_0(s))\}. \tag{3.7}$$

A proof of this assertion may be found in [**10**, §5].

Since we have now placed Problem 1.2 into the framework of Problem 2.1, and have shown that Assumptions 2.1—2.4 are satisfied, we can make use of Theorem 2.1 to obtain necessary conditions for solutions of Problem 1.2.

4. **The maximum principle.** Let the pair x_0, u_0 be a solution of Problem 1.2 (we shall first consider the problem without the constraint (1.4), and later the problem with this additional constraint), so that u_0 is an admissible control and $x_0 \in S$ satisfies Equation (3.4). By virtue of Theorem 2.1 and Equations (3.2), we can conclude (after making an obvious change of notation) that there exist multipliers λ_i, $i = 1, \cdots, m_1$, and $\bar{\lambda}_i$, $i = 0, 1, \cdots, m_2$, not all zero, such that

$$\sum_{i=1}^{m_1} \lambda_i \gamma_i'(x_0(t_0)) \cdot x(t_0) + \sum_{i=0}^{m_2} \bar{\lambda}_i \bar{\gamma}_i'(x_0(t_1)) \cdot x(t_1) \leqq 0 \tag{4.1}$$

for all $x \in S$ of the form of (3.3)—where $\xi \in R^n$ and the admissible control u are arbitrary—and such that $\bar{\lambda}_0 \leqq 0$.

If we choose functions $x \in S$ defined by (3.3) with $u = u_0$, then relation (4.1) gives rise to the inequality

$$\left[\sum_{i=1}^{m_1} \lambda_i \gamma_i'(x_0(t_0))\, \Phi(t_0) + \sum_{i=0}^{m_2} \bar{\lambda}_i \bar{\gamma}_i'(x_0(t_1))\, \Phi(t_1)\right] \cdot \xi \leqq 0 \qquad \text{for all } \xi \in R^n,$$

which is possible only if

$$\sum_{i=1}^{m_1} \lambda_i \gamma_i'(x_0(t_0))\, \Phi(t_0) + \sum_{i=0}^{m_2} \bar{\lambda}_i \bar{\gamma}_i'(x_0(t_1))\, \Phi(t_1) = 0. \tag{4.2}$$

On the other hand, if we choose functions $x \in S$ defined by (3.3) with $\xi = 0$, then (4.1) implies that

$$\sum_{i=0}^{m_2} \bar{\lambda}_i \bar{\gamma}_i'(x_0(t_1))\, \Phi(t_1) \times \int_{t_0}^{t_1} \Phi^{-1}(\tau) \left[f(x_0(\tau), u(\tau), \tau) - f(x_0(\tau), u_0(\tau), \tau)\right] d\tau \leqq 0 \qquad \text{for every admissible control } u. \tag{4.3}$$

Let the function $\Psi \in S$ be defined as follows (it is convenient to consider the values of Ψ to be row-vectors)

$$\Psi(t) = \sum_{i=0}^{m_2} \bar{\lambda}_i \bar{\gamma}_i'(x_0(t_1))\, \Phi(t_1)\, \Phi^{-1}(t). \tag{4.4}$$

Then (4.3) can be rewritten in the form

$$\int_{t_0}^{t_1} \Psi(t)\, f(x_0(t), u(t), t)\, dt \leqq \int_{t_0}^{t_1} \Psi(t)\, f(x_0(t), u_0(t), t)\, dt \qquad \text{for every admissible control } u. \tag{4.5}$$

Let us suppose that the vectors $\bar{\gamma}_i'(x_0(t_1))$, $i = 0, 1, \cdots, m_2$, are linearly independent, and that the same is true of the vectors $\gamma_i'(x_0(t_0))$, $i = 1, \cdots, m_1$. It then follows from (4.2), (4.4) and (3.5) that

$$\dot{\Psi}(t) = -\Psi(t)\, f_x(x_0(t), u_0(t), t) \quad \text{for almost all } t \in [t_0, t_1], \tag{4.6}$$

$$\Psi(t) \not\equiv 0, \tag{4.7}$$

$$(4.8)\qquad \Psi(t_1) = \sum_{i=0}^{m_2} \bar{\lambda}_i \bar{\gamma}_i'(x_0(t_1)), \quad \bar{\lambda}_0 \leqq 0,$$

$$(4.9)\qquad \Psi(t_0) = -\sum_{i=1}^{m_1} \lambda_i \gamma_i'(x_0(t_0)).$$

Relation (4.5), in conjunction with (4.6) and (4.7), is the maximum principle in integral form which was first obtained in [**11**]; Equations (4.8) and (4.9) are transversality conditions.

It is not difficult to show, on the basis of our hypotheses regarding f and the definition of the class of admissible controls (see, e.g., [**11**, p. 125]), that (4.5) can be rewritten in the form

$$(4.10)\qquad \Psi(t) f(x_0(t), u_0(t), t) = \max_{u \in U} \Psi(t) f(x_0(t), u, t) \quad \text{for almost all } t \in [t_0, t_1].$$

Relation (4.10) (in conjunction with (4.6) and (4.7)) is essentially the celebrated Pontryagin maximum principle [**1**].

Thus, in summary, we have shown that if x_0, u_0 is a solution of Problem 1.2, and if the functions γ_i and $\bar{\gamma}_i$ satisfy certain smoothness and linear independence relations, then there exists an absolutely continuous function $\Psi \in S$ such that relations (4.6)—(4.10) are satisfied.

As was pointed out in §3, the set M, defined as the convex hull of all $x \in S$ of the form of (3.3), approximates L in the sense of Assumption 2.2 even if we considerably modify our definition of the set of admissible controls. Thus, our basic result—relations (4.5)—(4.9) as necessary conditions—remains in force for such problems, although it will not, in general, be possible to conclude that (4.10) also holds.

Let us now pass to Problem 1.2 with the additional constraint (1.4), a so-called optimal control problem with restricted phase coordinates. By virtue of Theorem 2.1 and Equations (3.2), (3.6) and (3.7), if the pair x_0, u_0 is a solution of this problem, then there exist multipliers λ_i, $i = 1, \cdots, m_1$, $\bar{\lambda}_i$, $i = 0, 1, \cdots, m_2$, and λ^*, not all zero, and a linear functional l defined on S such that

$$(4.11)\qquad \lambda^* l(x) + \sum_{i=1}^{m_1} \lambda_i \gamma_i'(x_0(t_0)) \cdot x(t_0) + \sum_{i=0}^{m_2} \bar{\lambda}_i \bar{\lambda}_i'(x_0(t_1)) \cdot x(t_1) \leqq 0$$

for all $x \in S$ of the form of (3.3)—where $\xi \in R^n$ and the admissible control u are arbitrary—and such that $\lambda_0 \leqq 0$, $\lambda^* \leqq 0$. Further, the functional l satisfies the inequality

$$l(x) \leqq \sup_{t \in I_e} \gamma'(x_0(t)) \cdot x(t) \quad \text{for all } x \in S, \tag{4.12}$$

where I_e is given by (3.7).

If $\lambda^* = 0$, the necessary conditions for this problem reduce to the form previously discussed. Therefore, we shall suppose that $\lambda^* < 0$, or, without loss of generality, that $\lambda^* = -1$.

Using standard arguments regarding the representation of linear functionals (see, e.g., [**10**, pp. 128-130]), we can show that (4.12) implies that there exists a scalar-valued, nondecreasing function $\nu(\cdot)$ defined on $[t_0, t_1]$ and continuous from the right in (t_0, t_1), such that $\nu(t_1) = 0$, ν is constant on every subinterval of $[t_0, t_1]$ which does not meet I_e, and

$$l(x) = \int_{t_0}^{t_1} \gamma'(x_0(t)) \cdot x(t)\, d\nu(t) \quad \text{for all } x \in S. \tag{4.13}$$

Arguing as before, and performing some additional manipulations (these are carried out in detail in [**10**, pp. 130-133]), we finally arrive at the following necessary conditions:

If x_0, u_0 is a solution of Problem 1.2 with the additional constraint (1.4), then there exist an absolutely continuous function $\Psi \in S$ and a function ν possessing the properties outlined above, such that

(1) $$\dot{\Psi}(t) = -\Psi(t) f_x(x_0(t), u_0(t), t) - \nu(t)\, p_x(x_0(t), t),$$

where

$$p(x,t) = \gamma'(x) f(x, u_0(t), t),$$

(2) $$[\Psi(t) + \nu(t)\gamma'(x_0(t))] \cdot f(x_0(t), u_0(t), t) = \max_{u \in U} [\Psi(t) + \nu(t)\gamma'(x_0(t))] \cdot f(x_0(t), u, t)$$

for almost all $t \in [t_0, t_1]$,

(3) $$\Psi(t_0) = -\sum_{i=1}^{m_1} \lambda_i \gamma_i'(x_0(t_0)) - \nu(t_0)\gamma'(x_0(t_0)),$$

(4) $$\Psi(t_1) = \sum_{i=0}^{m_2} \bar{\lambda}_i \bar{\gamma}_i'(x_0(t_1)), \quad \bar{\lambda}_0 \leqq 0,$$

(5) $\Psi(t) + \nu(t)\gamma'(x_0(t)) \neq 0$ for t in a subset of $[t_0, t_1]$ of positive measure.

Conditions (1)—(5) are the modified form of the maximum principle for optimal control problems with restricted phase coordinates. Under special assumptions, they were first obtained in [**12**].

5. **Conclusion.** We have defined a general class of optimal control problems, reduced this class to a certain class of mathematical programming problems, and, on the basis of general necessary conditions for the latter problems, have obtained particular necessary conditions for the optimal control problems.

References

1. L. S. Pontryagin, V. G. Boltyanskiĭ, R. V. Gamkrelidze and E. F. Mishchenko, *The mathematical theory of optimal processes,* English transl., Wiley, New York, 1962.

2. L. Cesari, *Existence theorems for optimal solutions in Pontryagin and Lagrange problems,* J. Soc. Indust. Appl. Math. Control **3** (1965), 475-498.

3. A. V. Balakrishnan and L. W. Neustadt, Editors, *Computing methods in optimization problems,* Academic Press, New York, 1964.

4. B. N. Pshenichniy, *Linear optimal control problems,* J. Soc. Indust. Appl. Math. Control **4** (1966),577-593.

5. B. Paiewonsky, "Synthesis of optimal controls", in *Topics in optimization,* Academic Press, New York, 1967, pp. 391-416.

6. E. B. Lee, *A sufficient condition in the theory of optimal control,* J. Soc. Indust. Appl. Math. Control **1** (1963) 241-245.

7. M. Canon, C. Cullum and E. Polak, *Constrained minimization problems in finite-dimensional spaces,* Ibid. **4** (1966), 528-547.

8. H. Halkin and L. W. Neustadt, *General necessary conditions for optimization problems,* Proc. Nat. Acad. Sci. U. S. A. **56** (1966), 1066-1071.

9. L. W. Neustadt, *An abstract variational theory with applications to a broad class of optimization problems.* I: *General theory,* J. Soc. Indust. Appl. Math. Control **4** (1966), 505-527.

10. ———, *An abstract variational theory with applications to a broad class of optimization problems.* II: Applications, Ibid. **5** (1967), 90-137.

11. R. V. Gamkrelidze, *On some extremal problems in the theory of differential equations with applications to the theory of optimal control,* Ibid. **3** (1965) 106-128.

12. J. Warga, *Minimizing variational curves restricted to a preassigned set,* Trans. Amer. Math. Soc. **112** (1964), 432-455.

UNIVERSITY OF SOUTHERN CALIFORNIA

E. Polak[1]

Necessary Conditions of Optimality in Control and Programming[2]

Introduction. Until quite recently, the calculus of variations, nonlinear programming, and optimal control were considered to be loosely related fields. It was only in the past few years that a new approach has evolved in which all constrained minimization problems are considered to be special cases of a canonical, or basic, optimization problem. This has resulted in very general theorems of first order necessary conditions from which classical as well as new necessary conditions of optimality can be obtained as corollaries, by simply invoking the structure of a particular problem of interest.

The purpose of this paper is to sketch out these new developments by first considering constrained minimization problems in finite dimensional spaces and then indicating the straight-forward generalizations which are required to obtain an extension of the results to linear topological spaces. Among the applications presented are such well-known results as the Kuhn-Tucker [**4**] conditions and the Maximum Principle of Pontryagin [**3**]. For further details the reader is referred to literature cited at the end of this paper.

[1] Department of Electrical Engineering and Computer Sciences and the Electronics Research Laboratory, University of California, Berkeley, California.
[2] Research reported herein was supported in part by the National Aeronautics and Space Administration under Grant NsG-354, Supplement 3.

I. Finite dimensional problems.

(1) *The Basic Problem.* Given the continuously differentiable functions $f: E^n \to E^1$ and $r: E^n \to E^m$, and Ω, a subset of E^n, find a vector $\hat{z} \in E^n$ satisfying

(2) $$\hat{z} \in \Omega, \qquad r(\hat{z}) = 0,$$

such that

(3) $$\text{for all } z \in \Omega \text{ with } r(z) = 0, \;\; f(\hat{z}) \leqq f(z).$$

We shall call a vector $\hat{z}$ satisfying (2) and (3) an *optimal solution* to the Basic Problem.

For the problem defined above, we shall state a necessary condition of optimality in the form of an inequality which will be valid for all vectors δz in a convex cone which is a "linearization" of the set Ω. We shall make use of two kinds of "linearizations" of the set Ω at a point $\hat{z}$. We begin by introducing the simpler one.

(4) DEFINITION [1]. A convex cone[3] $C(\hat{z}, \Omega) \subset E^n$ will be called a *linearization of the first kind* of the constraint set Ω at $\hat{z}$ if for any finite collection $\{\delta z_1, \delta z_2, \cdots, \delta z_k\}$ of linearly independent vectors in $C(\hat{z}, \Omega)$ there exists an $\epsilon > 0$, possibly depending on $\hat{z}$, $\delta z_1, \delta z_2, \cdots, \delta z_k$, such that $\mathrm{co}\{\hat{z}, \hat{z} + \epsilon\delta z_1, \cdots, \hat{z} + \epsilon \delta z_k\}$[4] $\subset \Omega$.

Clearly, all the linearizations of the first kind must be contained in the cone defined below.

(5) DEFINITION. The *radial cone* to the set Ω at a point $\hat{z} \in \Omega$ will be denoted by $\mathrm{RC}(\hat{z}, \Omega)$ and is defined by

$$\mathrm{RC}(\hat{z}, \Omega) = \{\delta z | \, (\hat{z} + \epsilon \delta z) \in \Omega \text{ for all } \epsilon \in [0, \epsilon_1(z, \delta z)], \text{ where } \epsilon_1(\hat{z}, \delta z) > 0\}.$$

Consequently, in the various theorems to follow, the radial cone $\mathrm{RC}(\hat{z}, \Omega)$ should always be used if possible, since this will result in stronger necessary conditions.

We now define the continuously differentiable map $F: E^n \to E^{m+1}$ to be

---

[3] A set C is a cone with vertex x_0 if for every $x \in C$, $x \neq x_0$, $x_0 + \lambda(x - x_0) \in C$ for all $\lambda \geqq 0$. Since the vertex x_0 of the cone C will normally be obvious, we shall omit mentioning it.

[4] $\mathrm{co}\{\hat{z}, \hat{z} + \epsilon\delta z_1, \cdots, \hat{z} + \epsilon\delta z_k\}$ is the convex hull of $\hat{z}$, $\hat{z} + \epsilon \delta z_1, \cdots, z + \epsilon \delta z_k$, i.e., the set of all points, y, of the form $y = \mu^0 \hat{z} + \mu^1(\hat{z} + \epsilon\delta z_1) + \cdots + \mu^k(\hat{z} + \epsilon \delta z_k)$, where $\sum_{i=0}^k \mu^i = 1$, $\mu^i \geqq 0$ for all i.

(6) $$F(z) = (f(z), r(z)).$$

We shall number the components of E^{m+1} from 0 to m, i.e., $y \in E^{m+1}$ is given by $y = (y^0, y^1, \cdots, y^m)$ and we shall denote the Jacobian matrix of the map F by $\partial F(z)/\partial z$.

For the Basic Problem (1) the following theorem gives a necessary condition for optimality.

(7) THEOREM [1]. *If $\hat{z}$ is an optimal solution to the basic problem and $C(\hat{z}, \Omega)$ is a linearization of the first kind of Ω at $\hat{z}$, then there exists a nonzero multiplier vector $\psi = (\psi^0, \psi^1, \cdots, \psi^m) \in E^{m+1}$, with $\psi^0 \leqq 0$, such that for all $\delta z \in \overline{C}(\hat{z}, \Omega)$ (the closure of $C(\hat{z}, \Omega)$ in E^n)*

(8) $$\langle \psi, (\partial F(\hat{z})/\partial z)\, \delta z \rangle \leqq 0.$$

PROOF. Let $K(\hat{z}) \subset E^{m+1}$ be the cone defined by

(9) $$K(\hat{z}) = (\partial F(\hat{z})/\partial z)\, C(\hat{z}, \Omega),$$

and let $\hat{y} = F(\hat{z})$, then it is seen that the condition (8) simply states that the convex cone $\hat{y} + K(\hat{z})$ must be separated from the ray

(10) $$R = \{y \mid y = \hat{y} + \beta(-1, 0, \cdots, 0), \quad \beta > 0\},$$

or, in expanded form, that there must exist a nonzero vector $\psi \in E^{m+1}$ satisfying

(11) $$\langle \psi, y - \hat{y} \rangle \leqq 0 \quad \text{for every } y \in \hat{y} + K(\hat{z}),$$

(12) $$\langle \psi, y - \hat{y} \rangle \geqq 0 \quad \text{for every } y \in R.$$

To obtain a contradiction, let us suppose that the cone $\hat{y} + K(\hat{z})$ and the ray R are not separated. Then the cone $\hat{y} + K(\hat{z})$ must be of dimension $m + 1$ and R must be an interior ray of $\hat{y} + K(\hat{z})$ (i.e., all points of R are interior points of $\hat{y} + K(\hat{z})$).

We can therefore construct in the cone $\hat{y} + K(\hat{z})$ a simplex Σ with vertices $\hat{y}, \hat{y} + \delta y_1, \hat{y} + \delta y_2, \cdots, \hat{y} + \delta y_{m+1}$ such that

(13) the ray R passes through its interior, i.e., there is a point $y = (\hat{y} + \delta y_0) \in R$, with $\delta y_0 = \gamma(-1, 0, \cdots, 0)$ and $\gamma > 0$, which lies in the interior of Σ;

(14) to each of the vectors δy_i, $i = 1, 2, \cdots, m$, there corresponds a vector $\delta z_i \in C(\hat{z}, \Omega)$ such that

(15) $$\delta y_i = (\partial F(\hat{z})/\partial z)\, \delta z_i, \qquad i = 1, \cdots, m + 1$$

and

$$\text{(16)} \qquad \operatorname{co}\{\hat{z}, \hat{z}+\delta z_1, \cdots, \hat{z}+\delta z_{m+1}\} \subset \Omega.$$

Note that the vectors δz_i, $i = 1, 2, \cdots, m+1$ are linearly independent because the vectors $\delta y_1, \delta y_2, \cdots, \delta y_{m+1}$ are linearly independent.

For $0 < \alpha \leqq 1$, we define $S_\alpha \subset \Sigma$ to be a closed ball with center $\hat{y} + \alpha \delta y_0$ and radius αr, where $r > 0$, i.e., $S_\alpha = \{y \in E^{m+1} | \, \|y - \hat{y} - \alpha\delta y_0\| \leqq \alpha r\}$. Since $\hat{y} + \delta y_0$ is an interior point of Σ we can always find an $r > 0$ which will make this construction possible. Let $\alpha \in (0, 1]$ be arbitrary. We now define the map G_α from the ball $S_\alpha - (\hat{y} + \alpha\delta y_0)$ with center at the origin into E^{m+1} as follows. For any $x \in S_\alpha - (\hat{y} + \alpha \delta y_0)$, let

$$\text{(17)} \qquad G_\alpha(x) = -\,[F(\hat{z} + ZY^{-1}(\alpha\delta y_0 + x)) - (\hat{y} + \alpha\delta y_0)] + x$$

where Y is a $(m+1) \times (m+1)$ matrix whose ith column is δy_i, $i = 1, \cdots, m+1$, and Z is a $n \times (m+1)$ matrix whose ith column is δz_i. The matrix Y is invertible because the vectors δy_i, $i = 1, 2, \cdots, m+1$ are linearly independent by construction.

Expanding the right-hand side of (17) about $\hat{z}$, we get

$$\text{(18)} \qquad \begin{aligned} G_\alpha(x) = -\,[\hat{y} + (\partial F(\hat{z})/\partial z)\, ZY^{-1}(\alpha\delta y_0 + x) - (\hat{y} + \alpha\delta y_0) \\ + o(ZY^{-1}(\alpha\delta y_0 + x))] + x \end{aligned}$$

where $o(\cdot)$ is a continuous function such that $\lim_{\|y\| \to 0} \|o(y)\|/\|y\| = 0$. But by definition, $(\partial F(\hat{z})/\partial z)\, Z = Y$, and hence (18) becomes

$$G_\alpha(x) = -\,o(ZY^{-1}(\alpha\delta y_0 + x)).$$

Since in the above $\|x\| \leqq \alpha r$, there exists a constant $M > 0$ such that

$$\text{(20)} \qquad \|\alpha\delta y_0 + x\| \leqq \alpha M$$

for all x satisfying $\|x\| \leqq \alpha r$. Now $o(\alpha z)/\alpha \to 0$ as $\alpha \to 0$ for all z such that $\|z\| \leqq M$ and hence there exists a $\alpha^* \in (0, 1]$ for which

$$\text{(21)} \qquad \|o(\alpha^* z Y^{-1}(\delta y_0 + x))\| < \alpha^* r$$

with $x \in S_{\alpha^*} - (y + \alpha^*\delta y_0)$ arbitrary. Thus, G_{α^*} maps $S_{\alpha^*} - (\hat{y} + \alpha^*\delta y_0)$ into itself and it is continuous. We therefore conclude from Brouwer's Fixed Point Theorem that there exists a $\tilde{x} \in S_{\alpha^*} - y + \alpha^*\delta y_0$ such that

(22) $$G_{\alpha^*}(\tilde{x}) = \tilde{x},$$

i.e.,

(23) $$F(\hat{z} + ZY^{-1}(\alpha^*\delta y_0 + \tilde{x})) = \hat{y} + \alpha^*\delta y_0.$$

Expanding (23), we get

(24) $$r(\hat{z} + ZY^{-1}(\alpha^*\delta y_0 + \tilde{x})) = 0$$

and

(25) $$f(\hat{z} + ZY^{-1}(\alpha^*\delta y_0 + \tilde{x})) = f(\hat{z}) - f(\hat{z}) - \alpha^*\gamma < f(\hat{z}).$$

Finally, since for any δy in the simplex Σ the vector $z = \hat{z} + ZY^{-1}\delta y$ belongs to $\mathrm{co}\{\hat{z}, \hat{z} + \delta z_1, \cdots, \hat{z} + \delta z_{m+1}\}$ and (16) holds, we have

(26) $$(\hat{z} + ZY^{-1}(\alpha^*\delta y_0 + \tilde{x})) \in \Omega.$$

Hence $\hat{z}$ is not optimal, which is a contradiction. We therefore conclude that the cone $\hat{y} + K(\hat{z})$ and the ray R must be separated. But, if $\hat{y} + K(\hat{z})$ and R are separated, then so are $\hat{y} + \overline{K}(\hat{z})$ and R and hence there must exist a nonzero vector ψ such that

(27) $$\begin{aligned} \langle \psi, (\partial F(\hat{z})/\partial z)\delta z\rangle &\leqq 0 \text{ for every } \delta z \in \overline{C}(\hat{z}, \Omega), \text{ and} \\ \langle \psi, y\rangle &\geqq 0 \text{ for every } y \in \hat{y} + \overline{k}(\hat{z}). \end{aligned}$$

Substituting for y from (10) in (12) we have

(28) $$\langle \psi, (-1, 0, \cdots, 0)\rangle = -\psi^0 \geqq 0.$$

This completes the proof.

Theorem (7) can be extended to problems for which linearizations of the first kind do not exist (for example suppose Ω is a piece of a smooth, nonplanar surface) by introducing a more complex linearization for the set Ω.

(29) DEFINITION [1], [2]. A convex cone $C(\hat{z}, \Omega) \subset E^n$ will be called a *linearization of the second kind* of the constraint set Ω at $\hat{z}$, if for any finite collection $\{\delta z_1, \delta z_2, \cdots, \delta z_k\}$ of linearly independent vectors in $C(\hat{z}, \Omega)$, there exists an $\epsilon > 0$, possibly depending on $\hat{z}, \delta z_1, \cdots, \delta z_k$, and a continuous map ζ from $\mathrm{co}\{\hat{z}, \hat{z} + \delta z_1, \cdots, \hat{z} + \delta z_k\}$ into Ω, such that $\zeta(\hat{z} + \delta z) = \hat{z} + \epsilon\delta z + o(\epsilon\delta z)$, where $\lim_{\beta\to 0}\|o(\beta\delta z)\|/\beta \to 0$ uniformly for all $\delta z \in \mathrm{co}\{0, \delta z_1, \delta z_2, \cdots, \delta z_k\}$.

(30) REMARK. Note that any linearization of the first kind of Ω at $\hat{z}$, is also a linearization of the second kind of Ω at $\hat{z}$, with the map ζ defined by $\zeta(\hat{z} + \delta z) = \hat{z} + \epsilon\delta z$, i.e., ζ is an affine map.

Consequently, unless this fact is pertinent, we shall omit reference to the kind of a linearization in our statements and shall simply speak of linearizations. Theorem (7) thus becomes the

(31) FUNDAMENTAL THEOREM. *If $\hat{z}$ is an optimal solution to the Basic Problem* (1) *and* $C(\hat{z}, \Omega)$ *is a linearization of* Ω *at* $\hat{z}$, *then there exists a nonzero vector* $\psi = (\psi^0, \psi^1, \cdots, \psi^m) \in E^{m+1}$ *with* $\psi^0 \leqq 0$, *such that for all* $\delta z \in \overline{C}(\hat{z}, \Omega)$, (*the closure of* $C(\hat{z}, \Omega)$ *in* E^n), $\langle \psi, (\partial F(\hat{z})/\partial z)\, \delta z \rangle \leqq 0$.

We shall now show how a number of classical optimization problems can be recast in the form of the Basic Problem (1) and we shall then apply Theorem (31) to rederive several classical conditions for optimality, as well as to obtain some new ones.

Classical theory of Lagrange multipliers. Consider the problem $\min \{ f(z) | \, r(z) = 0, \ z \subset E^n \}$, where f, r are continuously differentiable functions from E^n into E^1 and E^n into E^m, respectively.

This problem is readily recognized to be the Basic Problem (1), with $\Omega = E^n$. Since E^n is a linearization for E^n at any $z \in E^n$, we conclude from Theorem (31) that if $\hat{z}$ is an optimal solution to the above problem, then there exists a nonzero vector $\psi \in E^{m+1}$ such that

(32) $$\langle \psi, (\partial F(\hat{z})/\partial z)\, \delta z \rangle \leqq 0 \quad \text{for all } \delta z \in E^n$$

and $\psi^0 \leqq 0$.

The inequality (32) may be rewritten as

(33) $$\langle (\partial F(\hat{z})^T/\partial z) \psi, \delta z \rangle \leqq 0 \quad \text{for all } \delta z \in E^n$$

and since for any $\delta z \in E^n$, $-\delta z$ is also in E^n, we conclude from (32) that

(34) $$(\partial F(\hat{z})^T/\partial z)\, \psi = 0.$$

Now, $\partial F(z)^T/\partial z$ is an $n \times (m-1)$ matrix with columns $\nabla f(\hat{z})$, $\nabla r^1(\hat{z}), \cdots, \nabla r^m(\hat{z})$, where $\nabla f(\hat{z}) = (\partial f(\hat{z})/\partial z^1, \cdots, \partial f(\hat{z})/\partial z^n)$, $\nabla r^i(\hat{z}) = (\partial r^i(\hat{z})/\partial z^1, \cdots, \partial r^i(\hat{z})/\partial z^n)$. We may therefore expand (34) as follows

(35) $$\psi^0 \nabla f(\hat{z}) + \sum_{i=1}^{m} \psi^i \nabla r^i(\hat{z}) = 0.$$

We have thus established the following well-known result.

(36) THEOREM. *Let* $f(\cdot), r^1(\cdot), r^2(\cdot), \cdots, r^m(\cdot)$ *be real valued, continuously differentiable functions on* E^n. *If* $z \in E^n$ *minimizes* $f(z)$

subject to the constraints $r^i(z) = 0,\ i = 1, 2, \cdots, m$, *then there exist scalar multipliers*, $\psi^0, \psi^1, \cdots, \psi^m$, *not all zero, such that* (35) *is satisfied.*

We now state an important special case.

(37) COROLLARY. *If the gradient vectors* $\nabla r^i(\hat{z}),\ i = 1, 2, \cdots, m$, *are linearly independent, then the multiplier* ψ^0 *in* (35) *cannot be zero (and hence can be taken to be* 1).

Nonlinear programming. Consider the problem $\min\{f(z) \mid r(z) = 0,\ q(z) \leqq 0,\ z \in E^n\}$ where f is a real valued continuously differentiable function on E^n and r, q are continuously differentiable functions from E^n into E^m and E^k, respectively.

This is the standard nonlinear programming problem which we also recongized to be a special case of the Basic Problem (1), with $\Omega = \{z : q(z) \leqq 0\}$. We shall now show how Theorem (31) can be used to establish commonly known necessary conditions for z to be an optimal solution.

It will often be necessary for us to differentiate between active and inactive inequality constraints at a given point. We shall do this by means of the following index set.

(38) DEFINITION. For any $\hat{z} \in \Omega$, let the index set $I(\hat{z})$ be defined by

$$I(\hat{z}) = \{i \mid q^i(\hat{z}) = 0,\ i \in \{1, 2, \cdots, k\}\}. \tag{39}$$

The constraints q^i, $i \in I(\hat{z})$ will be called *active* at $\hat{z}$. We shall denote by $\bar{I}(\hat{z})$ the complement of $I(\hat{z})$ in $\{1, \cdots, k\}$, and we shall call the constraints q^i, $i \in \bar{I}(\hat{z})$, *inactive*.

We now state a condition on the set $\Omega = \{z : q(z) \leqq 0\}$ without which the necessary conditions to be derived can be satisfied trivially.

(40) ASSUMPTION. Let $\hat{z} \in \Omega$ be an optimal solution of our nonlinear programming problem. Then there exists a vector $h \in E^n$ such that

$$\langle \nabla q^i(\hat{z}), h \rangle < 0 \quad \text{for all } i \in I(\hat{z}).$$

(41) DEFINITION. For any $\hat{z} \in \Omega$, the internal cone of Ω at $\hat{z}$, denoted by $\mathrm{IC}(\hat{z}, \Omega)$ is defined by

$$\mathrm{IC}(z, \Omega) = \{\delta z \mid \langle \nabla q^i(\hat{z}), \delta z \rangle < 0 \quad \text{for all } i \in I(\hat{z})\}.$$

Thus, the gist of assumption (40), is that the convex cone $\mathrm{IC}(\hat{z}, \Omega)$

is nonempty. The following lemma can be proved by expanding the appropriate functions into a first order Taylor expansion with remainder.

(42) LEMMA. *If* $\mathrm{IC}(\hat{z},\Omega) \neq \emptyset$, *the empty set, then*

(43) $\mathrm{IC}(\hat{z},\Omega)$ *is a linearization of the first kind of* Ω *at* $\hat{z}$,

(44) $\overline{\mathrm{IC}}(\hat{z},\Omega) = \{\delta z \mid \langle \nabla q^i(\hat{z}), \delta z\rangle \leqq 0 \text{ for all } i \in I(\hat{z})\}$.

When specialized to our nonlinear programming problem, Theorem (7) assumes the following form.

(45) THEOREM [1]. *If* $\hat{z}$ *is an optimal solution to our nonlinear programming problem and* (40) *is satisfied, then there exists a nonzero vector* $\psi \in E^{m+1}$, *with* $\psi^0 \leqq 0$, *such that*

$$\left\langle \psi^0 \nabla f(z) + \sum_{i=1}^{m} \psi^i \nabla r^i(z), z \right\rangle \leqq 0 \tag{46}$$

for all δz *satisfying*

$$\langle \nabla q^i(\hat{z}), \delta z\rangle \leqq 0, \quad i \in I(\hat{z}). \tag{47}$$

Combining Theorem (45) and Farkas' Lemma we obtain the following possibly more familiar necessary condition for optimality.

(48) THEOREM [1]. *If* $\hat{z}$ *is an optimal solution to the Nonlinear Programming Problem, and* (40) *is satisfied, then there exists a nonzero vector* $\psi \in E^{m+1}$, *with* $\psi^0 \leqq 0$, *and a vector* $\mu \leqq 0$, *such that*

$$\psi^0 \nabla f(\hat{z}) + \sum_{i=1}^{m} \psi^i \nabla r^i(\hat{z}) + \sum_{i=1}^{k} \mu^i \nabla q^i(\hat{z}) = 0 \tag{49}$$

and

$$\mu^i q^i(\hat{z}) = 0 \quad \text{for } i = 1, 2, \cdots, k. \tag{50}$$

Most of the other well-known necessary conditions for nonlinear programming problems can be obtained from Theorem (45) by making additional assumptions on the functions r and q. For example, the following corollaries to Theorem (45) are immediate consequences of that theorem.

(51) COROLLARY [1]. *If assumption* (40) *is satisfied and the vectors* $\nabla r^i(\hat{z})$, $i = 1, \cdots, m$, *are linearly independent, then there exist vectors* $\psi \in E^{m+1}$, $\mu \in E^k$ *which satisfy the conditions of Theorem* (48) *and* $(\psi^0, \mu) \neq 0$.

(52) COROLLARY. *If* $\nabla r^i(\hat{z})$, $i = 1, \cdots, m$, *together with* $\nabla q^i(\hat{z})$, $i \in I(\hat{z})$, *are linearly independent vectors, there exists a vector* $\psi \in E^{m+1}$

satisfying the conditions of Theorem (45) *with* $\psi^0 < 0$.

The assumptions in Corollary (52) ensure that the Kuhn-Tucker constraint qualification [**4**] is satisfied (see [**1**]).

(53) COROLLARY. *If there exists a vector* $h \in E^n$ *such that* $\langle \nabla q^i(\hat{z}), h \rangle < 0$ *for all* $i \in I(\hat{z})$, $\langle \nabla r^i(\hat{z}), h \rangle = 0$ *for* $i = 1, \cdots, m$, *and the vectors* $\nabla r^i(\hat{z})$, $i = 1, \cdots, m$, *are linearly independent, then there exists a vector* $\psi \in E^{m+1}$ *satisfying the conditions of Theorem* (45) *with* $\psi^0 < 0$.

The assumption in this corollary is sufficient to guarantee that the weakened constraint qualification be satisfied (see [**1**]).

II. **Infinite dimensional problems.** We shall now show how one may proceed in order to extend the Fundamental Theorem (31) to problems in infinite dimensional spaces. We shall then indicate how the Pontryagin Maximum Principle, [**3**] for fixed time optimal control problems can be obtained as a particular case of the generalized theorem.

First let us reformulate the Basic Problem (1) in an infinite dimensional space.

(54) BASIC PROBLEM. Given a function $f(\cdot)$ mapping a linear topological space L into the reals, a function $r(\cdot)$ mapping L into E^m, and a subset $\Omega \subset L$, find a vector $\hat{x} \in \Omega$ satisfying $r(\hat{x}) = 0$, such that for all $x \in \Omega$ satisfying $r(x) = 0$,

$$f(\hat{x}) \leqq f(x). \tag{55}$$

We shall call any $\hat{x}$ with the above properties an *optimal solution*.

Observe that in the above formulation we did not specify that the functions f and r are differentiable, as we did in (1), since differentiability is not a well defined concept in a general linear topological space.

The simplest way [**7**], [**2**], [**6**] to obtain an extension of the Fundamental Theorem (31) is to postulate the existence of a linear function from L into E^{m+1} which will take the place of the Jacobian matrix $\partial F(\hat{z})/\partial z$ in (31) and a suitable continuous function from L into E^{m+1} which will take the place of the function $o(\cdot)$ in (18). We take care of this by incorporating the required functions into the definition of a linearization. As we shall see later, this is not a restrictive practice.

(56) **Definition** [7]. A convex cone $C(\hat{x}, \Omega) \subset L$ will be called a linearization of the set Ω at $\hat{x} \in \Omega$, with respect to the map

$$F \triangleq (f, r),$$

if there exists a linear function $F'(\hat{x})(\cdot)$ from L into E^{m+1} such that for any finite collection $\{\delta x_1, \delta x_2, \cdots, \delta x_k\}$ of linearly independent vectors in $C(\hat{x}, \Omega)$ there exists an $\epsilon > 0$, a continuous map $\zeta(\cdot)$ from $\mathrm{co}\{\hat{x}, \hat{x} + \delta\hat{x}_1, x + \delta x_2, \cdots, x + \delta x_k\}$ into Ω, and a continuous map $o(\cdot)$ from L into E^{m+1}, with ϵ, ζ, and o possibly depending on $\hat{x}, \delta x_1, \delta x_2, \cdots, \delta x_k$ which satisfy

(57) $$\lim_{\beta \to 0} \frac{\| o(\beta y) \|}{\beta} \to 0$$

uniformly for all $y \in \mathrm{co}\{0, \delta x_1, \delta x_2, \cdots, \delta x_k\}$, and,

(58) $$F(\zeta(x)) = F(\hat{x}) + \epsilon F'(\hat{x})(x - \hat{x}) + o(\epsilon(x - \hat{x}))$$

for all $x \in \mathrm{co}\{\hat{x}, \hat{x} + \delta x_1, \cdots, \hat{x} + \delta x_k\}$.

Proceeding essentially as in the previous section, we now obtain the following result.

(59) **Theorem** [2], [6], [7]. *If $\hat{x}$ is an optimal solution to the Basic Problem* (54) *and $C(\hat{x}, \Omega)$ is a linearization of Ω at $\hat{x} \in \Omega$ with respect to the map $F = (f)$, then there exists a nonzero vector $\psi = (\psi^0, \psi^1, \cdots, \psi^m)$ in E^{m+1}, with $\psi^0 \leqq 0$, such that $\langle \psi, F'(\hat{x})(\delta x) \rangle \leqq 0$ for all $\delta x \in \overline{C}(\hat{x}, \Omega)$, where $\overline{C}(\hat{x}, \Omega)$ is the closure of $C(\hat{x}, \Omega)$ in L.*

III. **The maximum principle** [2], [3], [7]. Consider a dynamical system described by the differential equation

(60) $$dx/dt = \mathbf{h}(x, u)$$

for all t in the compact interval $I = [0, T]$, where $x(t) \in E^n$ is the state of the system at time t, $u(t) \in E^m$ is the input or control of the system at time t, and $\mathbf{h}$ is a function continuous in u and continuously differentiable in x which maps $E^n \times E^m$ into E^n.

(61) The *Fixed time optimal control problem* is that of finding a control $\hat{u}$, and a corresponding trajectory $\hat{x}$, both defined on I and determined by (60), which satisfies the following conditions:

(62) For $t \in I$, $\hat{u}$ is a measurable, essentially bounded function whose range is contained in a given subset U of E^m.

(63) At time $t = 0$, $x(0) = \hat{x}_0$ a given vector in E^n, and at $t = T$, the terminal state $\hat{x}(T) \in S$, where $S = \{x \in E^n | g(x) = 0\}$, and

g is a differentiable map from E^n into E^l $(l \leqq n)$, whose Jacobian $\partial g(x)/\partial x$ is of rank l for all $x \in S$.

For every control u, and corresponding trajectory x, satisfying the conditions (62), and (63),

$$\int_0^T f^0(x(t), u(t))\, dt \geqq \int_0^T f^0(\hat{x}(t), \hat{u}(t))\, dt \tag{64}$$

where $f^0(\cdot)$ is a cost function continuous in u and continuously differentiable in x, mapping $E^n \times E^m$ into the reals.

Let P_1 be a $1 \times (n+1)$ projection matrix of the form

$$P_1 = (1, 0, 0, \cdots, 0) \tag{65}$$

and let P_2 be a $n \times (n+1)$ projection matrix as shown below

$$P_2 = \left(\begin{array}{c|cccc} 0 & 1 & 0 & \cdot & 0 \\ 0 & 0 & 1 & 0 & 0 \\ \cdot & \cdot & \cdot & 1 & \cdot \\ 0 & 0 & 0 & 0 & 1 \end{array} \right) . \tag{66}$$

Finally, let $h\colon E^{n+1} \times E^m \to E^{n+1}$ be the function defined by

$$h(z, u) = (f^0(P_2 z, u), \underline{h}(P_2 z, u)), \quad z \in E^{n+1},\ u \in E^m. \tag{67}$$

It is clear that the optimal control problem (61) becomes that of finding a control $\hat{u}$ defined for $t \in I$ and a corresponding trajectory $\hat{z}$, determined by the differential equation

$$dz(t)/dt = h(z(t), u(t)) \tag{68}$$

such that for $t \in I$, $\hat{u}$ is a measurable, essentially bounded function, whose range is contained in the given subset U of E^m;

$$\hat{z}(0) = (0, \hat{x}_0) = \hat{z}_0, \tag{69}$$

where $\hat{x}_0$ is the given initial condition,

$$\hat{z}(T) \in \{ z \in E^{n+1} \mid g(P_2 z) = 0 \}, \tag{70}$$

where g maps E^n into E^l, and for every control u defined on I and corresponding trajectory z satisfying (68), (69), (70),
z satisfying (68), (69), (70),

$$P_1 \hat{z}(T) \leqq P_1 z(T). \tag{71}$$

To complete the transcription of the optimal control problem (61) into the form of the basic problem (54) we define

(72) $$f(z) = P_1(z(T)),$$

(73) $$r(z) = g(P_2 z(T)),$$

(74) and we let Ω be the set of all absolutely continuous functions z from I into E^{n+1} which, for some measurable, essentially bounded function u from I into U, satisfy the differential equation (68) for almost all t in I, with $z(0) = (0, \hat{x}_0)$.

We must still define the linear topological vector space L. From (74) it is clear that Ω is a subset of the linear space of all absolutely continuous functions from I into E^{n+1}. However, it is convenient to use an available linearization first constructed by Pontryagin et al. [**3**] and which does not consist of absolutely continuous functions. We therefore imbed Ω into a larger linear topological space which we define below.

Let $\mathscr{U}$ be the set of all upper semicontinuous real valued functions[5] defined on I, and let $\mathscr{S} = \mathscr{U} - \mathscr{U}$. The set $\mathscr{S}$ is a linear vector space. We now define L to be the Cartesian product $\mathscr{S}^{n+1} = \mathscr{S} \times \mathscr{S} \times \cdots \times \mathscr{S}$, with the pointwise topology.

It is easy to show that f and r, defined respectively by (72) and (73) are continuous on L.

Let $\hat{z}$, corresponding to the control $\hat{u}$, be an optimal solution to the optimal control problem (in Basic Problem form). We now present the linearization for the constraint set Ω at $\hat{z}$, which was used by Pontryagin et al. [**3**] (see also [**2**] and [**7**]).

Let $I_1 \subset I$ be the set of all points t at which the control $\hat{u}$ is regular, i.e.

(75) $$I_1 = \left\{ t \,|\, 0 \leqq t \leqq T \quad \text{and} \quad \lim_{\text{meas}(J) \to 0} \frac{\text{meas}(\hat{u}^{-1}(N) \cap J)}{\text{meas}(J)} = 1, \right.$$
$$\left. \text{for every neighborhood } N \text{ of } u(t), \ J \subset I \text{ and } t \in J \right\}.$$

Let $\Phi(t, \tau)$ be the $(n+1) \times (n+1)$ matrix which satisfies the linear differential equation

(76) $$\frac{d}{dt}\Phi(t, \tau) = \frac{\partial h(\hat{z}(t), \hat{u}(t))}{\partial t}\Phi(t, \tau)$$

for almost all $t \in I$, with $\Phi(\tau, \tau) = I_{n+1}$, the $(n+1)$ identity matrix.

---

[5] DEFINITION. A real valued function $f: E^1 \to E^1$ is called *upper semicontinuous* at a point t_0 in E^1, if $\limsup_{t \to t_0} f(t) \leqq f(t_0)$. And it is called *lower semicontinuous* if $-f$ is upper semicontinuous.

Finally, for any $s \in I_1$ and $v \in U$ let

$$(77)\qquad \begin{aligned} \delta z_{s,v}(t) &= 0 \text{ for } 0 \leqq t < s, \\ &= \Phi(t,s)\,[h(\hat{z}(s),v) - h(\hat{z}(s),\hat{u}(s))],\; s \leqq t \leqq T. \end{aligned}$$

It can then be shown [2] that the cone

$$(78)\qquad \begin{aligned} C(\hat{z},\Omega) = \Big\{ \delta z \in L \mid \delta z(t) = \sum_{i=1}^{k} \alpha_i \delta z_{s_i,v_i}(t), \{s_1, s_2, \cdots, s_k\} \subset I_1, \\ \{v_1, v_2, \cdots, v_k\} \subset U, \alpha_i \geqq 0, \text{ for } i = 1, 2, \cdots, k, \\ k \text{ arbitrary finite} \Big\}, \end{aligned}$$

is a linearization of Ω at $\hat{z}$. However, the proof of this fact is rather involved and quite long. The linear maps $f'(\hat{z})$ and $r'(\hat{z})$, which are used with this linearization are defined as follows. For every $\delta z \in L$

$$(79)\qquad f'(\hat{z})(\delta z) = P_1 \delta z(T)$$

and

$$(80)\qquad r'(\hat{z})(\delta z) = \frac{\partial g(P_2\hat{z}(T))}{\partial x} P_2 \delta z(T).$$

Applying Theorem (59) we conclude that if $\hat{z}$ is optimal, then there exists a nonzero vector $\psi = (\psi^0, \eta)$ in E^{n+1}, with $\psi^0 \leqq 0$ such that

$$(81)\qquad \psi^0 P_1 \delta z(T) + \left\langle \eta, \frac{\partial g(P_2\hat{z}(T))}{\partial x} P_2 \delta z(T) \right\rangle \leqq 0 \text{ for all } \delta z \in \overline{C(\hat{z},\Omega)}.$$

Since every $\delta z_{s,v}$, defined in (77), is in $C(\hat{z},\Omega)$, (81) implies that

$$(82)\qquad \begin{aligned} &\psi^0 P_1 \Phi(T,s)(h(\hat{z}(s),v) - h(\hat{z}(s),\hat{u}(s))) \\ &+ \langle \eta, (\partial g(P_2\hat{z}(T))/\partial x) P_2 \Phi(T,s)(h(\hat{z}(s),v) - h(\hat{z}(s),\hat{u}(s)))\rangle \leqq 0 \end{aligned}$$

for every $s \in I_1$ and $v \in U$.

Hence

$$(83)\qquad \begin{aligned} \langle \Phi^T(T,t)\,[\psi^0 P_1^T + P_2^T(\partial g^T(P_2\hat{z}(T))/\partial x)\eta], \\ h(\hat{z}(t),v) - h(\hat{z}(t),\hat{u}(t))\rangle \leqq 0 \end{aligned}$$

for every $t \in I_1$, and $v \in U$. Let $\psi(t) = (\psi^0(t), \psi^1(t), \cdots, \psi^n(t))$ be defined by

$$(84)\qquad \psi(t) = \Phi^T(T,t)(\psi^0 P_1^T + P_2^T(\partial g(P_2\hat{z}(T))/\partial x)^T \eta),$$

i.e. for almost all t in I, $\psi(t)$ satisfies the differential equation

$$\frac{d}{dt}\psi(t) = -\left(\frac{\partial h(\hat{z}(t), \hat{u}(t))}{\partial z}\right)^T \psi(t); \tag{85}$$

with

$$\psi(T) = \psi^0 P_1^T + P_2^T (\partial g(P_2\hat{z}(T))/\partial x)^T \eta. \tag{86}$$

Combining (83) and (84), we obtain

$$\langle \psi(t), h(\hat{z}(t), \hat{u}(t)) \rangle = \text{Maximum}\{ \langle \psi(t), h(\hat{z}(t), v) \rangle \mid v \in U \}, \text{ a.e.} \tag{87}$$

Thus we have proved the following theorem:

(88) MAXIMUM PRINCIPLE. *If $\hat{u}$ is an optimal control for the optimal control problem* (61) (*in basic problem form* (72), (73), (74)) *and $\hat{z}$ is a corresponding optimal trajectory, then there exists a function $\psi: I \to E^{n+1}$, not identically zero such that $\psi(t) = (\psi^0(t), \psi^1(t), \cdots, \psi^n(t))$ satisfies* (85) *and* (86), $\psi^0(t) = \psi^0 \leqq 0$, *and* (87) *is satisfied.*

This concludes our presentation of the unification which has recently taken place in the approach to constrained minimization problems.

References

1. M. Canon, C. Cullum and E. Polak, *Constrained minimization problems in finite dimensional spaces,* J. Soc. Indust. Appl. Math. Ser. A. Control **4** (1966), 528-547.
2. L. W. Neustadt, *An abstract variational theory with applications to a broad class of optimization problems.* Part I: *General theory,* J. Soc. Indust. Appl. Math. **4** (1966), 505-527; Part II: *Applications,* J. Soc. Indust. Appl. Math. **5** (1967), 90-137.
3. Pontryagin et al, *The mathematical theory of optimal processes,* Interscience, New York, 1962.
4. H. W. Kuhn and A. W. Tucker, "Nonlinear programming" in *Proc. of the second Berkeley symposium on mathematical statistics and probability,* Univ. of California Press, Berkeley, Calif., 1951, pp. 481-492.
5. F. John, "Extremum problems with inequalities as side conditions" in *Studies and essays,* Courant Anniversary Volume edited by K. O. Friedrichs, O. E. Neugebauer and J. J. Stoker, Wiley, New York, 1948, pp. 187-204.
6. H. Halkin and L. W Neustadt, *General necessary conditions for optimization problems,* USCEE Report 173, 1966.
7. N. O. Da Cunha and E. Polak, *Constrained minimization under vector valued criteria in linear topological spaces,* ERL-M191, Electronics Research Laboratory, University of California, Berkeley, California.

UNIVERSITY OF CALIFORNIA, BERKELEY

George B. Dantzig[1]

Linear Control Processes and Mathematical Programming[2]

Linear control process defined [8], [14]. We shall consider an "object" defined by its $n+1$ coordinates $X = (x_0, x_1, \cdots, x_n)$, whose "motion" described as a function of a parameter, "time" (t), can be written as a linear system of differential equations

$$dX/dt = A^t X + B^t u, \tag{1}$$

where A^t, B^t are known matrices that may depend on t and

$$u = (u_1, u_2, \cdots, u_n)$$

is a control vector that must be chosen from a convex set, $u \in U(t)$ for every $0 \leqq t \leqq T$. The time period $0 \leqq t \leqq T$ is fixed and known in advance. The coordinate $x_0 = x_0(t)$ represents the "cost" of moving the object from its initial position to $x_0(t)$. For this purpose it may be assumed that $x_0(0) = 0$. Defining

$$\overline{X} = (0, x_1, x_2, \cdots, x_n), \tag{2}$$

the object is required to start somewhere in a convex domain

[1] Operations Research House, Stanford University, Stanford, California. Research on this paper was partially supported by the National Science Foundation under Grant GP-2633 with the University of California.

[2] Reprinted with permission from SIAM J. Control, 4 (1966), pp. 56-60.

$\bar{X}(0) \in S_0$ and to terminate at $t = T$ somewhere on another convex domain $\bar{X}(T) \in S_T$.

Problem. Find $u \in U(t)$ and boundary values $\bar{X}(0) \in S_0$, $\bar{X}(T) \in S_T$, such that $x_0(T)$, the *objective*, is minimized.

Assuming $u \in U(t)$ is known, the system of differential equations can be integrated to yield an expression for $X(T)$ in terms of $X(0)$ and $u \in U(t)$. This is true in general but will be illustrated for the case when A^t and B^t do not depend on t; in this case

$$X(T) = e^{TA}X(0) + \int_0^T e^{(T-t)A}Bu(t)\,dt, \tag{3}$$

where $u(t) \in U(t)$ is a convex set and where we assume the integral exists whatever be the choice of the $u(t) \in U(t)$ for $0 \leqq t \leqq T$.

Generalized linear program. Our general objective is to see how *mathematical programming* and, in particular, how the *decomposition principle* in the form of the generalized linear program can be applied to this class of problems. An elegant constructive theory emerges, [**2**], [**10**], [**13**].

A *generalized linear program* differs from a standard linear program in that the vector of coefficients, say P, associated with any variable μ need not be constant but can be selected from a convex set C. For example:

Problem. Find $\max \lambda, \mu \geqq 0$ such that

$$U_0\lambda + P\mu = Q_0, \qquad \mu = 1, \tag{4}$$

where U_0, Q_0 are specified vectors and $P \in C$ convex.

It is assumed that the elements of C are only known implicitly (for example, as some solution to a linear program) but that particular choices of P can be easily obtained which minimize any given linear form in the components of P.

The method of solution assumes we have initially[3] on hand m particular choices $P_i \in C$ with the property that

$$\begin{aligned} U_0\lambda + P_1\mu_1 + P_2\mu_2 + \cdots + P_m\mu_m &= Q_0, \\ \mu_1 + \mu_2 + \cdots + \mu_m &= 1, \end{aligned} \tag{5}$$

[3] This is not a restrictive assumption since there is an analogous method for obtaining such a starting solution, see [**2**].

has a unique "feasible" solution; that is to say, $\lambda = \lambda^0$, $\mu_i = \mu_i^0 \geqq 0$ and the matrix

$$B^0 = \begin{bmatrix} U_0 & P_1 & \cdots & P_m \\ 0 & 1 & \cdots & 1 \end{bmatrix} \tag{6}$$

is nonsingular (i.e. the columns of B^0 form a basis). Because $P_i \in C$, the vector $P^0 = \sum P_i \mu_i^0$ constitutes a solution $P = P^0$ for (4) except that $\lambda = \lambda^0$ may not yield the maximal λ.

To test whether or not P^0 is an optimal solution, one determines a row vector $\bar{\pi} = \bar{\pi}^0$ such that

$$\bar{\pi}^0 B^0 = (1, 0, \cdots, 0), \tag{7}$$

and then a value δ and a vector $P_{m+1} \in C$ such that

$$\delta = \bar{\pi}^0 \bar{P}_{m+1} = \min_{P \in C} \bar{\pi}^0 \bar{P}, \tag{8}$$

where we denote

$$\bar{P} = \begin{bmatrix} P \\ 1 \end{bmatrix}. \tag{9}$$

If it turns out that $\delta = 0$, then $P = P^0$ is an optimal solution.

If P^0 is not optimal, system (5) is augmented by P_{m+1}. After one or several iterations k the augmented system takes the form of a linear program:

Problem. Find $\max \lambda, \mu_i \geqq 0$,

$$U_0 \lambda + \sum_1^{m+k} P_i \mu_i = Q_0, \qquad \sum_1^{m+k} \mu_i = 1. \tag{10}$$

Letting B^k denote the basis associated with an optimal basic feasible solution $\mu_i = \mu_i^k$ to (10), π^k is defined analogous to (7) and δ^{k+1} and P_{m+k+1} analogous to (8). If it turns out that $\delta = 0$, the solution

$$P^k = \sum_1^{m+k} P_i \mu_i^k \tag{11}$$

is optimal. If not the system is augmented by P_{m+k+1} and the iterative process is repeated.

It is known under certain general conditions, such as C bounded and the initial solution nondegenerate (i.e. $\mu_i^0 > 0$), that $\bar{\pi}^k \rightarrow \bar{\pi}^*$

and $P^k \to P^*$ on some subsequence k and that $P = P^*$ is optimal. The two fundamental properties of $\bar{\pi}^*$ are

$$\bar{\pi}^* \neq 0 \quad \text{and } \bar{\pi}^* \bar{P} \geqq \bar{\pi}^* \bar{P}^* = 0 \quad \text{for all } P \in C. \tag{12}$$

The entire process can be considered as constructive providing it is not difficult to compute the various P_{m+k+1} from (8) with $\bar{\pi} = \bar{\pi}^{m+k}$. For example, if C is a parallelepiped or more generally a convex polyhedral set, then $\min \bar{\pi}\bar{P}$ constitutes the minimization of a linear form with known coefficients $\bar{\pi} = \bar{\pi}^{m+k}$ subject to linear inequality constraints in the unknown components of $\bar{P}$, i.e. a linear program. In this case the iterative process terminates in a finite number of steps and P_{m+k} constitute extreme solutions from it. In all cases an estimate is available on how close the kth solution is to an optimal value of λ.

Application of the generalized program to the linear control process. Let us denote

$$P = \int_0^T e^{(T-t)A} Bu(t)\, dt, \tag{13}$$

and note that P is an element of a convex set C_μ generated by choosing all possible $u(t) \in U(t)$. We specify that $U_0 = (1, 0, \cdots, 0)$, and denote by $\lambda = -X_0(T)$, where $X_0(T)$ is the coordinate of $X(T)$ to be minimized. Then

$$X(T) = -U_0\lambda + \bar{X}(T). \tag{14}$$

We further define Q_0 by

$$\bar{X}(T) = e^{TA}X(0) + Q_0. \tag{15}$$

Substitution of these into (3) formally converts[4] the integrated form of the control problem into a generalized linear program (4).

Each cycle of the iterative process yields a known row vector, which we partition

$$\bar{\pi}^{k+1} = [\pi, \theta], \tag{16}$$

where π represents its first $n+1$ components corresponding to P and θ its last component. Since π is known, our choice for P_{m+k+1} is

[4] Actually Q_0 is not given but is an element of a convex set. To simplify the discussion which follows we assume Q_0 is a fixed vector.

(17)
$$\pi P_{m+k+1} = \min_{u \in U(t)} \left\{ \int_0^T \pi e^{(T-t)A} Bu(t)\, dt \right\} = \int_0^T \left\{ \min_{u \in U(t)} \pi e^{(T-t)A} Bu(t) \right\} dt,$$

where clearly the minimum is obtained when, in (17), the integrand for each t is selected to be minimum.

Note that

(18)
$$\phi_{t,\pi} = \pi e^{(T-t)A} B$$

is a row vector that can be computed for each t. For example, $\phi_{t,\pi}$ can be represented by a finite sum of vectors whose weights depend on t and the eigenvalues of A.[5] The new extremal solution P_{m+k+1} is obtained by choosing the control which minimizes the linear form in u for each t; i.e. find

(19)
$$\min(\phi_{t,\pi} u), \qquad u \in U(t).$$

For example, if $U(t)$ is a polyhedral set then (19) is a linear program. If $U(t)$ is the same for all t, then only the objective form, $\phi_{t,\pi} u$, varies for different t; except for the objective form the linear programs are the same for all t.

If optimal π^* is used, then the optimal control u (except for a set of measure zero) satisfies

(20)
$$\min[\phi^*(t) u], \qquad u \in U(t),$$

where $\phi^*(t) = \pi^* e^{(T-t)A} B$. Pontryagin refers to this as the *maximal principle*. It is, as we have just shown, also a consequence of the decomposition principle of linear programming [**9**].

Conclusion. In our approach the general control obtained for each cycle is a linear combination of exactly $n + 1$ special controls obtained by minimizing for each t, the linear expression (19) in u for $n + 1$ choices of π. These special controls may be referred to as extreme controls. The latter each in themselves do not maintain feasibility, that is to say, guarantee that the object will move from $\overline{X}(0)$ to $\overline{X}(T)$. Each new linear combination of these special

[5] If the roots λ_i are real and distinct, $e^{tA} = \sum_0^n M_i e^{\lambda_i t}$ where M_i is independent of t. [F. T. Smith, RAND RM-3526-PR, February, 1963].

controls will, however, generate a new feasible control with a lower value[6] for the total cost $X_0(T)$. Under the conditions stated this iterative process is known to converge [**2**, p. 471].

References

1. R. Bellman, *Adaptive control processes; A guided tour,* Princeton Univ. Press, Princeton, N. J., 1961.

2. G. B. Dantzig, *Linear programming and extensions,* Princeton Univ. Press, Princeton, N. J., 1963.

3. A. F. Filippov, *On certain questions in the theory of optimal control,* English transl., J. Siam Control **1** (1962), 76-84.

4. H. Halkin, *On the necessary conditions for optimal control of nonlinear systems,* J. Analyse Math. **12** (1964), 1-82.

5. J. P. LaSalle, "The time optimal control problem" in *Contributions to the theory of nonlinear oscillations,* Vol. V, Princeton Univ. Press, Princeton, N. J., 1958.

6. J. P. LaSalle and S. Lefschetz, *Stability by Liapunov's direct method,* Academic Press, New York, 1961.

7. G. Leitmann, ed., *Optimization techniques,* Academic Press, New York, 1962.

8. L. W. Neustadt, "Discrete time optimal control systems" in *Nonlinear differential equations and nonlinear mechanics,* edited by J. P. LaSalle and S. Lefschetz, Academic Press, New York, 1963.

9. L. S. Pontryagin, V. G. Boltyanskii, R. V. Gamkrelidze and E. F. Mishchenko, *The mathematical theory of optimum processes,* Interscience, New York, 1962.

10. B. H. Whalen, *Linear programming for optimal control,* Ph. D. dissertation, University of California, Berkeley, 1962.

11. ———, *On linear programming and optimal control,* Correspondence to IRE Trans. on Automatic Control, AC-7 (1962), p. 46.

12. R. M. Van Slyke, *Mathematical programming,* Ph.D. dissertation, University of California, Berkeley, 1965.

13. L. A. Zadeh, *A note on linear programming and optimal control,* Correspondence to IRE Trans. on Automatic Control, AC-7 (1962), p. 46.

14. L. A. Zadeh and C. A. Desoer, *Linear system theory, the state-space approach,* McGraw-Hill, New York, 1963.

Stanford University

[6] If basic solution is nondegenerate.

J. B. Rosen

Numerical Solution of Optimal Control Problems[1]

Summary.

I. The discrete optimal control problem to be considered is as follows:

Let $x_i \in E^n$ denote the state vector at time t_i, and $u_i \in E^r$ the corresponding control vector. The system dynamics are given by

$$x_{i+1} = x_i + f(x_i, u_i), \quad i = 0, \cdots, m-1, \tag{1.1}$$

where the controls u_i must be selected so that

$$\begin{aligned} u_i &\in U_i \subset E^r, \quad i = 0, 1, \cdots, m-1, \\ x_i &\in X_i \subset E^n, \quad i = 0, 1, \cdots, m. \end{aligned} \tag{1.2}$$

It is assumed that the sets X_i and U_i are compact and convex, and that f is a real valued function, continuous on $X_i \times U_i$ for such i. We call the sequence $\{x_i, i = 0, \cdots, m\}$ the state *trajectory*, the sequence $\{u_i, i = 0, \cdots, m-1\}$ the *control*, and we denote by $z = \{x_i, u_i\}$ the direct product of these two sequences. We say that z is admissible if $\{x_i\}$ and $\{u_i\}$ satisfy (1.1) and (1.2). Note that we can specify the initial and terminal values x_0 and x_m by setting $X_0 = x_0$ and $X_m = x_m$.

[1] This research was sponsored in part by NSF Research Grant GP 6070.

We are given a real valued function σ, assumed to be continuous on $U_i \times X_i$, for each i, and we define

$$\phi(z) \equiv \sum_{i=0}^{m-1} \sigma(x_i, u_i). \tag{1.3}$$

We wish to find an admissible z^* such that $\phi(z)$ attains its minimum, over all admissible z, at $z = z^*$.

Now let us consider the following continuous optimal control problem. Let $x(t)$ and $u(t)$ satisfy

$$\begin{aligned} &\dot{x} \equiv dx/dt = \bar{f}(x,u) \\ &u(t) \in U(t) \qquad\qquad t \in [0, T]. \\ &x(t) \in X(t) \end{aligned} \tag{1.4}$$

Find $x^*(t)$ satisfying (1.4) such that

$$\phi[u] = \int_0^T \bar{\sigma}(x(t), u(t))\, dt \tag{1.5}$$

attains its minimum over all $x(t)$ and $u(t)$ which satisfy (1.4). Suppose that we choose a finite difference step $\Delta t = T/m$, and use the simplest approximation $\dot{x}(i\Delta t) = (x_{i+1} - x_i)/\Delta t$. We also evaluate the integral (1.5) by the trapezoidal rule and let $f = \Delta t \bar{f}$ and $\sigma = \Delta t \bar{\sigma}$. We then formally obtain the equivalent discrete problem (1.1), (1.2) and (1.3).

The terminal time T is assumed to be specified in the continuous problem as given by (1.4) and (1.5). We can however put a problem with variable terminal time into this fixed time formulation by introducing an additional state variable. To illustrate this, suppose the variable time problem is given by

$$dy/d\tau = g(y,u), \quad y,g \in E^{n-1},$$

$$\int_0^{\bar{\tau}} \eta(y,u)\, d\tau = \text{min}, \quad \bar{\tau} \text{ variable}.$$

We introduce a new state variable $\xi > 0$, and let $\tau = \xi t$, $t \in [0, 1]$. We require that ξ satisfy $\dot{\xi} = 0$ (i.e., $\xi = \text{const.}$), with its initial value $\xi(0)$ to be determined. If we define vectors in E^n by

$$x = \begin{pmatrix} y \\ \xi \end{pmatrix}, \qquad \bar{f} = \begin{pmatrix} \xi g \\ 0 \end{pmatrix}$$

and let $\bar{\sigma}(x,u) = \xi\eta(y,u)$, the resulting problem given by (1.4) and (1.5) with $T = 1$ is equivalent to the variable time problem.

It should also be remarked that explicit dependence on t of $\bar{\sigma}$ and $\bar{f}$ can be handled with no essential difficulty. Such dependence leads to functions σ_i and f_i in (1.1) and (1.3) which depend explicitly on the index i. To simplify the presentation, we will not consider such dependence.

II. We will now show that the discrete optimal control problem can be considered as a mathematical programming problem (in general, nonlinear) with a special structure [1]. We let $s = mr + (m+1)n$, and consider the vector z in the product space E^s. We denote by $Z \subset E^s$, the compact, convex subset

$$Z = \left\{ z \;\middle|\; \begin{matrix} u_i \in U_i, & i = 0, 1, \cdots, m-1 \\ x_i \in X_i, & i = 0, 1, \cdots, m \end{matrix} \right\}. \tag{2.1}$$

We also define a vector mapping v: $E^s \to E^{mn}$, so that the recursion relations (1.1) are all given by $v(z) = 0$. That is, we define $v_{i+1} \in E^n$, by

$$v_{i+1} \equiv f(x_i, u_i) + x_i - x_{i+1}, \quad i = 0, 1, \cdots, m-1 \tag{2.2}$$

and we let the components $in + 1, in + 2, \cdots, (i+1)n$ of v be given by v_{i+1}. The discrete optimal control problem can now be stated as that of finding a z^* which solves the mathematical programming problem

$$\min_z \left\{ \phi(z) \;\middle|\; \begin{matrix} z \in Z \\ v(z) = 0 \end{matrix} \right\} \tag{2.3}$$

where $\phi(z)$ is given by (1.3).

The admissible (feasible) set $S \subset E^s$ is given by

$$S = \left\{ z \;\middle|\; \begin{matrix} z \in Z \\ v(z) = 0 \end{matrix} \right\}. \tag{2.4}$$

The set S may be empty, in which case no control and corresponding trajectory exist which satisfy (1.1) and (1.2). In many practical situations the existence of an admissible control and trajectory with the given dynamics and imposed constraints is the primary question. If no admissible solution exists, it is necessary to relax the control constraints (by increasing the allowable range on some of the controls, for example) or relax the state constraints (by increasing the size of the target set X_m, for example) before the determination of an optimum solution can be considered. In some cases an admissible solution may also be achieved by an appropriate modification of the system dynamics. In any event, the determination of whether or not an admissible solution exists has been reduced to finding any feasible solution to the problem (2.3).

If S is not empty, it is a compact set, so that ϕ attains its minimum on S. If the null space of v is convex, then S is also convex. If ϕ is convex on Z (σ convex on $X_i \times U_i$), then (2.3) is a convex programming problem for which both necessary and sufficient optimality conditions can be stated, and for which efficient computational methods of solution are available. It follows from (2.2) that for linear f, $f = Ax + Bu + q$, where A and B are matrices and q is a constant vector, the null space of v will be a linear manifold (and therefore convex). Thus linear dynamics and a convex functional lead to a reasonably well understood convex problem. In general, however, if f is not linear, the set S will not be convex. Necessary optimality conditions will be given for S nonconvex, but in general, conditions which are also sufficient are not known for such problems. Furthermore, for nonconvex S a problem may have many constrained local minima even with ϕ linear. Thus even if a method finds a local minimum such a minimum may be far from the desired global minimum.

III. We will now consider optimality conditions for the problem (2.3), and use these to obtain the adjoint equations and a "minimum principle" for the discrete optimal control problem. We assume that ϕ and v are continuously differentiable in Z, and that S is nonempty. We denote by v_z the Jacobian matrix of a function v, and the transpose of p by p', so that for example, $p'p$ denotes an inner product.

SUFFICIENCY THEOREM. *Let ϕ be convex and v be linear on Z. A sufficient condition that $z^* \in S$ solves* (2.3) *is that there exists $p \in E^{mn}$ such that*

$$[\phi_z(z^*) + p'v_z(z^*)](z - z^*) \geqq 0, \quad \forall\, z \in Z. \tag{3.1}$$

PROOF. Since ϕ is convex and v linear, the function $\psi = \phi + p'v$ is convex on Z. Then for every $z \in Z$,

$$\psi(z) - \psi(z^*) \geqq \psi_z(z^*)\,[z - z^*] \geqq 0,$$

by (3.1). Since $z^* \in S$, $v(z^*) = 0$, so that we have

$$\phi(z) - \phi(z^*) \geqq -p'v(z) = 0, \quad \forall\, z \in S.$$

Thus, z^* is a global minimum on S.

NECESSITY THEOREM. *Assume that the compact, convex set* Z *has a nonempty interior. Let* z^* *solve* (2.3). *Then there exists a scalar* $\mu \geqq 0$, *and* $p \in E^{mn}$, *not both zero, such that*

$$\Phi_z(z^*)(z - z^*) \geqq 0, \quad \forall\, z \in Z, \tag{3.2}$$

where

$$\Phi(z) \equiv \mu\phi(z) + p'v(z). \tag{3.3}$$

The proof of this theorem is too long to be included here, and is given in [2]. It should also be noted that if an appropriate constraint qualification is satisfied we can choose $\mu = 1$.

If we restrict the functions ϕ and v as in the sufficiency theorem, we obtain a *Minimum principle*.

Let ϕ *be convex and* v *linear on* Z, *and let* z^* *solve* (2.3). *Then there exist multipliers* $\mu \geqq 0$, *and* p, *not both zero, such that* $\Phi(z)$ *attains its minimum over* $z \in Z$ *at* z^*, *where* $\Phi(z)$ *is given by* (3.3). *That is*

$$\Phi(z^*) \leqq \Phi(z), \quad \forall\, z \in Z. \tag{3.4}$$

The proof follows immediately from the convexity of Φ and (3.2).

We are now in a position to apply these results directly to the discrete optimal control problem. If we denote by $p_i \in E^n$, the multipliers corresponding to v_i, we obtain from (1.3) and (3.3)

$$\Phi(z) = \mu \sum_{i=0}^{m-1} \sigma(x_i, u_i) + \sum_{i=0}^{m-1} p'_{i+1}[f(x_i, u_i) + x_i - x_{i+1}]. \tag{3.5}$$

We will let $f^*_{xi} \equiv f_x(x^*_i, u^*_i)$, etc., $p_0 \equiv 0$, and I be the $n \times n$ unit matrix. Then the necessary condition (3.2) can be written

$$[\mu\sigma^*_{xi} + p'_{i+1}(I + f^*_{xi}) - p'_i](x_i - x^*_i) \geqq 0, \quad \forall\, x_i \in X_i; \quad i = 0, 1, \cdots, m-1, \tag{3.6}$$

$$-p'_m(x_m - x^*_m) \geqq 0, \quad \forall\, x_m \in X_m, \tag{3.7}$$

$$[\mu\sigma^*_{ui} + p'_{i+1}f^*_{ui}](u_i - u^*_i) \geqq 0, \quad \forall\, u_i \in U_i; \quad i = 0, 1, \cdots, m-1. \tag{3.8}$$

If we assume that the initial state is specified as a point, i.e., $X_0 = \bar{x}_0$, then $x^*_0 = \bar{x}_0$ and (3.6) is satisfied for $i = 0$. Now suppose further that no state constraints are active except for the terminal constraint, i.e., $x^*_i \in \operatorname{int} X_i$, $i = 1, \cdots, m-1$. Then the expression $[\cdot]$ in (3.6) must vanish, since otherwise we could choose some

$x_i \in X_i$ such that (3.6) would be violated. That is

(3.9) $\quad p_i' - p_{i+1}' = p_{i+1}' f_{xi}^* + \mu \sigma_{xi}^*, \quad i = m-1, m-2, \cdots, 1.$

The multipliers p_i are therefore determined by the recursion relation (3.9) starting with a vector p_m satisfying (3.7). For x_m^* on the boundary of X_m, p_m must be parallel to the (outward) normal vector to a supporting hyperplane at x_m^*. The recursion relation (3.9) is seen to be a finite difference approximation to the adjoint differential equation

$$-\dot{p}' = p'\bar{f}_x + \mu\bar{\sigma}_x \tag{3.10}$$

for the continuous optimal control problem.

Let us now define

$$H(x, u, p, q) = p' f(x, u) + (p - q)' x + \mu \sigma(x, u). \tag{3.11}$$

By rearranging terms we obtain from (3.5)

$$\Phi(z) = \sum_{i=0}^{m-1} H(x_i, u_i, p_{i+1}, p_i) - p_m x_m. \tag{3.12}$$

For σ convex and f linear on $X_i \times U_i$, the previous minimum principle applies. Therefore, by (3.4), the optimal trajectory $\{x_i^*\}$ and control $\{u_i^*\}$ must satisfy the following:

Discrete minimum principle. If σ is convex and f is linear on $X_i \times U_i$, and $\{x_i^*\}$ and $\{u_i^*\}$ are optimal, then

$$H(x_i^*, u_i^*, p_{i+1}, p_i) \leqq H(x_i, u_i, p_{i+1}, p_i), \quad \forall\ x_i \in X_i,\ u_i \in U_i, \quad i = 0, 1, \cdots, m-1 \tag{3.13}$$

where the adjoint vectors p_i are determined by (3.7) and (3.9).

IV. In order to describe the computational solution of the discrete optimal control problem we first consider a linear recursion relation with $f = Ax + Bu$, and also give a more explicit statement of the constraint sets X_i and U_i. To simplify the discussion we will assume that $X_0 = \bar{x}_0$, and that the X_i and U_i are polyhedral sets defined by the linear inequalities

$$\begin{aligned} X_i &= \{x \mid G_i' x - \bar{g}_i \geqq 0\}, \quad i = 1, \cdots, m \\ U_i &= \{u \mid H' u - \bar{h} \geqq 0\}, \quad i = 0, \cdots, m-1 \end{aligned} \tag{4.1}$$

where G_i and H are constant matrices and $\bar{g}_i$ and $\bar{h}$ are constant vectors. If these sets are just upper and lower bounds we have $G_i = (I_n \ -I_n)$ and $H = (I_r \ -I_r)$, where I_n and I_r are unit matrices of order n and r, respectively.

The problem (1.1)—(1.3) now becomes

$$(4.2)\quad \min_{x_i, u_i} \left\{ \sum_{i=0}^{m-1} \sigma(x_i, u_i) \;\middle|\; \begin{array}{l} x_0 = \bar{x}_0 \\ x_{i+1} = (I + A)x_i + Bu_i, \ i = 0, \cdots, m-1 \\ G_i' x_i - \bar{g}_i \geqq 0, \ i = 1, \cdots, m \\ H' u_i - \bar{h} \geqq 0, \ i = 0, \cdots, m-1 \end{array} \right\}.$$

If σ is convex, this is a convex programming problem with linear equality and inequality constraints. It consists of $m(n + r)$ variables, mn equality constraints, and $2m(n + r)$ inequality constraints if they are bounds. This could be solved directly, but will be a large problem if m is large. A more efficient method of solution is to use the linear equations to eliminate the state vectors x_i and map the entire problem into the control space. This is done by means of the following

Lemma. *Let x_i satisfy the recursion relation*

$$(4.3)\qquad x_{i+1} = K_i x_i + b_i, \quad i = 0, 1, \cdots, m-1$$

with x_0 specified and K_i nonsingular. Then

$$(4.4)\qquad x_i = Y_i x_0 + Y_i \sum_{j=0}^{i-1} \Lambda_{j+1}' b_j, \quad i = 1, \cdots, m$$

where the matrices Y_i and Λ_i satisfy

$$(4.5)\qquad Y_{i+1} = K_i Y_i, \quad Y_0 = I, \quad i = 0, 1, \cdots, m-1$$

and

$$(4.6)\qquad \Lambda_i' = \Lambda_{i+1}' K_i, \quad \Lambda_m' = Y_m^{-1}, \quad i = m-1, \cdots, 1.$$

If we apply this to the equality constraints with $K_i = I + A$, and $b_i = Bu_i$ we obtain

$$(4.7)\qquad x_i = Y_i x_0 + \sum_{j=0}^{i-1} Q_{ij}' u_j, \quad i = 1, \cdots, m$$

where $Q_{ij}' = Y_i \Lambda_{j+1}' B$. The important point to note is that given the matrices A and B, we can explicitly compute the matrices Y_i and Q_{ij} by operations with $(n \times n)$ matrices. Since n (the dimensionality of the state vector) is usually small compared to m, this is an efficient computation.

We now use (4.7) to eliminate the x_i from the state constraints and $\sigma(x_i, u_i)$ in (4.2). The state constraints can then be written

$$\sum_{j=0}^{i-1} D'_{ij} u_j - d_i \geqq 0, \quad i = 1, \cdots, m \tag{4.8}$$

where $D'_{ij} = G'_i Q'_{ij}$ and $d_i = \bar{g}_i - G'_i Y_i x_0$. A similar substitution in $\sigma(x_i, u_i)$ gives an objective function $\rho(u)$ depending on the controls $u = \{u_i\}$, only. If σ is convex, $\rho(u)$ will also be convex because of the linearity of (4.7). The problem (4.2) has therefore been reduced to

$$\min_u \left\{ \rho(u) \;\middle|\; \begin{array}{ll} \sum_{j=0}^{i-1} D'_{ij} u_j \geqq d_i, & i = 1, \cdots, m \\ H' u_i \geqq \bar{h}, & i = 0, \cdots, m-1 \end{array} \right\}. \tag{4.9}$$

This reduced problem has mr variables, no equalities, and the same number of inequalities as (4.2). In most control problems $r < n$, so that we have cut the problem size at least in half. Since there are more inequality constraints than variables, and no nonnegativity requirements, this is best solved by a convex method in the dual space (such as gradient projection). Computational efficiency is also improved by taking advantage of the upper triangular structure of the first $2mn$ constraints and the block diagonal structure of the remaining $2mr$ constraints.

In the important case where σ is linear, (4.9) should be considered as the unsymmetric dual problem. The equivalent primal, with mr rows is then efficiently solved by any standard LP routine, which of course gives the desired dual variables (the controls) as the elements of the pricing vector. If we let ν_m denote the primal activity vector corresponding to the cost vector d_m, then the terminal adjoint vector p_m is given by $p_m = G_m \nu_m$. All the remaining adjoint variables are then given by (3.9). For a more complete discussion of the recursion relation problem, see [**1**].

V. The method of the previous section can only be applied directly when the system dynamics are described by a linear recursion relation. However, with appropriate convexity requirements on f, the nonlinear problem can be solved by an iterative solution of linearized problems. At each iteration the function f is linearized about the previous state and control, and a linearized problem of the kind discussed above is solved. This method is fully described in [**3**] and [**4**].

References

1. J. B. Rosen, *Optimal control and convex programming,* Proc. IBM Sympos. on Control Theory and Appl., Yorktown Heights, N. Y., 1964, pp. 223-237.

2. O. L. Mangasarian, *Nonlinear programming,* McGraw-Hill, New York, 1969, pp. 168-169.

3. J. B. Rosen, *Iterative solution of nonlinear optimal control problems,* J. Soc. Indust. Appl. Math. Ser. A. Control 4 (1966), 223-244.

4. Robert Meyer, *The validity of a family of optimization methods,* J. Soc. Indust. Appl. Math. Ser. A. Control 8 (1970).

UNIVERSITY OF WISCONSIN

VII
MATHEMATICAL ECONOMICS

H. W. Kuhn

Lectures on Mathematical Economics

Introduction. The lectures which follow were presented at the American Mathematical Society Summer Institute on the Mathematics of the Decision Sciences at Stanford University on July 13, 14, 17, 18, and 19, 1967. The first two lectures deal with some duality concepts of mathematical programming and their application to various problems in economic theory. The last three lectures, which stand as an independent unit, treat a selection of problems from the mathematical theory of international trade. A cumulative bibliography appears at the end of the five lectures.

With the exception of the addition of details that were not given in the lectures due to time constraints, the manuscript follows the spoken version quite closely. I am indebted to Mr. Alan Kirman, whose notes made this possible.

Lecture 1. The theory of optimization has a long and rich history within mathematics, reaching back at least to the 27th Theorem of the 6th Book of Euclid. However, for our purposes, it is useful to oversimplify this history by contrasting two principal trends. The first trend begins with the discovery of the calculus in the 17th century and continues to the present with applications to problems drawn largely from the physical sciences and engineering. The second trend is the intense development of the theory of constrained optimization during the past 30 years and the associated growth of the subject of Mathematical Programming. The problems falling in this area may be interpreted as problems in economics since mathematical programming is concerned with *the allocation of scarce resources so as to fulfill certain requirements while optimizing some objective function*. It has had an impressive record of practical success in solving real problems.

The pattern of development of mathematical programming has been typical of much current applied mathematics, namely, the bulk of the theory was done quite quickly (or was available through the prior work of pure mathematicians) and the subsequent emphasis has been on computational methods directed at applications. This is only partly true. The theory of mathematical programming is largely the theory of duality and it is true that the foundations had been laid by the early 1950's. However, to say that the theory of duality is complete is to ignore the important work of Rockafellar [**27**] and others, who are extending its range and hence its applicability.

Duality has two faces. On the one hand, it has contributed significantly to computational methods. For instance, the full power of the Simplex Method can only be understood if one keeps the pair of dual programs in evidence. The class of techniques known as primal-dual methods, beginning with the Hungarian Method [**19**], including the work of Ford and Fulkerson [**10**] on the Transportation Problem, and culminating with the elegant work of Balinski and Gomory [**2**], all originated in explicit dual formulations. On the other hand, duality has contributed almost all of the significant economic interpretations.

Therefore, my main theme in these first two lectures will be duality in various forms, with particular emphasis on economic applications. The mathematical level has been chosen to emphasize the ideas behind some well-known results and to apply them in unusual contexts, rather than to extend their generality mathematically.

The various conditions for a constrained optimum that can be derived from Lagrange multiplier techniques (including the so-called Kuhn-Tucker conditions) go hand-in-hand with duality. The use of these conditions in the solution of specific computational problems is relatively direct. Indeed, with the difference that we are solving systems of equations and inequalities instead of equations alone, it parallels the classical applications of Lagrange multipliers.

The application of the conditions to qualitative economics is much less direct. Here, they enable us to derive general conditions about the nature of the solutions to optimization problems, even when we deal with rather general functions, about which we may only have qualitative information. *This may be the most powerful technique provided to economic theory by mathematical programming.*

To understand this somewhat better, it should be recalled that there is no closed analytic expression for the solution of a general programming problem. This may be contrasted with the situation in traditional calculus problems, where there is often an explicit solution in terms of equations, possibly involving derivatives. (A typical example of this is the derivation of the so-called Slutsky equations in the theory of consumer demand [**15**].) However, we can use the multiplier conditions to derive general conclusions about the behavior of the solutions to nonnumerical problems. After we have developed the necessary machinery, several examples will illustrate this.

The simple setting of linear programming gives us a natural introduction to the derivation of necessary conditions for a constrained maximum. Let X denote an n-vector; then the *feasible set* for a linear program can be given as

$$S = \{X \mid N_k \cdot X \geqq n_k \text{ for } k \in K\}$$

where the "normals" N_k are n-vectors, n_k are scalars, $N_k \cdot X$ denotes the inner product of N_k and X, and K is a finite set of indices. Let the objective function be $g(X)$, a linear function of X. Thus, we have a prototype linear program:

$$\text{Maximize } g(X) \text{ subject to } X \in S.$$

For making local comparisons of the values of g near a point $\overline{X} \in S$, the following definition is useful.

DEFINITION 1.1. The vector E *enters* S *from* $\overline{X}$ if $\overline{X} + tE \in S$ for $0 \leqq t < \epsilon$ and some ϵ.

For the linear case, we have the following simple characterization of entering vectors:

LEMMA 1.1. *The vector E enters S from $\overline{X}$ if and only if E makes a nonnegative inner product with all of the inward pointing normals to the boundary of the feasible set at $\overline{X}$.*

PROOF. Let $\overline{X} \in S$ and let $N_k \cdot \overline{X} = n_k$ for $k \in \overline{K}$ (possibly empty). Let E enter S from $\overline{X}$. Then there exists an $\epsilon > 0$ such that $N_k \cdot (\overline{X} + tE) \geqq n_k$ for $k \in K$ and $0 \leqq t < \epsilon$. If $k \in \overline{K}$ then $N_k \cdot \overline{X} = n_k$ and hence

$$N_k \cdot E \geqq 0 \quad \text{for all } k \in \overline{K}.$$

On the other hand, suppose $N_k \cdot E \geqq 0$ for all $k \in \overline{K}$. Then, for $k \in K - \overline{K}$, $N_k \cdot \overline{X} > n_k$ and hence

$$N_k \cdot (\overline{X} + tE) = N_k \cdot \overline{X} + tN_k \cdot E \geqq n_k \quad \text{for } t \text{ small enough.}$$

For $k \in \overline{K}$, $N_k \cdot \overline{X} = n_k$ and hence

$$N_k \cdot (\overline{X} + tE) = n_k + tN_k \cdot E \geqq n_k \quad \text{for all } t \geqq 0. \qquad \text{Q.E.D.}$$

The other important ingredient for the linear case is given by the following lemma, in which $\nabla g(\overline{X})$ denotes the gradient of g, evaluated at $\overline{X}$.

LEMMA 1.2. *If $\overline{X}$ maximizes $g(X)$ for $X \in S$, then $-\nabla g(\overline{X}) \cdot E \geqq 0$ for all vectors E entering S from $\overline{X}$.*

PROOF. This follows directly from the fact that $g(\overline{X} + tE) - g(\overline{X}) \leqq 0$ for $0 \leqq t < \epsilon$ for some ϵ and hence

$$\nabla g(\overline{X}) \cdot E = \lim_{t \to 0+} \frac{g(\overline{X} + tE) - g(\overline{X})}{t} \leqq 0. \qquad \text{Q.E.D.}$$

Lemmas 1.1 and 1.2 provide exactly the hypotheses of Farkas' Lemma [9]. Namely, if $\overline{X}$ maximizes $g(X)$ for $X \in S$, then $-\nabla g(\overline{X}) \cdot E \geqq 0$ for all E such that $E \cdot N_k \geqq 0$ for $k \in \overline{K}$. We may conclude that

$$-\nabla g(\overline{X}) = \sum_{k \in \overline{K}} \overline{p}_k N_k = \sum_{k \in K} \overline{p}_k N_k$$

where $\overline{p}_k \geqq 0$ for $k \in \overline{K}$ and $\overline{p}_k = 0$ for $k \in K - \overline{K}$. We restate this result as a theorem, setting $f_k(X) = N_k \cdot N - n_k$ and noting $N_k = \nabla f_k(X)$.

THEOREM 1.1. *Suppose $\overline{X}$ maximizes $g(X)$ subject to $X \in S = \{X \mid f_k(X) \geqq 0 \text{ for } k \in K\}$. Then there exist $\overline{p}_k \geqq 0$ such that*

$$\nabla g(\overline{X}) + \sum_{k \in K} \overline{p}_k \nabla f_k(\overline{X}) = 0, \tag{1.1}$$

$$\sum_{k \in K} \overline{p}_k f_k(\overline{X}) = 0. \tag{1.2}$$

It must be emphasized that we have proved this theorem only for the linear case at this point. We may restate the theorem in terms of a "Lagrangian" function as follows:

Form $l(X) = g(X) + \sum_{k \in K} p_k f_k(X)$. If $\overline{X}$ maximizes $g(X)$ subject to $f_k(X) \geqq 0$ for $k \in K$ then there exist $\overline{p}_k \geqq 0$ such that

$$\nabla l(\overline{X}) = 0, \tag{1.3}$$

$$\sum_{k \in K} \overline{p}_k f_k(\overline{X}) = 0. \tag{1.4}$$

For the case of inequality constraints, it is the "orthogonality condition" (1.4) that carries the main content in most applications. Since both the multipliers $\overline{p}_k$ and the constraint function values $f_k(\overline{X})$ are nonnegative, this condition is a compact statement of the requirement that, of each pair, at least one must be zero.

As simple as this argument has been, it is strong enough to show the principal duality theorem of linear programming. Namely, consider the linear program

$$\text{Maximize } C \cdot X \text{ subject to } B - AX \geqq 0, \quad X \geqq 0.$$

(Here C and X are n-vectors, B is an m-vector, and A is an m by n matrix.) Suppose $\overline{X}$ solves this problem. By our previous result, we should form

$$l(X) = C \cdot X + P \cdot (B - AX) + Q \cdot X$$

and conditions (1.3) and (1.4) become

$$C - \overline{P}A + \overline{Q} = 0, \tag{1.5}$$

$$\overline{P}(B - A\overline{X}) + \overline{Q} \cdot \overline{X} = 0, \tag{1.6}$$

for some $\overline{P} \geqq 0$ and $\overline{Q} \geqq 0$. Consider the dual program:

$$\text{Minimize } P \cdot B \text{ subject to } PA - C \geqq 0, \ P \geqq 0.$$

Clearly $\overline{P}$ is feasible for this program by (1.5). For any pair of feasible vectors, X and P, for the dual programs, we have

$$P \cdot B - C \cdot X = P \cdot (B - AX) + (PA - C) \cdot X \geqq 0,$$

while, for the particular feasible pair, $\overline{X}$ and $\overline{P}$,

$$\overline{P} \cdot B - C \cdot \overline{X} = \overline{P} \cdot (B - A\overline{X}) + \overline{Q} \cdot \overline{X} = 0,$$

by (1.5) and (1.6). Hence $\overline{P}$ solves the minimum program and $\overline{P} \cdot B = C \cdot \overline{X}$ establishes the equality of the objective functions in the solutions to the dual programs. This well-known result is the heart of duality theory for linear programming.

Our treatment of the linear case suggests the conjecture that Theorem 1.1 might extend, without change, to the case in which

g and f_k are general nonlinear differentiable functions. The following example (drawn from [22]) illustrates the difficulty that is encountered.

EXAMPLE 1.1. Maximize $g(x_1, x_2) = x_1$ subject to $x_1 \geqq 0$, $x_2 \geqq 0$, $x_3 = (1 - x_1)^3 - x_2 \geqq 0$.

The Lagrangian is

$$l(x_1, x_2) = x_1 + p_1 x_1 + p_2 x_2 + p_3[(1 - x_1)^3 - x_2]$$

and conditions (1.3) are

$$l_{x_1} = 1 + p_1 - 3p_3(1 - x_1)^2 = 0, \qquad l_{x_2} = p_2 - p_3 = 0.$$

If $p_3 > 0$ then $p_2 > 0$ and hence $x_2 = x_3 = 0$. This implies $x_1 = 1$ and hence $1 + p_1 = 0$ which is a contradiction. If $p_3 = 0$ we have $1 + p_1 = 0$, the same contradiction. Hence, the necessary conditions for a maximum cannot be satisfied but the problem has the obvious solution $\overline{x}_1 = 1$, $\overline{x}_2 = 0$, $\overline{x}_3 = 0$ as illustrated by the following graph:

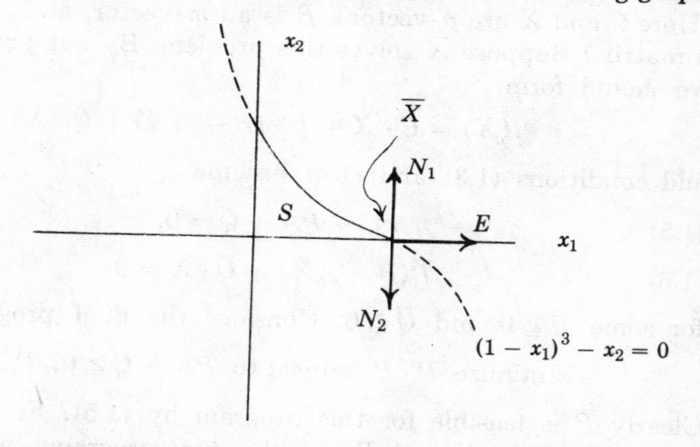

Of course, the difficulty arises in this nonlinear case because the vectors entering S from $\overline{X}$ do not coincide with the vectors E such that $E \cdot N_k \geqq 0$ for all inward pointing normals N_k. The constraints defining S give $N_1 = (0, 1)$ and $N_2 = (0, -1)$ as the inward pointing normals at $\overline{X}$. $E = (1, 0)$ makes a zero inner product with both and yet does not enter S from $\overline{X}$.

The difficulty uncovered by this example is removed in a rather direct (and, perhaps, too obvious) way by the following sequence of definitions. We first weaken the conditions on an entering vector with the nonlinear case in mind.

DEFINITION 1.2. The vector E *enters* S from $\overline{X}$ if there is a function $X(\cdot)$ such that $X(0) = \overline{X}$, $X(t) \in S$ for $0 \leqq t < \epsilon$ and $(dX/dt)_{t=0+} = E$.

This definition coincides with the previous definition if S is defined by linear inequalities and, if $\overline{X}$ maximizes $g(X)$ for $X \in S$, it preserves the property that $-\nabla g(\overline{X}) \cdot E \geqq 0$ for all vectors E entering S from $\overline{X}$. We next make precise what is meant by the inward pointing normals at $\overline{X}$.

DEFINITION 1.3. If $S = \{X \mid f_k(X) \geqq 0 \text{ for } k \in K\}$, where the f_k are differentiable functions, and if $\overline{X} \in S$, then the *inward normals at* $\overline{X}$ are the vectors $N_k = \nabla f_k(\overline{X})$ for those k with $f_k(\overline{X}) = 0$.

Of course, this definition coincides with the definition in the linear case. We now rule out disagreeable examples by fiat.

DEFINITION 1.4. Let $S = \{X \mid f_k(X) \geqq 0 \text{ for } k \in K\}$ and let $\overline{X} \in S$. Then the feasible set S (as presented by the constraints $f_k(X) \geqq 0$) satisfies the *constraint qualification* at $\overline{X}$ if $N_k \cdot E \geqq 0$ for all inward normals N_k at $\overline{X}$ implies E enters S from $\overline{X}$.

With these transparent changes, Theorem 1.1 can now be modified so that it covers the nonlinear case. The proof given before carries through *without change*.

THEOREM 1.2. *Suppose $\overline{X}$ maximizes $g(X)$ subject to $X \in S = \{X \mid f_k(X) \geqq 0 \text{ for } k \in K\}$ where g and the f_k are differentiable functions. Then, if S satisfies the constraint qualification at $\overline{X}$, there exist $\bar{p}_k \geqq 0$ such that*

$$\nabla g(\overline{X}) + \sum_{k \in K} \bar{p}_k \nabla f_k(\overline{X}) = 0, \tag{1.1}$$

$$\sum_{k \in K} \bar{p}_k f_k(\overline{X}) = 0. \tag{1.2}$$

We shall conclude this lecture with a simple example of an economic application suggested by W. J. Baumol [3].

EXAMPLE 1.2. Consider a revenue maximizing firm manufacturing a single product. The revenue R depends on the quantity q manufactured and the amount a of advertising expenditure, while the costs are manufacturing costs $c(q)$ plus advertising expenditure. The profit of the firm is not to fall below a lower limit m. We assume, as is normal, that $\partial R/\partial a > 0$ and $\partial c/\partial q > 0$.

We shall explore the consequences of assuming the existence of a revenue maximum at a quantity $q > 0$. Assuming the constraint qualification satisfied, form

$$l(q,a) = R(q,a) + uq + va + w[R(q,a) - c(q) - a - m].$$

The necessary conditions for the maximum are

$$\partial R/\partial q + u + w\,\partial R/\partial q - w\,\partial c/\partial q = 0 \tag{1.7}$$

$$\partial R/\partial a + v + w\,\partial R/\partial a - w = 0 \tag{1.8}$$

$$uq + va + w\,[R(q,a) - c(q) - a - m] = 0 \tag{1.9}$$

for some $u \geqq 0$, $v \geqq 0$, $w \geqq 0$. By assumption and using (1.8),

$$0 < \partial R/\partial a \leqq (1 + w)\,\partial R/\partial a = w - v.$$

Hence $w > v \geqq 0$ and we have

$$\text{profit} = m. \tag{1.10}$$

On the other hand $q > 0$ implies $u = 0$ and hence by (1.7) and the assumption that costs rise with the quantity manufactured,

$$(1 + w)\,\partial R/\partial q = w\,\partial c/\partial q - u = w\,\partial c/\partial q > 0.$$

Hence

(1.11) marginal revenue is positive.

Finally, if profit is denoted by Π,

$$w\partial\Pi/\partial q = w(\partial R/\partial q - \partial c/\partial q) = -\,\partial R/\partial q - u < 0,$$

and hence

(1.12) marginal profit is negative.

We may, of course, interpret the multiplier w as the marginal loss in maximum revenue due to the limit on profit.

Lecture 2. As is often the case, the original motivation for the Lagrangian conditions had special features that made the resulting generalization somewhat unnatural. It may be instructive to trace this motivation to lead to a comparison of the various forms that the conditions can take.

Consider the following pair of dual linear programs:

$$\text{Maximize } C \cdot X \text{ subject to } AX \leqq B,\ X \geqq 0.$$

$$\text{Minimize } U \cdot B \text{ subject to } UA \geqq C,\ U \geqq 0.$$

In this framework, and in view of parallels with matrix game theory, it is natural to form the "saddle-value function"

$$s(X, U) = C \cdot X + U \cdot (B - AX) = U \cdot B - (UA - C) \cdot X.$$

This stands in contrast to the Lagrangian function

$$l(X, U, V) = C \cdot X + U \cdot (B - AX) + V \cdot X$$

which would follow from the maximum program by the rules of Lecture 1. The absence of the multipliers V lead to changes in the form of the conditions derived. However, before making these comparisons, we state the principal theorem for the saddle-value function (for the linear case, as before):

THEOREM 2.1. *The vector $\overline{X}$ solves the maximum program if and only if there exists $\overline{U} \geqq 0$ such that*

$$s(X, \overline{U}) \leqq s(\overline{X}, \overline{U}) \leqq s(\overline{X}, U)$$

for all $X \geqq 0$, $U \geqq 0$.

PROOF. If $\overline{X}$ solves the maximum program then (as in Lecture 1) there exists $\overline{U} \geqq 0$ such that $\overline{U}A \geqq C$ and with $\overline{U} \cdot B = C \cdot \overline{X} = s(\overline{X}, \overline{U})$. Hence, for any $X \geqq 0$,

$$s(X, \overline{U}) = \overline{U} \cdot B - (\overline{U}A - C) \cdot X \leqq s(\overline{X}, \overline{U}),$$

and, for any $U \geqq 0$,

$$s(\overline{X}, U) = C \cdot \overline{X} + U \cdot (B - A\overline{X}) \geqq s(\overline{X}, \overline{U}).$$

This proves the necessity of the conditions.

In proving the sufficiency, we shall state the problem more generally to emphasize that some of our special assumptions are not needed in the proof. As in Lecture 1, consider the (possibly nonlinear) maximum problem: Maximize $g(X)$ subject to $F(X) \geqq 0$ (that is, $f_k(X) \geqq 0$ for $k \in K$) and $X \geqq 0$. (Note that the conditions $X \geqq 0$ are now present explicitly.) Then, without further hypotheses, the following theorem holds.

THEOREM 2.2. *Form the saddle-value function*

$$s(X, U) = g(X) + U \cdot F(X)$$

associated with the maximum problem. If there exist $\overline{X} \geqq 0$ and $\overline{U} \geqq 0$ such that

$$s(X, \overline{U}) \leqq s(\overline{X}, \overline{U}) \leqq s(\overline{X}, U)$$

for all $X \geqq 0$ and $U \geqq 0$, then $\overline{X}$ solves the maximum problem.

PROOF. Since

$$s(\overline{X}, \overline{U}) = g(\overline{X}) + \overline{U} \cdot F(\overline{X}) \leqq g(\overline{X}) + U \cdot F(\overline{X}) = s(\overline{X}, U)$$

for all $U \geqq 0$ we must have $F(\overline{X}) \geqq 0$. If we take $U = 0$ in this inequality, we have $\overline{U} \cdot F(\overline{X}) \leqq 0$ and hence, since $\overline{U} \geqq 0$ and $F(\overline{X}) \geqq 0$,

$$\overline{U} \cdot F(\overline{X}) = 0.$$

On the other hand,

$$s(X, \overline{U}) = g(X) + \overline{U} \cdot F(X) \leqq g(\overline{X}) = s(\overline{X}, \overline{U})$$

for all $X \geqq 0$. Hence for any $X \geqq 0$ which is feasible for the maximum problem (that is, $F(X) \geqq 0$),

$$g(X) \leqq g(\overline{X})$$

which completes the proof of Theorem 2.2 (and also the sufficiency part of Theorem 2.1).

As before our success with the linear case suggests attempting the generalization of the necessity of Theorem 2.1 to the nonlinear case. The following simple example illustrates a difficulty that is encountered.

EXAMPLE 2.1. Maximize $g(x_1) = x_1$ subject to $-x_1^2 \geqq 0$ and $x_1 \geqq 0$. The saddle-value function is

$$s(x_1, u_1) = x_1 - u_1 x_1^2.$$

It is clear that $\overline{x}_1 = 0$ is the unique solution to the maximum problem. Moreover, any $\overline{u}_1$ achieves the $\min_{u_1} s(\overline{x}_1, u_1)$. However, the only $\overline{u}_1$ for which $\max_{x_1} s(x_1, \overline{u}_1)$ exists are $\overline{u}_1 > 0$ and are achieved at $\overline{x}_1 = (2\overline{u}_1)^{-1} > 0$. Hence $s(x_1, u_1)$ has no saddle-point.

Seemingly the most natural hypothesis under which a saddle-point is a necessary condition for a solution to the maximum problem is given in the following theorem [**18**].

THEOREM 2.3. *Let g and f_k be concave functions and suppose that there exists a vector $X' \geqq 0$ for which $F(X') > 0$. If $\overline{X}$ maximizes $g(X)$ subject to $F(X) \geqq 0$, $X \geqq 0$, then there exists a $\overline{U} \geqq 0$ such that $(\overline{X}, \overline{U})$ is a saddle-point for $s(X, U) = g(X) + U \cdot F(X)$.*

We may call a maximum program *regular* if there exists $X' \geqq 0$ with $F(X') > 0$. The implications connecting the three problems are then summarized in the following diagram. Although we have only proved the implications represented by solid arrows in this diagram, proofs of the remaining implications are readily available [**22**], [**18**], [**14**]. Throughout the diagram, differentiability is assumed where needed when the Lagrangian conditions are in question.

LAGRANGIAN CONDITIONS

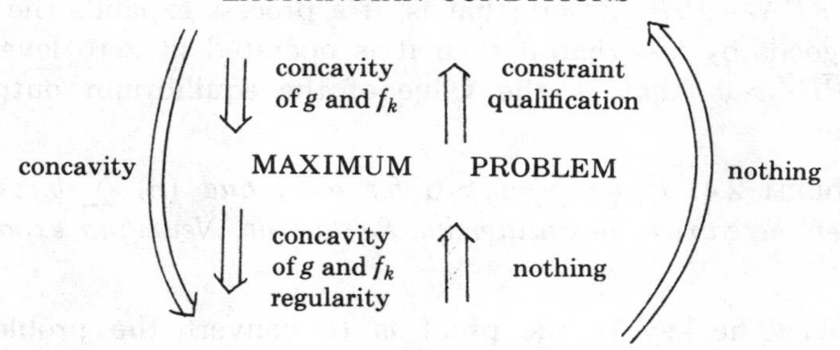

SADDLE-VALUE PROBLEM

Returning now to economic applications of the Lagrangian conditions, we shall show a typical application to establishing the existence of equilibrium in an economic model. The model we shall use to illustrate this type of application is a familiar one, namely, the von Neumann expanding economy, and is the subject of many papers in the literature [**26**]. However, the proof of the existence of equilibrium is thought to be new. In view of the many sources in which the economic implications of the model are discussed at length, we shall only describe the framework in detail sufficient for the existence proof.

Consider an economy in which m goods $(i = 1, \cdots, m)$ are made by n processes $(j = 1, \cdots, n)$. The technology is entirely described by two m by n nonnegative matrices $A = (a_{ij})$ and $B = (b_{ij})$, which specify, respectively, how many units of good i are used as inputs (a_{ij}) or produced as outputs (b_{ij}) when process j is run at a unit level. If the processes are run at nonnegative levels $Z = (z_j)$ then $X = AZ$ is the vector of goods input to production at the beginning of a period and $Y = BZ$ is the vector of goods output from production at the end of a period.

The nonnegative vector P will denote prices for the goods and the nonnegative scalars $\alpha \geqq 0$ and $\beta \geqq 0$ will denote the technical and economic expansion factors respectively. With these definitions, equilibrium for the economy is defined to be $\overline{Z} \geqq 0$, $\overline{P} \geqq 0$, $\overline{\alpha} \geqq 0$, and $\overline{\beta} \geqq 0$ such that

(a) $B\overline{Z} \geqq \overline{\alpha} A\overline{Z}$ (that is, output of each good is at least $\overline{\alpha}$ times input);

(b) $\overline{P} \cdot (B\overline{Z} - \overline{\alpha} A\overline{Z}) = 0$ (that is, if good i is expanded by more than $\overline{\alpha}$ then its price is zero);

(c) $\overline{P}B \leqq \overline{\beta}\,\overline{P}A$ (that is, no process expands the value of the goods by more than $\overline{\beta}$ or, equivalently, each activity is "profitless" at the interest rate $\overline{r} = \overline{\beta} - 1$);

(d) $(\bar{\beta}\bar{P}A - \bar{P}B) \cdot \bar{Z} = 0$ (that is, if a process expands the value of the goods by less than $\bar{\beta}$ then it is operated at zero level);

(e) $\bar{P}B\bar{Z} > 0$ (that is, the value of the equilibrium output is positive).

THEOREM 2.4. *If* (i) $\sum_i a_{ij} > 0$ *for all* j *and* (ii) $\sum_j b_{ij} > 0$ *for all* i, *then there exists an equilibrium for the von Neumann expanding economy.*

PROOF. The key to the proof is to convert the problem of equilibrium into a maximum problem. From well-known properties of the von Neumann model, the following formulation is appropriate.

Maximize α subject to

$$BZ - \alpha AZ \geqq 0, \tag{2.1}$$

$$SAZ - 1 \geqq 0, \tag{2.2}$$

$$Z \geqq 0, \tag{2.3}$$

$$\alpha \geqq 0, \tag{2.4}$$

where the m-vector $S = (1, \cdots, 1)$. Following the treatment of Lecture 1, we form

$$l(Z, \alpha) = \alpha + P \cdot (BZ - \alpha AZ) + u(SAZ - 1) + V \cdot Z + w\alpha.$$

Then, setting $\nabla l = 0$, we obtain

$$PB - \alpha PA + uSA + V = 0, \tag{2.5}$$

$$1 - PAZ + w = 0, \tag{2.6}$$

and

$$P \cdot (BZ - \alpha AZ) = 0, \tag{2.7}$$

$$u(SAZ - 1) = 0, \tag{2.8}$$

$$V \cdot Z = 0, \tag{2.9}$$

$$w\alpha = 0. \tag{2.10}$$

Of course, (2.1) is (a) and (2.7) is (b) of the equilibrium conditions. On the other hand, (2.5) asserts

$$\alpha PA = PB + uSA + V \geqq PB$$

which is (c) with $\beta = \alpha$. This also verifies (d) as (2.7). Finally, from (2.6), $PAS = 1 + w > 0$ and hence

$$PBZ = \alpha PAZ = \alpha(1 + w) > 0$$

if $\alpha > 0$. Thus (e) is verified and the existence of equilibrium is established provided the maximum problem has a solution α that is positive. The traditional conditions (i) and (ii) are designed precisely to show that α has a maximum and that this maximum is finite. We leave to the reader the verification of these facts; we also point out that it is necessary to confirm the Constraint Qualification to complete the proof.

We conclude this lecture with an example of an economic application of the Lagrangian conditions, brought to my attention by W. J. Baumol.

EXAMPLE 2.2 (AVERCH AND JOHNSON [1]). Suppose a firm has a single output $y = f(x_1, x_2)$, where x_1 and x_2 are the respective quantities of capital and labor used. The unit costs (interest and wages) of capital and labor are assumed to be the constants c_1 and c_2. Assume a regulatory constraint which restricts the ratio of net revenue to capital to be not greater than a fixed proportion s. One can show that the marginal revenue product of capital cannot equal its marginal cost at a profit maximum. Thus, the regulatory constraint distorts the "efficient" relative proportions of capital and labor that would be used in the absence of the constraint.

Lecture 3. The subject of the next three lectures is the pure theory of international trade. We shall treat a selection of mathematical problems chosen from this area. This sampling is rather special and personal although we shall draw heavily on topics treated by Chipman in his superb survey of the subject [**5**], [**6**], [**7**].

What type of questions will concern us? In the main they will not differ in kind from the problems of the two previous lectures although the setting will be different. We shall set up a model of international trade and will be interested in the existence of equilibrium for such a model. We shall be interested in which countries produce which goods under various definitions of equilibrium efficiency. Since these are basically problems of constrained optimization, the tools of the two previous lectures will reappear. The reader with a modest training in economics will recognize the initial model to be the simple formulation of Ricardo.

We shall first provide a generalization of the law of International Value, stated by J. S. Mill [**25**] for two countries and two commodities. Our treatment was stimulated by Chipman's observations in [**5**].

It is assumed that there are m countries $(i = 1, \cdots, m)$ and n goods $(j = 1, \cdots, n)$. There is but one primary resource (say, labor) and the fixed endowment of country i is b_i. To manufacture a unit of good j in country i requires a_{ij} units of labor in country i. Hence, if x_{ij} is the quantity of good j made in country i, and if

$$x_j = \sum_i x_{ij} \tag{3.1}$$

defines the total world output of good j, then this output is subject to the constraints:

$$\sum_j a_{ij} x_{ij} \leqq b_i, \qquad (i = 1, \cdots, m), \tag{3.2}$$

$$x_{ij} \geqq 0, \qquad (i = 1, \cdots, m; j = 1, \cdots, n). \tag{3.3}$$

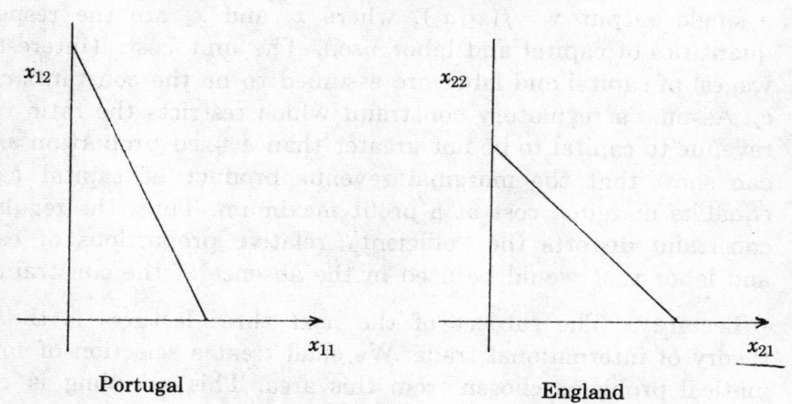

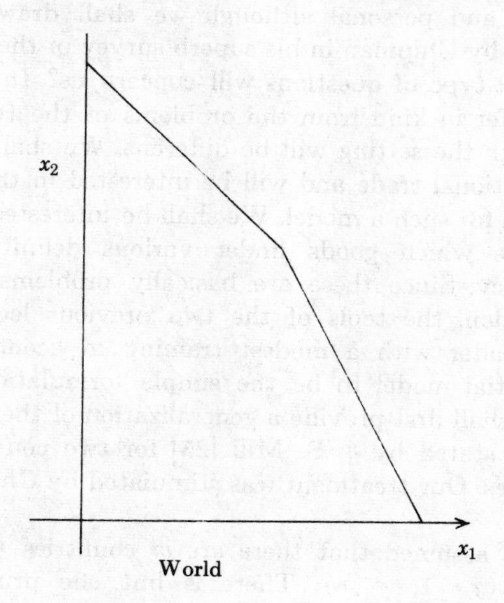

For those who are unfamiliar with this formulation, we shall provide an example in the Ricardian tradition. Let countries 1 and 2 be Portugal and England, respectively. Let goods 1 and 2 be cloth and wine, respectively. The single factor is labor and 30 units are available in Portugal and 20 units in England. The manufacture of a unit of cloth requires 2 units of labor in Portugal and 1 unit in England. The manufacture of a unit of wine uses 1 unit of labor in both countries.

The figures on p. 62 show the production possibility sets in Portugal, England, and the "world" of these two countries combined.

It is further assumed, following Mill, that the quantities produced maximize

$$U(x_1, x_2, \cdots, x_n) = x_1 x_2 \cdots x_n \tag{3.4}$$

subject to (3.1), (3.2), and (3.3). For a *pattern of specialization* in which country i produces good $j(i)$ only, we shall use the symbol $\langle j(1), j(2), \cdots, j(n)\rangle$. Our problem is to find conditions which are necessary and sufficient for this pattern of specialization to maximize the utility function (3.4) subject to the technological constraints (3.1), (3.2), and (3.3). This is provided by the following theorem.

Theorem 3.1. *A pattern of specialization maximizes the utility function subject to the technological constraints if and only if every good is produced and no country can produce the total world output of any good with less labor than that required for the total world output of the good in which it specializes.*

Proof. The objective function

$$\begin{aligned} U(x_1, \cdots, x_n) &= x_1 \cdots x_n \\ &= (x_{11} + \cdots + x_{m1}) \cdots (x_{1n} + \cdots + x_{mn}) = V(x_{ij}) \end{aligned}$$

may be regarded either as a function U of the world outputs x_j or as a function V of the outputs x_{ij} in the individual countries. It is easily verified that the quasi-concavity of U implies the quasi-concavity of V. Furthermore, the "world production possibility set" $\mathfrak{X}$ is clearly a closed bounded polyhedral set with an interior. Hence we may apply the Kuhn-Tucker theorem [**24**] to provide necessary and sufficient conditions. Before stating these, it is convenient to point out that it is necessary that every good be produced (and hence necessarily $m \geqq n$). This follows from the fact that U takes on positive values in the interior of $\mathfrak{X}$ while, if some good is not produced, $U = 0$.

Hence, a necessary and sufficient condition for $\langle j(1), j(2), \cdots, j(m)\rangle$ to maximize V subject to (3.1), (3.2), and (3.3) is that there exist $\bar{w}_1 \geqq 0, \cdots, \bar{w}_m \geqq 0$ such that

$$\bar{w}_i a_{ij} \geqq \Pi_{k \neq j} \bar{x}_k \quad \text{for all } i \text{ and } j, \tag{3.5}$$

where strict inequality implies $\bar{x}_{ij} = 0$. (In these conditions, $\bar{x}_{ij} = b_i/a_{ij}$ if $j(i) = j$ and is zero otherwise, while $\bar{x}_j = \sum_i \bar{x}_{ij}$.) Multiplying each inequality in (3.5) by $\bar{x}_j$, and letting $\bar{V} = \bar{x}_1 \cdots \bar{x}_n$, the conditions become

$$\bar{w}_i a_{ij} \bar{x}_j \geqq \bar{V} \quad \text{for all } i \text{ and } j, \tag{3.6}$$

where strict inequality implies $\bar{x}_{ij} = 0$. Noting that equality holds for $j = j(i)$ and that $\bar{w}_i > 0$ for all i, the conditions may be written:

$$a_{ij(i)} \bar{x}_{j(i)} \leqq a_{ij} \bar{x}_j \quad \text{for all } i \text{ and } j. \tag{3.7}$$

This is exactly the statement of the theorem.

To show that this generalizes Mill's statement, let $q_{ij} = b_i/a_{ij}$ denote the maximum quantity that country i can produce of good j. Then, for $m = n = 2$, $\langle 2, 1\rangle$ is a solution if and only if

$$a_{12}\bar{x}_2 \leqq a_{11}\bar{x}_1 \quad \text{and} \quad a_{21}\bar{x}_1 \leqq a_{22}\bar{x}_2 \tag{3.8}$$

by (3.7). Noting that $\bar{x}_1 = b_2/a_{21}$ and $\bar{x}_2 = b_1/a_{12}$, these become

$$b_1 \leqq b_2 a_{11}/a_{21} \quad \text{and} \quad b_2 \leqq b_1 a_{22}/a_{12} \tag{3.9}$$

or

$$q_{11} < q_{21} \quad \text{and} \quad q_{22} \leqq q_{12}. \tag{3.10}$$

These are exactly Mill's conditions as derived by Chipman (see [**6**, p. 487]). The reader should apply these conditions to the example introduced above.

We shall next give a concise proof of the existence of equilibrium in Graham's model of world trade (first proved by McKenzie in [**23**]). We shall use a general technique developed by Gale [**11**] for a simple but extremely flexible model of a competitive market. In a previous paper [**20**], Gale's result has been proved along lines which parallel the verbal explanation of economic equilibrium in a competitive market.

The basic model is the same as that used above. The "world production possibility set," $\mathfrak{X}$, consisting of all $X = (x_1, \cdots, x_n)$ satisfying (3.1), (3.2), and (3.3), is a closed polyhedral bounded convex set. Moreover, it is assumed that all $a_{ij} > 0$ and all $b_i > 0$. Hence, if we denote by $\mathscr{P}$ the $(n-1)$-dimensional simplex of relative prices $P = (p_1, \cdots, p_n)$ such that all $p_j \geqq 0$ and $\sum_j p_j = 1$, then for each $P \in \mathscr{P}$ there exists an X such that $P \cdot X > 0$. Furthermore, it is clear that the function

$$r(P) = \max_{X \in \mathfrak{X}} P \cdot X \tag{3.11}$$

is a continuous real-valued function defined on $\mathscr{P}$ which only assumes positive values.

It may be useful to point out that the fact that $r(P)$ is a continuous function is easily verified and does not need the general results invoked by McKenzie (see [**23**, p. 156 and p. 160]). We must show that, if P^k is a sequence of price vectors that converges to P^o then $r(P^k)$ converges to $r(P^o)$. For arbitrary $\epsilon > 0$, choose an integer K such that $|P^k - P^o| \leqq \epsilon / C$ for $k \geqq K$ where C is a positive constant such that $|X| \leqq C$ for $X \in \mathfrak{X}$. Then

$$|(P^k - P^o) \cdot X| \leqq \epsilon \quad \text{for all } X \in \mathfrak{X} \quad \text{and all } k \in K. \tag{3.12}$$

Now let $\overline{X}^k$ and $\overline{X}^o$ be such that $P^k \cdot \overline{X}^k = r(P^k)$ for all k and $P^o \cdot \overline{X}^o = r(P^o)$. Then

$$P^o \cdot \overline{X}^o - \epsilon \leqq P^k \cdot \overline{X}^o \leqq P^k \cdot \overline{X}^k \leqq P^o \cdot \overline{X}^k + \epsilon \leqq P^o \cdot \overline{X}^o + \epsilon \tag{3.13}$$

for all $k \geqq K$. Hence $r(P^k)$ converges to $r(P^o)$.

The competitive equilibrium in world trade is expressed in terms of wages w_i in country i and world prices p_j:

$$w_i a_{ij} \geqq p_j, \tag{3.14}$$

$$w_i \geqq 0. \tag{3.15}$$

These express the condition that the manufacture of product j should be profitless. Clearly, in addition,

$$\text{if } w_i a_{ij} > p_j, \quad \text{then } x_{ij} = 0, \tag{3.16}$$

while

$$\text{if } \sum_j a_{ij} x_{ij} < b_i, \quad \text{then } w_i = 0. \tag{3.17}$$

Thus, we are seeking the solutions to a pair of dual linear programs. The primal program has variables x_{ij} and objective function $P \cdot X$ to be maximized; the dual program has variables w_i and objective function $W \cdot B$ to be maximized.

The demand functions introduced by Graham, following the lead of J. S. Mill, have unit price and income elasticity and are defined by

(3.18) $$y_j = f_j r(P)/p_j \quad \text{where } f_j > 0 \quad \text{and} \quad \sum_j f_j = 1.$$

These demand functions imply that the same fraction of world income is spent on the jth good regardless of its price.

THEOREM 3.2. *Given* $a_{ij} > 0$, $b_j > 0$, $f_j > 0$, $\sum_j f_j = 1$, *there exist* x_{ij}, x_j, p_j, w_i *satisfying* (3.1), (3.2), (3.3), (3.11), (3.14), (3.15), (3.16) *and* (3.17) *such that* $x_j = y_j$ *for all* y_j *defined by* (3.18). *Moreover,* $P = (p_j) > 0$ *and* $X = (x_j) > 0$.

To prove this theorem we shall make use of the model developed by Gale [11]. This model makes provision for j goods at relative prices $p_1, \cdots, p_n$ which are nonnegative and sum to one. The activity of the various economic units in the market is summarized by a net supply function $\mathscr{F}$ which assigns to each price vector $P = (p_1, \cdots, p_n)$ a set $\mathscr{F}(P)$ of commodity bundles $Z = (z_1, \cdots, z_n)$. In each bundle $Z \in \mathscr{F}(P)$, the component z_j measures the net supply of good j (i.e. the total amount supplied by producing units diminished by the total amount demanded by consuming units) that might occur at prices P. Each bundle Z associated with P is assumed to satisfy the budget inequality $P \cdot Z = p_1 z_1 + \cdots + p_n z_n \geqq 0$; this holds necessarily if each economic unit receives at least as much income from the goods that it produces as it pays for the goods it consumes. Equilibrium for this model consists of a price vector P and a bundle $Z \in \mathscr{F}(P)$ such that $Z \geqq 0$. With these identifications, Gale's theorem is the following:

Let $\mathscr{F}$ be a bounded continuous set-valued function from the unit $(n-1)$-simplex $\mathscr{P}$ into R_n such that (a) $\mathscr{F}(P)$ is nonempty and convex for all $P \in \mathscr{P}$, and (b) if $Z \in \mathscr{F}(P)$ then $P \cdot Z \geqq 0$. Then there exists $\overline{P} \in \mathscr{P}$ and $\overline{Z} \in \mathscr{F}(\overline{P})$ such that $\overline{Z} \geqq 0$.

PROOF. Given $P \in \mathscr{P}$, let $\mathscr{S}(P)$ be the set of commodity bundles X that maximize $P \cdot X$ subject to (3.1), (3.2), and (3.3). To define a demand function that is continuous on $\mathscr{P}$ we follow McKenzie [23] in modifying (3.18). Let $q_j > \sum_i q_{ij}$ where $q_{ij} = b_i/a_{ij}$ for all i and j. Then define

(3.18′) $$\begin{aligned} y_j(P) &= q_j \quad \text{if } f_j r(P)/p_j \geqq q_j \quad \text{or } p_j = 0 \\ &= f_j r(P)/p_j \quad \text{otherwise.} \end{aligned}$$

Clearly $D(P) = (y_1(P), \cdots, y_n(P))$ is a continuous function defined on $\mathscr{P}$ and satisfies $P \cdot D(P) \leqq \sum_j f_j r(P)$ for all $P \in \mathscr{P}$.

The "net supply function" $\mathscr{F}$ is now defined by

(3.19) $\mathscr{F}(P) = \{Z | Z = X - Y \text{ for } X \in \mathscr{S}(P) \text{ and } Y = D(P)\}.$

The hypotheses of Gale's theorem are now easily verified. For example,

$$P \cdot Z = P \cdot X - P \cdot D(P) \geqq r(P) - \sum_j f_j r(P) = 0,$$

shows that the budget constraint is satisfied.

Hence there exists $\overline{P}$ such that $\overline{Z} \in \mathscr{F}(\overline{P})$ with $\overline{Z} \geqq 0$. This means $\overline{Z} = \overline{X} - \overline{Y} \geqq 0$ or $\overline{X} \geqq \overline{Y}$ with $\overline{X} \in \mathscr{S}(\overline{P})$ and $\overline{Y} = D(\overline{P})$. Note that $\overline{x}_j \geqq \overline{y}_j = q_j$ is impossible and hence $\overline{P} > 0$ and $\overline{y}_j = f_j r(\overline{P}) / \overline{p}_j$ for all j. Therefore,

$$\overline{P} \cdot \overline{Y} = \sum_j f_j r(\overline{P}) = r(\overline{P}) = \overline{P} \cdot \overline{X}.$$

Since $\overline{X} \leqq \overline{Y}$ and $\overline{P} > 0$, this equation implies $\overline{X} = \overline{Y}$. The other prices needed to establish the equilibrium (namely, the wages $\overline{W} = (\overline{w}_i)$) are provided by the duality theory of linear programming or by direct and obvious solutions of (3.14) and (3.16).

Lecture 4. The setting of the last lecture, and the two country two-good example contained in it, provide a background for a discussion of the classical concept of "comparative advantage" or "comparative costs." Recalling the world production possibility set for the example (shown again below), we note that the (negative) slopes of the northeast boundary segments are the "natural" price ratios for the two goods in the two countries in a world without trade.

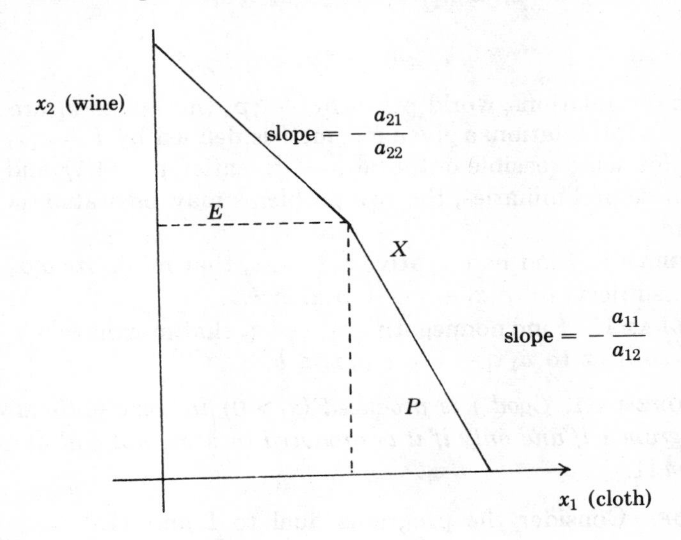

Thus, it is said in classical theory, that if the world price ratio lies between these limits, i.e. $a_{21}/a_{22} < p_1/p_2 < a_{11}/a_{12}$, then world production will lie at X. At this point each country specializes in the good in which it has the comparative advantage, with England specializing in cloth and Portugal in wine. In reaching this result, the objectives of the countries involved are often left unstated or are ambiguous at best. Very few authors have been explicit in stating an objective function.

Our first goal in this lecture is to resolve an apparent difference in the objective functions used by various authors when they are discussing the theory of comparative costs. Thus, for example, Viner [**30**] distinguishes "three different methods of dealing with the question of 'gain' from trade" and identifies the theory of comparative costs as that in which "economy in cost of obtaining a given income was the criterion of gain." On the other hand, when Dorfman, Samuelson, and Solow [**9**] discuss the theory of comparative advantage, they assert that "the problem is to maximize National Product (NP) subject to the technical production-possibility curve."

We shall show a sense in which these points of view are equivalent in the Ricardian setting. Both refer to a single country. As in Lecture 3, there is but one primary resource (say, labor) with fixed endowment and it is used to make n goods, $j = 1, \cdots, n$. To manufacture a unit of good j requires $a_j > 0$ units of labor. Hence if x_j is the quantity of good j made, these outputs are subject to the constraints:

$$a_1 x_1 + \cdots + a_n x_n \leqq b, \tag{4.1}$$

$$x_1 \geqq 0, \cdots, x_n \geqq 0. \tag{4.2}$$

In both formulations, world prices $p_1, \cdots, p_n$ (not all zero) are given. In Viner's formulation, a given income I is defined by $I = p_1\bar{x}_1 + \cdots + p_n\bar{x}_n$ for some feasible output $\bar{x}_1, \cdots, \bar{x}_n$ satisfying (4.1) and (4.2). With these preliminaries, the two problems may be stated as linear programs:

PROGRAM I. Find nonnegative $x_1, \cdots, x_n$ that minimize $a_1x_1 + \cdots + a_nx_n$ subject to $p_1x_1 + \cdots + p_nx_n \geqq I$.

PROGRAM II. Find nonnegative $x_1, \cdots, x_n$ that maximize $p_1x_1 + \cdots + p_nx_n$ subject to $a_1x_1 + \cdots + a_nx_n \leqq b$.

THEOREM 4.1. *Good j is produced ($x_j > 0$) in some optimal output for Program* I *if and only if it is produced in some optimal output for Program* II.

PROOF. Consider the programs dual to I and II.

DUAL PROGRAM I. Find $u \geqq 0$ so as to maximize uI subject to

$$up_j \leqq a_j \quad (j = 1, \cdots, n). \tag{4.3}$$

This program has an obvious solution, namely,

$$\bar{u} = \min_j \frac{a_j}{p_j}. \tag{4.4}$$

Furthermore, $x_j > 0$ in some optimal output for Program I if and only if $a_j/p_j = \bar{u}$.

DUAL PROGRAM II. Find $v \geqq 0$ so as to minimize vb subject to

$$va_j \geqq p_j \quad (j = 1, \cdots, \mathrm{n}). \tag{4.5}$$

This program has an obvious solution, namely,

$$\bar{v} = \max_j \frac{p_j}{a_j}. \tag{4.6}$$

Furthermore, $x_j > 0$ in some optimal output for Program II if and only if $p_j/a_j = \bar{v}$. Since the indices for which $a_j p_j$ achieves the minimum $\bar{u}$ are the same as those for which p_j/a_j achieves the maximum $\bar{v}$, this completes the proof of the theorem.

We also note the following connections between the two programs:

(4.7) If $x_1, \cdots, x_n$ is optimal for I then $x_1, \cdots, x_n$ is feasible for II.

(4.8) If $x_1, \cdots, x_n$ is optimal for II then $x_1, \cdots, x_n$ is feasible for I.

To show (4.7), note that if $x_1, \cdots, x_n$ is optimal for I then

$$a_1x_1 + \cdots + a_nx_n \leqq a_1\bar{x}_1 + \cdots + a_n\bar{x}_n \leqq b.$$

To show (4.8) note that if $x_1, \cdots, x_n$ is optimal for II then

$$p_1x_1 + \cdots + p_nx_n \geqq p_1\bar{x}_1 + \cdots + p_n\bar{x}_n = I. \qquad \text{Q.E.D.}$$

The following figure further illustrates the relation between the two programs:

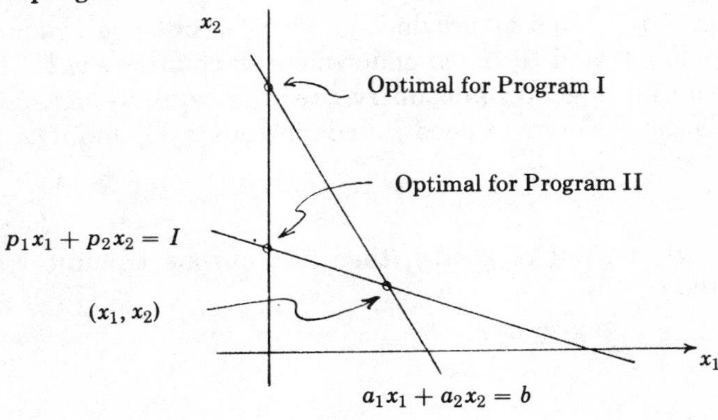

Returning to the two country two-good example with which this lecture began, the relation $a_{21}/a_{22} < a_{11}/a_{12}$ expresses the fact that Country 1 has a "comparative advantage" in Good 2, while Country 2 has a "comparative advantage" in Good 1. This is enough to determine the specialization of these countries in these goods. It is therefore quite natural to ask if, when more than two countries and two goods are involved, these comparisons suffice to determine the goods in which these countries specialize. Consider the following example, in which the entries of the matrix are our a_{ij}:

$$\begin{pmatrix} 2 & 1 & 5 \\ 5 & 2 & 1 \\ 1 & 5 & 2 \end{pmatrix}.$$

If we consider the specialization in which each Country i produces Good i (for $i = 1, 2, 3$) it is clear that, since $2/5 < 1/2$, the pairwise relations are satisfied. However, it is equally clear that the world as a whole will have twice as much of all goods if Country 1 makes Good 2, Country 2 makes Good 3, and Country 3 makes Good 1. Thus the validity of the pairwise comparisons is not enough to ensure world-wide efficiency.

We shall now attempt an explanation of this example by means of the McKenzie-Jones multiplicative formula (see [**24**, p. 171] and [**16**, p. 165]). In doing so we shall expand and, incidentally, correct Chipman's account [**6**]. We shall deal exclusively with an equal number of countries and goods. In this case, any one-to-one association of the goods and the countries to produce them will be called an *assignment*. Thus, if the goods are indexed by $j = 1, 2, \cdots, n$, an assignment is a permutation π of these integers. The country i that specializes in good j is $\pi(j)$; each country produces one good and each good is produced by one country.

The setting is the same as previously. There is but one primary resource (say, labor) and the fixed endowment of country i is b_i. To manufacture a unit of good j in country i requires a_{ij} units of labor. Hence, if x_{ij} is the quantity of good j made in country i, and if

(4.9) $$x_j = \sum_i x_{ij}, \quad (j = 1, \cdots, n),$$

defines the world output of good j, then this output is subject to the constraints:

(4.10) $$\sum_j a_{ij} x_{ij} \leqq b_i, \quad (i = 1, \cdots, n),$$

$$(4.11) \qquad x_{ij} \geqq 0, \quad (i = 1, \cdots, n;\ j = 1, \cdots, n).$$

The "world production possibility set," $\mathfrak{X}$, is the set of all $X = (x_1, \cdots, x_n)$ satisfying (4.9), (4.10), and (4.11).

For any assignment π, the world output $X^\pi = (x_1^\pi, \cdots, x_n^\pi)$ is defined by

$$(4.12) \qquad x_j^\pi = b_{\pi(j)}/a_{\pi(j),j}, \quad (j = 1, \cdots, n).$$

Since it is assumed that all $a_{ij} > 0$ and all $b_i > 0$, we have $X^\pi > 0$ for all assignments π.

DEFINITION. An output $X \in \mathfrak{X}$ is *efficient* if and only if there is no $X' \in \mathfrak{X}$ such that $X' \geq X$.[1] A nonzero output $X \in \mathfrak{X}$ is *extreme* if and only if there are no $X', X'' \in \mathfrak{X}$ such that $X' \neq X''$ and $X = tX' + (1-t)X''$ where $0 < t < 1$. An assignment π is said to be efficient (or extreme) if X^π is efficient (or extreme).

With these definitions, the problem studied by Jones was to find criteria that identify efficient and extreme assignments. These concepts are not independent as is shown by the following result.

PROPOSITION 4.1. *If X is extreme then X is efficient.*

PROOF. Suppose X is not efficient. Then there exists an $X' \in \mathfrak{X}$ such that $X' \geq X$. We may decompose the activities making X' into those making X and those making $X' - X$. Let the amount of labor used in the latter be b'; clearly, $b' > 0$ since $X' - X \geq 0$. With the labor b' we can make (at least) ϵX where $\epsilon > 0$. Hence by reassigning this labor we can make $(1+\epsilon)X$. Therefore

$$(4.13) \qquad X = \frac{1}{1+\epsilon}((1+\epsilon)X) + \frac{\epsilon}{1+\epsilon}0$$

is not extreme.

COROLLARY. *If the assignment π is extreme then π is efficient.*

It is clear that the converse of the proposition is false, since there are in general many efficient outputs which are not extreme. The converse of the corollary is also false, but requires "degeneracy" such as that illustrated in the following figure to counter it.

---

[1] We employ standard notation for vector inequalities: $A \geqq B$ means $a_k \geqq b_k$ for all k; $A \geq B$ means $A \geqq B$ but $A \neq B$; $A > B$ means $a_k > b_k$ for all k.

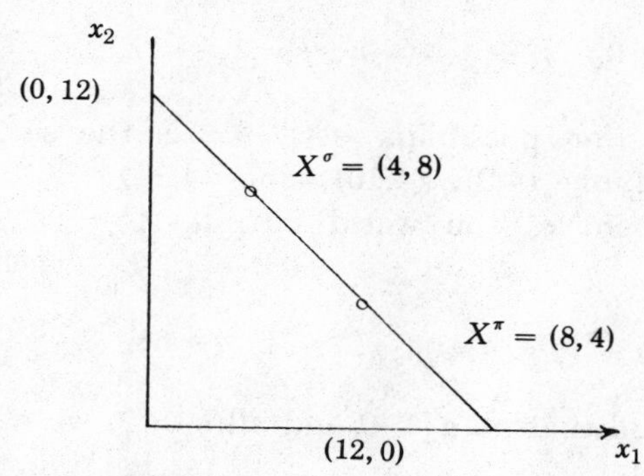

Here we suppose

$$A = \begin{pmatrix} 2 & 2 \\ 1 & 1 \end{pmatrix}, \quad B = \begin{pmatrix} 8 \\ 8 \end{pmatrix},$$

and $\pi(1) = 2,\ \pi(2) = 1$ while $\sigma(1) = 1,\ \sigma(2) = 2$.

THEOREM 4.2. *If the assignment π is extreme then $\Pi a_{\pi(j),j} < \Pi a_{\sigma(j),j}$ for all assignments $\sigma \neq \pi$.*

PROOF. Suppose π is extreme. Then $X^\pi > 0$ is extreme in $\mathfrak{X}$ and there exists a set of prices $p > 0$ such that X^π is the unique solution of the linear program: Maximize $P \cdot X$ subject to $X \in \mathfrak{X}$. The dual program may be stated: Minimize $W \cdot B$ subject to $w_i a_{ij} \geqq p_j$ and $w_i \geqq 0$. By the uniqueness of the solution, $w_i a_{ij} > p_j$ if and only if $x_{ij} = 0$. Hence

$$w_{\pi(j)} a_{\pi(j),j} = p_j, \quad (j = 1, \cdots, n), \tag{4.14}$$

$$w_i a_{ij} > p_j, \quad (i \neq \pi(j),\ j = 1, \cdots, n). \tag{4.15}$$

Let σ be any assignment different from π. Then

$$w_{\sigma(j)} a_{\sigma(j),j} \geqq p_j, \quad (j = 1, \cdots, n) \tag{4.16}$$

while

$$w_{\sigma(j)} a_{\sigma(j),j} > p_j, \quad \text{for at least one } j. \tag{4.17}$$

Hence

$$\frac{w_{\sigma(j)}}{w_{\pi(j)}} \geqq \frac{p_j / a_{\sigma(j),j}}{p_j / a_{\pi(j),j}} = \frac{a_{\pi(j),j}}{a_{\sigma(j),j}} \tag{4.18}$$

for all j and strict inequality holds for at least one j. Therefore

$$(4.19) \qquad 1 = \frac{w_1 \cdots w_n}{w_1 \cdots w_n} = \frac{w_{\sigma(1)} \cdots w_{\sigma(n)}}{w_{\pi(1)} \cdots w_{\pi(n)}} > \frac{a_{\pi(1),1} \cdots a_{\pi(n),n}}{a_{\sigma(1),1} \cdots a_{\sigma(n),n}}$$

and the theorem is proved.

THEOREM 4.3. *If the assignment* π *is efficient then* $\Pi a_{\pi(j),j} \leqq \Pi a_{\sigma(j),j}$ *for all assignments* σ.

PROOF. Suppose π is efficient. Then $X^\pi > 0$ is efficient and there exist prices $P > 0$ such that X^π solves the linear program: Maximize $P \cdot X$ subject to $X \in \mathfrak{X}$. The dual program may be stated: Minimize $W \cdot B$ subject to $w_i a_{ij} \geqq p_j$ and $w_i \geqq 0$. We have

$$(4.20) \qquad w_{\pi(j)} a_{\pi(j),j} = p_j, \quad (j = 1, \cdots, n),$$

while

$$(4.21) \qquad w_{\sigma(j)} a_{\sigma(j),j} \geqq p_j, \quad (j = 1, \cdots, n),$$

for all assignments σ. Thus, as before

$$(4.19) \qquad 1 = \frac{w_1 \cdots w_n}{w_1 \cdots w_n} = \frac{w_{\sigma(1)} \cdots w_{\sigma(n)}}{w_{\pi(1)} \cdots w_{\pi(n)}} \geqq \frac{a_{\pi(1),1} \cdots a_{\pi(n),n}}{a_{\sigma(1),1} \cdots a_{\sigma(n),n}}$$

and the theorem is proved.

THEOREM 4.4. *If* $\Pi a_{\pi(j),j} \leqq \Pi a_{\sigma(j),j}$ *for all assignments* σ *then the assignment* π *is efficient.*

PROOF. We shall show that, if π is not efficient then it is possible to find σ with $\Pi a_{\sigma(j),j} < \Pi a_{\pi(j),j}$. We have seen (in the proof of Proposition 4.1) that if X^π is not efficient then there exists $\epsilon > 0$ such that $(1 + \epsilon) X^\pi \in \mathfrak{X}$. If we choose the largest such ϵ then the resulting X' is efficient. Moreover, we can choose $P > 0$ such that X' solves the linear program: Maximize $P \cdot X$ subject to $X \in \mathfrak{X}$. Again the dual program is: Minimize $W \cdot B$ subject to $w_i a_{ij} \geqq p_j$ and $w_i \geqq 0$. Furthermore, if x'_{ij} are the outputs in the countries that give rise to the world output X',

$$(4.23) \qquad \text{if } x'_{ij} > 0 \quad \text{then } w_i a_{ij} = p_j.$$

Note that

$$(4.24) \qquad w_{\pi(j)} a_{\pi(j),j} \geqq p_j, \quad \text{for all } j$$

and that we must have

$$(4.25) \qquad w_{\pi(j)} a_{\pi(j),j} > p_j, \quad \text{for at least one } j.$$

Otherwise we would have

$$b_{\pi(j)} w_{\pi(j)} \frac{a_{\pi(j),j}}{b_{\pi(j)}} = b_{\pi(j)} w_{\pi(j)} \frac{1}{x_j^\pi} = p_j$$

for all j, and hence

$$W \cdot B = P \cdot X^\pi$$

which is impossible since X^π is not efficient. We now construct an assignment σ such that

(4.26) $\qquad \sigma(j) \neq \pi(j) \quad$ for some j with $w_{\pi(j)} a_{\pi(j),j} > p_j$

and

(4.27) $\qquad x'_{\sigma(j),j} > 0 \quad$ whenever $\sigma(j) \neq \pi(j)$.

For simplicity and without loss of generality we may assume that $\pi(j) = j$ for all j and that $w_1 a_{11} > p_1$. Note that, since $X' > X^\pi$, for each j, $x'_{ij} > 0$ for some $i \neq j$. Hence we can start at $(1, 1)$ and construct a cycle of the following type:

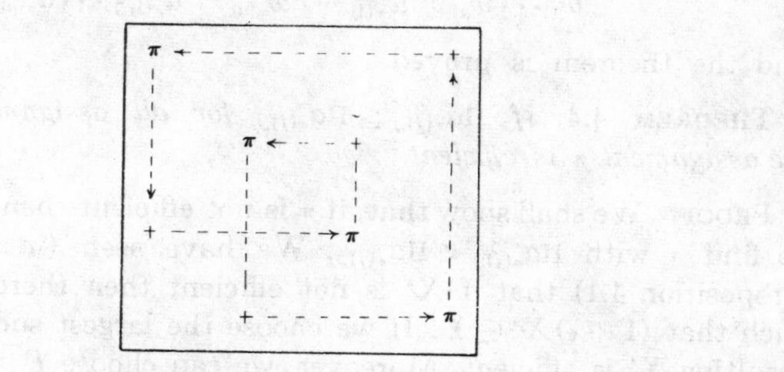

where each $+$ denotes a cell (i, j) with $x'_{ij} > 0$ and $i \neq j$ and each π denotes a cell (i, j) with $i = j$. Let σ differ from π only on the columns in the cycle and satisfy $\sigma(j) = i$ for each cell (i, j) marked with a $+$. Now, clearly

(4.28) $$1 = \frac{w_1 \cdots w_n}{w_1 \cdots w_n} = \frac{w_{\sigma(1)} \cdots w_{\sigma(n)}}{w_{\pi(1)} \cdots w_{\pi(n)}} < \frac{a_{\pi(1),1} \cdots a_{\pi(n),n}}{a_{\sigma(1),1} \cdots a_{\sigma(n),n}}$$

since, for $\sigma(j) = \pi(j)$ we have $w_{\sigma(j)}/w_{\pi(j)} = a_{\pi(j),j}/a_{\sigma(j),j} = 1$, while for $\sigma(j) \neq (j)$ we have $w_{\sigma(j)}/w_{\pi(j)} \leqq a_{\pi(j),j}/a_{\sigma(j),j}$ with strict inequality assured for $j = 1$. This completes the proof of the theorem.

COROLLARY. *There exists an efficient assignment* π.

THEOREM 4.5. *If* $\Pi a_{\pi(j),j} < \Pi a_{\sigma(j),j}$ *for all assignments* σ *then the assignment* π *is extreme.*

PROOF. To avoid making this note too long, the proof of this theorem will be left to the reader.

The confusion in Chipman's account (see [**6**, p. 508]) is caused by the possibility of no less than *five* concepts of the "efficiency" of an assignment. Let us list these:

(1) The assignment π is *extreme* if X^π is an extreme point of the world production possibility set $\mathfrak{X}$. (Chipman calls such an X^π a "vertex" of $\mathfrak{X}$.)

(2) The assignment π is *efficient* if $X \geq X^\pi$ for no $X \in \mathfrak{X}$.

(3) The assignment π is *strongly efficient* if $X > X^\pi$ for no $X \in \mathfrak{X}$. (Due to the particular geometry of $\mathfrak{X}$, with no faces distinct from and parallel to the coordinate hyperplanes, efficient and strongly efficient are the same property. This is essentially the first part of the proof of Proposition 4.1.)

To explain the other two concepts of "efficiency" which were introduced by McKenzie and used by Jones, we must define $C_{\pi,\sigma}$ the *consumption matrix* corresponding to a switch from the assignment π to the assignment σ,

$$C_{\pi,\sigma} = (c_{ij}) \text{ where } \begin{cases} c_{\pi^{-1}\sigma(j),j} = a_{\sigma(j),j}/a_{\sigma(j),\pi^{-1}\sigma(j)} \\ c_{ij} = 0 \quad \text{otherwise} \end{cases}$$

or

$$c_{kj} = \begin{cases} a_{\sigma(j),j}/a_{\sigma(j),k} & \text{for } \pi(k) = \sigma(j) \\ 0 & \text{otherwise.} \end{cases}$$

Then we have two additional concepts of efficiency.

(4) The assignment π can be *improved* by *switching* to σ if $C_{\pi,\sigma}$ is *semiproductive* (see Gale [**12**] for a definition).

(5) The assignment π can be *strictly improved* by *switching* to σ if $C_{\pi,\sigma}$ is *productive* (see Gale, *op. cit.*).

The relation to the previous results is established by the following two propositions (the proofs of both are left to the reader):

PROPOSITION 4.2. *The consumption matrix* $C_{\pi,\sigma}$ *is semiproductive if* $\Pi a_{\sigma(j),j} < \Pi a_{\pi(j),j}$.

PROPOSITION 4.3. *If* $C_{\pi,\sigma}$ *is productive then* $\Pi a_{\sigma(j),j} < \Pi a_{\pi(j),j}$.

The converse of Proposition 4.3 is not true as Chipman has suggested. A counterexample is provided as follows:

$$A = \begin{bmatrix} 2 & 1 & 10 & 10 \\ 1 & 1 & 10 & 10 \\ 10 & 10 & 2 & 1 \\ 10 & 10 & 1 & 1 \end{bmatrix}$$

$$\pi = \begin{pmatrix} 1 & 2 & 3 & 4 \\ 1 & 2 & 4 & 3 \end{pmatrix}$$

$$\sigma = \begin{pmatrix} 1 & 2 & 3 & 4 \\ 2 & 1 & 4 & 3 \end{pmatrix}$$

$$C_{\pi,\sigma} = \begin{pmatrix} 0 & \frac{1}{2} & 0 & 0 \\ 1 & 0 & 0 & 0 \\ 0 & 0 & 1 & 0 \\ 0 & 0 & 0 & 1 \end{pmatrix}$$

Here $C_{\pi,\sigma}$ is not productive but only semiproductive. Clearly, π is not efficient and yet it is not possible to shift resources to the assignment σ such that "the world output of each commodity is increased."

Lecture 5. Chipman has devoted a major portion of the third part of his survey, entitled "Modern Theory," to two theorems: the Samuelson factor price equalization theorem and the Stolper-Samuelson theorem on the effects of tariffs on real wages. The history of these two theorems is a complicated one and the literature that they have generated is voluminous. In this lecture, I shall focus my attention on some mathematical aspects which have been of interest to me; the limited time permits only a rapid outline of the economic background. Fortunately, the Chipman article will provide an easy entry to the subject for anyone who is interested in pursuing it further. Furthermore, the excellent bibliography to be found in Chipman's survey will allow us to cite articles by the year and refer the reader to his listing.

The Samuelson theorem (1948, 1949) has as its source the revision of classical international trade theory undertaken by Heckscher (1919) and Ohlin (1933). Ohlin, whose work was known in the English-reading world before that of Heckscher due to a thirty year delay in the translation of the 1919 paper, had asserted that *there is a tendency toward factor price equilization as a result of free trade*. However, he qualified his statement in a number of directions, even to the point of claiming that equalization would never be complete. Heckscher, on the other hand, made a clear claim for the equality of factor prices and it was this assertion that Samuelson attempted to prove. I shall attempt to sketch the final form that this proof has assumed and indicate the nature of the new mathematics which it has stimulated.

The background of the Stolper-Samuelson theorem is stated by Chipman as follows (italics are added): "While Heckscher had stated the factor price equalization theorem in terms of complete equalization, Ohlin (1933) stated it only in terms of tendencies. Stolper and Samuelson (1941) proved a very strong theorem concerning these tendencies, namely that *trade would lower the price of the 'scarce' factor expressed in terms of any commodity.* Thus Samuelson's 1948 and 1949 papers represented the evolution of what was an outgrowth of Ohlin's more cautious statement, back full circle to Heckscher's original position. So great has been the attention given to the conditions for full equalization, that *the problem originally posed by Stolper and Samuelson has remained relatively underdeveloped.*" Chipman goes on to sketch a proof of the Stolper-Samuelson theorem. With minor modifications, this proof is valid but only for the two good-two factor case; it is unusual that this restriction has not been more widely noted. The lecture concludes with some modest attempts to extend the result to more goods and factors.

Turning first to the Factor Price Equalization Theorem, let n goods be produced in amounts $y_1, \cdots, y_j, \cdots, y_n$ using m factors of production, indexed by $i = 1, \cdots, m$. The production functions

$$y_j = f_j(x_1, \cdots, x_m)$$

are assumed to be concave homogeneous functions of the first order defined for all nonnegative inputs. Let factor prices $w_1, \cdots, w_m$ be given and define

$$C_j(y_j; w_1, \cdots, w_m) = \min \left\{ \sum_i w_i x_i \,|\, y_j = f_j(x_1, \cdots, x_m) \right\}.$$

As Shephard [28] first proved, this minimum cost function factors as follows:

$$C_j(y_j; w_1, \cdots, w_m) = y_j \cdot g_j(w_1, \cdots, w_m);$$

the minimum unit cost functions can be shown to be continuous, homogeneous of the first order, and concave.

If we let x_{ij} be the quantity of factor i used in producing goods j, then a competitive equilibrium for this production economy may be defined as quantities of goods y_j and factors x_{ij} and prices of goods p_j and factors w_i such that

$$p_j \partial f_j / \partial x_{ij} \leqq w_i,$$

where if strict inequality holds then $x_{ij} = 0$, and

$$\sum_i w_i x_{ij} \leqq p_j,$$

where if strict inequality holds then $y_j = 0$.

It is known [22] that the first set of conditions are necessary and sufficient conditions for each set of factor inputs $(x_{1j}, \cdots, x_{mj})$ to achieve the minimum unit cost $g_j(w_1, \cdots, w_m)$ for the jth goods at factor prices $(w_1, \cdots, w_m)$. Thus, the existence of an equilibrium may be formulated in the following terms, which apply equally well to nondifferentiable production functions (arising, say, from a linear programming setting). Given the production functions f_j and positive prices p_j for the goods, there exist a set of nonnegative factor inputs $(x_{1j}, \cdots, x_{mj})$ such that

$$g_j(w_1, \cdots, w_m) = w_1 x_{1j} + \cdots + w_m x_{mj} \geqq p_j,$$

where if strict inequality holds then $y_j = 0$.

Up to now, we have dealt with only one country; however, since we assume identical production functions in each country, we have the *same* vector inequality

$$g(w) \geqq p$$

as the condition of equilibrium in each country. Therefore, as Chipman has pointed out, "factor price equalization will take place if (1) all n commodities are produced in each country (so that $y_j > 0$ and $g_j(w) = p_j$ for all j), and (2) the function g is globally invertible." For then the goods prices p_j will determine factor prices w_i, which will be the same in all countries.

Restating (2), we must show that, given $p_j > 0$, there are some factor endowments for which $g(w) = p$ has a solution and then we must show that the solution is unique. The existence of a solution was handled rigorously for the first time in 1959 [**21**], although Chipman has quite properly criticized the restrictive assumptions under which the theorem was proved. The proof is so short that we shall reproduce it here. The questionable hypothesis under which it was proved is

INTENSITY HYPOTHESIS. Suppose $m = n$ and that there is an indexing of goods and factors (associating with each goods the factor that is used intensively in its production) such that $w_i = 0$ implies

$$g_i(w_1, \cdots, w_m)/p_i < g_k(w_1, \cdots, w_m)/p_k$$

for some k. (Informally, this asserts that, if the factor that is used intensively in good i is free, then good i is not the most expensive to produce.)

PROOF. For each nonzero set of factor prices, define $h(w_1, \cdots, w_m) = \max_i g_i(w_1, \cdots, w_m)/p_i$. Then set

$$w_i' = w_i + h(w_1, \cdots, w_m) - g_i(w_1, \cdots, w_m)/p_i.$$

This defines a continuous mapping of nonnegative factor prices into themselves. By the Brouwer fixed-point theorem, there exist factor prices $\overline{w}_1, \cdots, \overline{w}_m$ and a constant $c > 0$ such that

$$c\overline{w}_i = \overline{w}_i + h(\overline{w}) - g_i(\overline{w})/p_i \quad \text{for all } i.$$

Choose k such that $g_k(\overline{w})/p_k = h(\overline{w})$. For this k, $c\overline{w}_k = \overline{w}_k$ and hence either $c = 1$ or $\overline{w}_k = 0$. The latter possibility is ruled out by the Intensity Hypothesis. Hence

$$g_i(\overline{w}) = h(\overline{w})p_i \quad \text{for all } i.$$

By the homogeneity of g_i,

$$g_i(\overline{w}/h(\overline{w})) = p_i \quad \text{for all } i,$$

and the existence of a solution is proved.

The question of uniqueness is delicate; sufficient conditions with a natural economic interpretation seem difficult to find. Samuelson's first attempt, which involved an incorrect application of the implicit function theorem, has led Nikaidô and Gale [13] to develop new results on the global univalence of mappings. Their principal theorem is that a sufficient condition for the global univalence of a differential mapping $g(w) = p$, where both w and p are m-dimensional is that the Jacobian matrix $[\partial g_i/\partial w_j]$ have all positive principle minors.

Now turning to the Stolper-Samuelson theorem, attempts to extend it to many commodities and factors from the two commodity and two factor case leads to an interesting and difficult technical problem. We shall follow the discussion of Chipman (see [8, pp. 37-39]) in stating the problem. As he has noted, "the argument of the Stolper-Samuelson theorem, as has been made clear by Bhagwati

[4], breaks into two distinct steps: (1) trade lowers the relative price of the commodity that employs the 'scarce' factor relatively intensively; (2) the fall in this relative price brings about a fall in the price of the 'scarce' factor relative to all commodity prices." We shall consider only the second step of the argument.

For the case of an equal number of commodities and factors, Chipman (see [8] or [5]) has isolated what is needed to validate (2) in a particularly elegant style. Namely, assuming that each factor has been associated with the commodity in which it is used intensively, then we must show $\partial \log w_i / \partial \log p_i > 1$, where p_i and w_i are the prices of the associated commodity i and factor i. That is, we must show that the elasticity of the ith factor price relative to the ith commodity price must exceed unity. These elasticities may be shown to be the diagonal elements of the matrix inverse to

$$A = (a_{ij}) = (x_{ij} w_j / p_i)$$

where x_{ij} is the quantity used of factor j in making one unit of commodity i at minimum cost when factor prices are $(w_1, \cdots, w_n)$. Clearly A is a stochastic matrix; that is, all $a_{ij} \geqq 0$ and $\sum_j a_{ij} = 1$ for $i = 1, \cdots, n$. Thus, the second step of the Stolper-Samuelson theorem is reduced to showing that the inverse of A has all of its diagonal elements greater than one.

At this point, Chipman has made a minor slip. We quote (see [8, p. 38]): "For the case $n = 2$ it is always true that the inverse of a stochastic matrix (when it exists) has its diagonal elements either greater than one or less than zero; so by appropriate permutations of rows and columns (that is, by suitable association of commodities with their 'intensive' factors—indeed, this provides us with a definition of 'intensity'), these diagonal elements will always exceed unity." Clearly, this cannot be true since

$$A = \begin{pmatrix} 1 & 0 \\ 0 & 1 \end{pmatrix}$$

is a stochastic matrix. A correct statement is provided by the following theorem.

THEOREM 5.1. *Let*

$$A = \begin{pmatrix} a_{11} & a_{12} \\ a_{21} & a_{22} \end{pmatrix}$$

be a stochastic matrix with positive determinant. Let

$$A^{-1} = \begin{pmatrix} b_{11} & b_{12} \\ b_{21} & b_{22} \end{pmatrix}.$$

Then $b_{11} \geqq 1$ *and* $b_{22} \geqq 1$. *Furthermore if* $a_{12} > 0$ *then* $b_{11} > 1$ *and if* $a_{21} > 0$ *then* $b_{22} > 1$.

PROOF. (Before proceeding to the proof itself, note that there is no restriction involved in the assumption that the determinant is positive; if the inverse exists then the determinant is nonzero and, with the possible permutation of the rows, can be assumed positive.) By direct calculation,

$$b_{11} = \frac{a_{22}}{a_{11}a_{22} - a_{12}a_{21}} \quad \text{and} \quad b_{12} = \frac{-a_{12}}{a_{11}a_{22} - a_{12}a_{21}}$$

and hence

$$b_{11} + b_{12} = \frac{a_{22} - a_{12}}{a_{11}a_{22} - a_{12}a_{21}} = \frac{a_{22} - a_{12}}{a_{22}(1 - a_{12}) - a_{12}(1 - a_{22})} = 1.$$

Hence $b_{11} \geqq 0$, and if $a_{12} > 0$ then $b_{11} > 1$. The proof for b_{22} is exactly analogous.

It should be noted that the assumption that A has a positive determinant is, in effect, an "intensity hypothesis." Namely, $a_{11}a_{22} - a_{12}a_{21} > 0$ implies

$$\frac{1}{p_1} x_{11} w_1 \frac{1}{p_2} x_{22} w_2 - \frac{1}{p_1} x_{12} w_2 \frac{1}{p_2} x_{21} w_1 > 0$$

and hence, if all of the prices are positive,

$$x_{11}x_{22} - x_{12}x_{21} > 0.$$

To insure $b_{11} > 1$ and $b_{22} > 1$, we must have all of $x_{ij} > 0$ and hence this may be written $x_{11}/x_{12} > x_{21}/x_{22}$. This says that relatively more of factor 1 is used in making one unit of commodity 1 than in making one unit of commodity 2.

Before proceeding to the case of three commodities and three factors we shall establish several general properties which might otherwise appear to be peculiar to the case just considered.

PROPOSITION 5.1. *If* A *has each of its row sums equal to one and if* $B = A^{-1}$, *then each of the rows of* B *has sum equal to one.*

PROOF. Let $A = (a_{ij})$ and $A^{-1} = B = (b_{ij})$. Then

$$\sum_j b_{ij} = \sum_j b_{ij} \left(\sum_k a_{jk} \right) = \sum_k \sum_j b_{ij} a_{jk} = 1.$$

Proposition 5.2. *If A has each of its row sums equal to one and $a_{kk}a_{ij} > a_{kj}a_{ik}$ for all k and all i, $j \neq k$, then*

$$a_{kk} > a_{ik} \quad \textit{for all } k \textit{ and all } i \neq k.$$

Proof. (Note that the conclusion of this proposition, namely, that the diagonal elements of A be larger than any other element in the same column, is the condition that Chipman has announced [8] as being sufficient for the case of $n = 3$ but not for $n \geqq 4$.) Assuming $a_{kk}a_{ij} > a_{kj}a_{ik}$ all k, all $i, j \neq k$ we have

$$\sum_{j \neq k} a_{kk}a_{ij} > \sum_{j \neq k} a_{kj}a_{ik}.$$

That is,

$$a_{kk}(1 - a_{ik}) > a_{ik}(1 - a_{kk})$$

and hence

$$a_{kk} > a_{ik} \quad \text{all } k, \quad \text{all } i \neq k.$$

Theorem 5.2 (**Chipman**). *Let $A = (a_{ij})$ be a 3 by 3 stochastic matrix and let $a_{kk} > a_{ik}$ for all k and all $i \neq k$. Then A is nonsingular. Let $A^{-1} = B = (b_{ij})$. Then $b_{kk} \geqq 1$ for all k and, if $a_{kj} > 0$ for some $j \neq k$ then $b_{kk} > 1$.*

Proof. If A can be shown to be nonsingular then

$$\begin{aligned}(b_{12} + b_{13}) \det A &= (a_{13}a_{32} - a_{12}a_{33}) + (a_{12}a_{23} - a_{13}a_{22}) \\ &\leqq (a_{13}a_{22} - a_{12}a_{33}) + (a_{12}a_{33} - a_{13}a_{22}) = 0,\end{aligned}$$

where strict inequality holds if $a_{12} > 0$ or $a_{21} > 0$. If $\det A$ can be shown to be positive then $b_{12} + b_{13} \leqq 0$ where strict inequality holds if $a_{13} > 0$ or $a_{12} > 0$. Since, by Proposition 5.1, $b_{11} + b_{12} + b_{13} = 1$, we have $b_{11} \geqq 1$ where strict inequality holds if $a_{12} > 0$ or $a_{13} > 0$. Thus the truth of the theorem depends on the

Lemma 5.1. $\det A > 0$.

Proof. We shall first prove A is nonsingular. If not, $XA = 0$ for some $X \neq 0$. Then $\sum_i x_i a_{ij} = 0$ and hence $0 = \sum_j \sum_i x_i a_{ij} = \sum_i x_i \sum_j a_{ij} = \sum_i x_i$. Since $X \neq 0$, we may assume $x_1 = 1$, $x_2 \leqq 0$, $x_3 \leqq 0$ and $x_2 + x_3 = -1$, that is, that the first row of A is a convex combination of the other rows of A. (Note that this may necessitate the simultaneous reordering of the rows and columns of A which will not change the column domination of the diagonal.) However, this implies

$$a_{11} = (-x_2)a_{21} + (-x_3)a_{31} < a_{11}(-x_2 - x_3) = a_{11},$$

which is a contradiction.

However, the set of stochastic matrices for which $a_{kk} > a_{ik}$ $(i \neq k)$ is a convex set. The function $\det A$ is continuous on this set and $\det I = 1$ for the matrix I which lies in the set. Since $\det A$ is never zero we must have $\det A > 0$ throughout. This completes the proof of the lemma and of the theorem.

The case of more than 3 goods is still largely open. First note that Chipman's condition (that the diagonal elements of A be larger than any other element in the same column) does not even imply that A is nonsingular. The following example shows this:

$$\begin{pmatrix} \frac{1}{3} & \frac{1}{6} & \frac{1}{4} & \frac{1}{4} \\ \frac{1}{6} & \frac{1}{3} & \frac{1}{4} & \frac{1}{4} \\ \frac{1}{4} & \frac{1}{4} & \frac{1}{3} & \frac{1}{6} \\ \frac{1}{4} & \frac{1}{4} & \frac{1}{6} & \frac{1}{3} \end{pmatrix}$$

However, no example is known of a 4 by 4 stochastic matrix satisfying the intensity hypothesis of Proposition 5.2 that is singular. Hence there is some (vague) hope that some intensity hypothesis may be strong enough to prove the Stolper-Samuelson theorem for four factors and four commodities.

Bibliography

1. Harvey Averch and L. L. Johnson, *Behavior of the firm under regulatory constraint,* Amer. Economic Rev. **52** (1952), 1052-1069.

2. M. L. Balinski and R. E. Gomory, "A mutual primal-dual simplex method" in *Recent advances in mathematical programming*, edited by Robert L. Graves and Philip Wolfe, McGraw-Hill, New York, 1963, pp. 17-26.

3. W. J. Baumol, *Economic theory and operations analysis,* 3rd ed., Prentice-Hall, Englewood Cliffs, N. J., 1968.

4. Jagdish Bhagwati, *Protection, real wages and real income,* Economic J. **69** (1959), 733-749.

5. J. S. Chipman, *Factor price equalization and the Stolper-Samuelson theorem,* (abstract) Econometrica **32** (1964), 682-683.

6. ———, *A survey of the theory of international trade.* Part 1: *The classical theory,* Econometrica **33** (1965), 477-519.

7. ———, *A survey of the theory of international trade.* Part 2: *The Neo-classical theory,* Econometrica **33** (1965), 685-760.

8. ———, *A survey of the theory of international trade.* Part 3: *The modern theory,* Econometrica **34** (1966), 18-76.

9. Robert Dorfman, P. A. Samuelson and R. M. Solow, *Linear programming and economic analysis,* McGraw-Hill, New York, 1958.

10. L. R. Ford, Jr. and D. R. Fulkerson, *Flows in networks,* Princeton Univ. Press, Princeton, N. J., 1962.

11. David Gale, *The law of supply and demand,* Math. Scand. **3** (1955), 155-169.

12. ———, *The theory of linear economic models,* McGraw-Hill, New York, 1960.

13. David Gale and Hukukane Nikaidô, *The Jacobian matrix and global univalence of mappings,* Math. Ann. **159** (1965), 81-93.

14. G. Hadley, *Nonlinear and dynamic programming,* Addison-Wesley, Reading, Mass., 1964.

15. J. M. Henderson and R. E. Quandt, *Microeconomic theory,* McGraw-Hill, New York, 1958.

16. R. W. Jones, *Comparative advantage and the theory of tariffs: A multi-country multi-commodity model,* Rev. Economic Studies **28** (1961), 161-175.

17. ———, *Factor proportions and the Heckscher-Ohlin theorem,* Rev. Economic Studies **24** (1956), 1-10.

18. Samuel Karlin, *Mathematical methods and theory in games, programming, and economics,* Addison-Wesley, Reading, Mass., 1959.

19. H. W. Kuhn, *The Hungarian method for solving the assignment problem,* Naval. Res. Logist. Quart. **2** (1955), 83-97.

20. ———, *A note on 'The law of supply and demand,'* Math. Scand. **4** (1956), 143-146.

21. ———, *Factor endowments and factor prices: Mathematical appendix,* Economica **26** (1959), 142-144.

22. H. W. Kuhn and A. W. Tucker, *Nonlinear programming,* Proc. Second Berkeley Sympos. Math. Statist. Probability., Univ. of California Press, Berkeley, Calif., 1951, pp. 481-492.

23. L. W. McKenzie, *On equilibrium in Graham's model of world trade and other competitive systems,* Econometrica **22** (1954), 147-161.

24. ———, *Specialization and efficiency in world production,* Rev. Economic Studies **21** (1954), 165-180.

25. J. S. Mill, *Principles of political economy with some of their applications to social philosophy,* 3rd ed., Parker, London, 1852. For the most recent scholarly edition, see *Collected works of John Stuart Mill.* Vol. III, Univ. of Toronto Press, Toronto, pp. 595-617.

26. Oskar Morgenstern and G. L. Thompson, *Private and public consumption and savings in the von Neumann model of an expanding economy,* Kyklos **20** (1967), 387-409.

27. R. T. Rockafellar, *Convex analysis,* Mimeographed lecture notes, Princeton University, 1966.

28. R. W. Shephard, *Cost and production functions,* Princeton Univ. Press, Princeton, N. J., 1953.

29. W. F. Stolper and P. A. Samuelson, *Protection and real wages,* Rev. Economic Studies **9** (1941), 51-73. Reprinted in Readings in the *Theory of international trade,* Blakiston, Philadelphia, Pa., 1949, pp. 333-357.

30. J. Viner, *Studies in the theory of international trade,* Harper and Brothers, New York, 1937, p. 437.

PRINCETON UNIVERSITY

Kenneth J. Arrow[1]

Applications of Control Theory to Economic Growth

1. **Review of the basic theorems of optimal control theory: finite horizon.** The basic criteria for optimization of dynamic processes in continuous times, as stated by L. S. Pontryagin and associates [**15**], will be restated in this and the following lecture. Some emphasis will be placed on special features appropriate to the use that will be made of these theorems in growth theory, in particular the assumption of an infinite horizon and the presence of constraints on the choice of control variables.

The object of study is a system, economic or other, evolving in time. At any moment, the system is in some *state*, which can be described by a finite-dimensional vector $x(t)$. For an economic system, the amount of capital goods of each type might constitute a suitable state description.

In an optimization problem there is some possibility of controlling the system. At any time, t, there is a vector $v(t)$ which can be chosen by a decision-maker from some set which may, in general, vary with both t and the state $x(t)$. The vector $v(t)$ is frequently referred to as the decision or control variable; following the terminology of Tinbergen [**18**, p. 7], the term *instrument* is used here. In an economic system the instruments are typically the allocations of resources to different productive uses and to consumption or

[1] This work was supported by the National Science Foundation under grant GS-1440 at Stanford University.

perhaps taxes and bond issues which at least partially determine allocations.

It is assumed that the state and the instrument variables at any point of time completely determine the rate of change of the state of the system. Thus, for a given technology and labor force, the capital structure (state), together with its allocation among different uses (by some of the instruments), determine the outputs of all goods. These in turn are allocated between consumption and capital accumulation (through other components of the instrument vector). In symbols, the evolution of the state of the system is governed by the differential equations

$$\dot{x} = T[x(t), v(t), t], \tag{1}$$

which will be referred to as *transition equations*. The time variable t may enter into T to allow for the possibility that the transition relations may vary over time due to technological progress, labor force growth or other exogenous factors.

Given, then, the state of the system at some time, say 0, and the choice of instruments as a function of time, $v(t)$, the whole course of the system is determined. To begin with, let us suppose that the analysis is carried out only until a finite horizon T, after which the process ceases.

By suitable choices of the values of instruments over time, alternative histories of the process can be achieved. As usual in economic analysis, we assume that these histories can be valued in some way, i.e. we can express preferences as between alternative histories, and these preferences can be given numerical value by a *utility functional* with arguments $x(t)$, $v(t)$, $(0 \leqq t \leqq T)$. The optimization problem is to choose the values of the instrument variables so as to maximize the utility functional subject to the constraints implied by (1), the constraints on the choices of the instruments, and the initial values of the state variables.

More specifically, it will be assumed that the utility functional is additive over time. That is, at each moment t there is a return or *felicity* (to use a term due to Gorman [5, p. 43]), which depends only on the values of the state variables and instruments at time t, such that the utility of a whole history is the sum or integral of the values of the felicities at each moment of time. Let

$$U(x, v, t) = \text{felicity at time } t \tag{2}$$

if the state is x and the instrument vector is v.

In addition to the felicity generated at each moment of time during the process, the decision-maker may also assign a value to the state achieved at the end of the process, T. In an industrial application the stock of machines may have a *scrap value*, and we will use this term generally. In a broader economic context, if T is not literally the end of the world but only the end of the planning period, the capital stock left over at T will have some use in the future. The scrap value will be denoted by $S[x(T)]$.

The general form of the optimization problem in time is then, with finite horizon,

$$\text{(3)} \qquad \text{Maximize} \int_0^T U[x(t), v(t), t]\,dt + S[x(T)]$$

with respect to choice of the instruments over time subject to (1), some constraints on the choice of instruments possibly depending on the current values of the state variables, and the initial conditions $x(0) = x$.

We assume that all functions satisfy sufficiently strong smoothness conditions, which will not be spelled out in detail. Then Pontryagin and associates [**15**] (see also Halkin [**6**]) have shown

PROPOSITION 1. *Let $v^*(t)$ be a choice of instruments* $(0 \leqq t \leqq T)$ *which maximizes* (3). *Then there exist auxiliary functions of time, $p(t)$, one for each state variable, such that, for each t,*

(a) *$v^*(t)$ maximizes $H[x(t), v(t), p(t), t]$, where $H(x, v, p, t) = U(x, v, t) + pT(x, v, t)$;*

the function $p(t)$ satisfies the differential equations

(b) *$\dot{p}_i = -\partial H/\partial x_i$, evaluated at $x = x(t), v = v^*(t)$, $p = p(t)$;*

and the transversality conditions

(c) *$p_i(T) = \partial S/\partial x_i$, evaluated at $x = x(T)$,*

hold.

The function H is known as the Hamiltonian. The auxiliary variables p can be given an economic interpretation. Consider the maximization of the utility function from any time t_0 to the horizon T; the history, before t_0, affects this problem only through the state at time t_0, as can easily be seen from (1) and the additive nature of the maximand in (3). Let this maximum be

$$\text{(4)} \qquad V(x, t_0) = \max \left\{ \int_0^T U[x(t), v(t), t]\,dt + S[x(T)] \right\}, \quad \text{where } x(t_0) = x.$$

Then the auxiliary variables are defined so that

(5) $$p_i = \partial V / \partial x_i.$$

An auxiliary variable measures the marginal contribution of the corresponding state variable to the utility functional at time t_0. Then $p_i \dot{x}_i = p_i T_i$ is the rate of increase of utility due to the current rate of increase of the state variable x_i, and therefore H is the current flow of utility from all sources, both enjoyed immediately, as expressed by U, and anticipated to be enjoyed in the future, as expressed by pT. The current instruments are chosen then to maximize H. The condition (b) is an equilibrium condition for holding the state variables constant (at an instant of time); the increment in utility plus speculative gain should be zero; if not, the individual would have wanted to have less or more of that state variable (read, "stock of a capital good in the economic context"). Finally, at time T, $V(x, T) = S(x)$; hence (c) holds by (5).

In the sequel, a slightly different formulation of end-of-period conditions will be useful. Instead of a scrap value, require simply that the end-of-period values of the state variables be nonnegative. Now approximate this condition by a scrap value function; that is, permit negative values but impose a very large penalty. Formally, let

$$S(x) = \sum_i P_i \min(x_i, 0),$$

where $\min(x_i, 0)$ means the smaller of x_i and 0, and the P_i's are chosen very large. For $x_i < 0$, $\partial S / \partial x_i = P_i$; for $x_i > 0$, $\partial S / \partial x_i = 0$. If $x_i = 0$, the right-hand derivative is zero and the left-hand derivative P_i, a fact which may be expressed loosely by the statement $0 \leqq \partial S / \partial x_i \leqq P_i$.

Now let the P_i's approach $+\infty$, so that we may be sure that the optimal policy will never lead to a final negative value, $x_i(T)$:

$$x(T) \geqq 0.$$

From the preceding discussion, $\partial S / \partial x_i \geqq 0$, and further if $x_i(T) > 0$, then $\partial S / \partial x_i = 0$. In view of Proposition 1(c), $p_i(T) \geqq 0$, $p_i(T) x_i(T) = 0$, all i, or

$$p(T) \geqq 0, \qquad p(T) x(T) = 0.$$

PROPOSITION 2. *Let $v^*(t)$ be a choice of instruments* $(0 \leqq t \leqq T)$ *which maximizes $\int_0^T U[x(t), v(t), t] \, dt$ subject to the conditions*

(a) $\dot{x} = T[x(t), v(t), t]$,

some constraints on the choices of instruments possibly involving current values of the state variables, and the terminal conditions,

$x(T) \geqq 0$. *Then there exist auxiliary variables*, $v(t)$, *such that* (a) *and* (b) *of Proposition* 1 *hold and for which*

(b) $p(T) \geqq 0, \quad p(T)x(T) = 0.$

The optimal path is a solution of the differential equations (1) and (b) of Proposition 1; the values of the instruments which enter into them are determined as functions of x, p, and t by Proposition 1(a). The number of these equations is twice that of the number of state variables. The solution is usually only determined when an equal number of initial conditions are specified. The values of the state variables at the beginning of the process, $x(0)$, are taken as known, but these constitute only half the needed conditions. The transversality conditions, Proposition 1(c) or Proposition 2(b), constitute the remaining conditions, but from a practical point of view they suffer from the severe difficulty of being defined at the end of the process, while the other initial conditions are defined at the beginning. The computation can proceed by guessing *initial* values, $p(0)$, solving the system of transition and auxiliary equations with the hope that the transversality conditions are satisfied, and correcting the initial guesses if not. It can also proceed by guessing the final state $x(T)$ and solving the equations backward in the hope that the initial conditions are satisfied.

Now consider more explicitly the constraints on the instruments. In general, they may depend on the values of the state variables. Thus, amounts of resources allocated to particular productive purposes are constrained by the total amounts available, which in turn are determined by the state variables. The following discussion is based on that in Pontryagin [**15**, Chapter VI] and on the theory of nonlinear programming due to Kuhn and Tucker [**8**].

Let the choice of instruments at any time t with state x satisfy a vector of inequality constraints,

$$F(x, v, t) \geqq 0. \tag{6}$$

For example, if output is a function of the stock of capital, $F(K)$, and if output is to be divided between consumption (C) and investment (I), then the instruments C and I satisfy the condition

$$F(K) - C - I \geqq 0,$$

which involves the state variable K. Some of the constraints in F might not include state variables; for example, nonnegativity conditions on the instruments.

It is well known from the general theory of nonlinear programming that if v^* maximizes H subject to the conditions (6), and if these constraints satisfy a certain condition known as the Constraint Qualification (see Kuhn and Tucker, [8, pp. 483-484]; Arrow, Hurwicz, and Uzawa [1]), then at any moment of time there exist multipliers q^0 such that

(7) $$q^0 \geqq 0, \qquad q^0 F(x, v^*, t) = 0,$$

and

(8) $$\partial L / \partial v_k = 0 \quad \text{at } v = v^*,\ q = q^0,$$

where

(9) $$L = H + qF.$$

It can be shown that $\partial H / \partial x_i = \partial L / \partial x_i$ when evaluated at $v = v^*$, $q = q^0$.

With the explicit formulation (6) of constraints, the conditions for optimization over time become

PROPOSITION 3. *Let $v^*(t)$ be a choice of instruments $(0 \leqq t \leqq T)$ which maximizes $\int_0^T U[x(t), v(t), t]\,dt$ subject to the conditions,*

(a) $\dot{x} = T[x(t), v(t), t]$,

a set of constraints,

(b) $F(x(t), v(t), t] \geqq 0$,

on the instruments possibly involving the state variables, initial conditions on the state variables, and the terminal conditions $x(T) \geqq 0$. If the Constraint Qualification holds, then there exist auxiliary variables $p(t)$, such that, for each t,

(c) *$v^*(t)$ maximizes $H[x(t), v, p(t), t]$ subject to the constraints (b), where $H(x, v, p, t) = U(x, v, t) + pT(x, v, t)$;*

(d) *$\dot{p}_i = -\partial L / \partial x_i$, evaluated at $x = x(t)$, $v = v^*(t)$, $p = p(t)$,*

where

(e) $L(x, v, q, t) = H(x, v, p, t) + qF(x, v, t)$,

and the Lagrange multipliers q are such that

(f) $\partial L / \partial v_k = 0$, for $x = x(t)$, $v = v^*(t)$, $p = p(t)$, $q \geqq 0$, $qF[x(t), v^*(t), t] = 0$,

and

(g) $p(T) \geqq 0$, $p(T)x(T) = 0$.

In many circumstances it is reasonable to consider, in addition, restrictions on the state variables in which the instruments do not enter. In particular, if the state variables are stocks of capital, negative values have no meaning. Here, nonnegativity conditions on the state variables,

(10) $$x(t) \geqq 0,$$

will be considered; the terminal condition $x(T) \geqq 0$ is implied.

For any i, if $x_i(t) > 0$, then the corresponding constraint (10) is ineffective and can be disregarded. Suppose that $x_i(t) = 0$ over some interval. Then, to avoid violation of (10), the instruments must be so constrained that $\dot{x}_i(t) \geqq 0$, and this constraint is clearly effective over that interval. But $\dot{x}_i = T_i$, so that the constraint $T_i(x, v, t) \geqq 0$ is effective over that interval. Then, in Proposition 3, this constraint can be regarded as added to the original set of constraints (b). Let q be the Lagrange multipliers associated with the original constraints (b), and let r_i be the multiplier associated with the new constraint $T_i \geqq 0$. As before, $r_i \geqq 0$. Define, in addition, $r_i = 0$ for each state variable for which $x_i(t) > 0$. Then, clearly, $r \geqq 0$, $rT = 0$, $rx = 0$.

PROPOSITION 4. *Let $v^*(t)$ be a choice of instruments* $(0 \leqq t \leqq T)$ *which maximizes $\int_0^T U[x(t), v(t), t]dt$ subject to the conditions,*

(a) $\dot{x} = T[x(t), v(t), t]$,

a set of constraints,

(b) $F[x(t), v(t), t] \geqq 0$,

involving the instruments and possibly the state variables, initial conditions on the state variables, and the nonnegativity conditions,

(c) $x(t) \geqq 0$,

on the state variables. If the Constraint Qualification holds, then there exist auxiliary variables $p(t)$ such that, for each t,

(d) *$v^*(t)$ maximizes $H[x(t), v, p(t), t]$ subject to the constraints* (b) *and the additional constraints $T_i[x(t), v, t] \geqq 0$ for all i for which $x_i(t) = 0$, where $H(x, v, p, t) = U(x, v, t) + pT(x, v, t)$;*

(e) *$\dot{p} = -\partial L / \partial x_i$, evaluated at $x = x(t)$, $v = v^*(t)$, $p = p(t)$, $q = q(t)$, $r = r(t)$, where*

(f) $L(x, v, p, q, r, t) = H(x, v, p, t) + qF(x, v, t) + rT(x, v, t)$,

and the Lagrange multipliers q and r are such that

(g) *$\partial L / \partial v_k = 0$, for $x = x(t)$, $v = v^*(t)$, $p = p(t)$, $q(t) \geqq 0$, $q(t) F[x(t), v^*(t), t] = 0$, $r(t) \geqq 0$, $r(t) x(t) = 0$, $r(t) T[x(t), v^*(t), t] = 0$;*

(h) $r(T) \geqq 0$, $p(T) x(T) = 0$.

So far, the propositions stated have been necessary conditions for the optimality of a policy. The situation is precisely analogous to the usual problem in calculus; the condition that a derivative be zero is necessary for a maximum but certainly not sufficient in general. However, the condition is sufficient if the function being maximized is concave. A basic property of concave functions is the following.

(11) If $f(x)$ is a concave function, then for any given point x^* and any other point x in the domain of definition, $f(x) \leqq f(x^*) + f_x^*(x - x^*)$, where f_x^* is the row vector with components $\partial f / \partial x_i$ evaluated at x^*.

Define the function

$$H^0(x,p,t) = \max_v H(x,v,p,t), \tag{12}$$

where v is constrained as in any of the Propositions 1—4. Then the concavity of H^0 as a function of x, for given p and t, implies that the Pontryagin conditions are sufficient for optimality. (This is a minor variation of a theorem of Mangasarian [**11**].)

PROPOSITION 5. *If H^0, as defined in* (12), *is concave in x for given p and t, then any choice of instruments satisfying the conditions of any of Propositions* 1—4 *is optimal for the corresponding problem.*

2. **Optimization with infinite horizon.** For many purposes it is more convenient to introduce the fiction that the horizon is infinite. Certainly, processes of capital accumulation for the economy as a whole have no natural stopping place in the definable future. At any given future date the state of the system (its capital structure) will have implications for the further future. If we choose to stop our analysis at any fixed date, it will be necessary, as already noted, to include in our utility functional some scrap value for the stock of capital at the end of the period. But the only logically consistent way of doing so is to determine the maximum utility attainable in the further future starting with any given stock of capital. Of course, the astronomers assure us that the world as we know it will come to an end in some few billions of years. But, as elsewhere in mathematical approximations to the real world, it is frequently more convenient and more revealing to proceed to the limit to make a mathematical infinity in the model correspond to the vast futurity of the real world.

Formally, the only change in the statement of the model is to let $T = +\infty$. But going to the limit, here as elsewhere, involves some risks. The utility functional, now an improper integral, might not converge at all; and even if it does, there might not exist an optimal policy. However, it is still possible to state necessary conditions and sufficient conditions for optimality, though existence of an optimal policy may be difficult to guarantee, and also it is not yet known how to state the appropriate transversality conditions. An extensive discussion of a case of nonexistence of an optimal path is given by Koopmans [7, pp. 251-253].

If an optimal policy exists, then it can be shown that the arguments for the necessity conditions of Propositions 1—4, except for the transversality conditions, are still valid. In the cases of interest in economics, the transversality conditions (Propositions 1(c), 2(b), 3(g), or 4(h)) are in fact valid, but so far it is necessary to verify this in each case. The infinite-horizon statement of the transversality conditions of Propositions 2—4 is

$$\lim_{t \to +\infty} p(t) \geqq 0, \qquad \lim_{t \to +\infty} p(t)x(t) = 0. \tag{13}$$

The sufficiency theorems remain completely valid, with the transversality condition (13).

It is customary and reasonable to assume that future felicities are discounted; i.e. the felicity obtained at time t is multiplied by a *discount factor* $\alpha(t)$, which is ordinarily taken to be a decreasing function of t. This corresponds to the intuitive idea that future pleasures are counted for less today. The utility functional is rewritten

$$\int_0^{+\infty} \alpha(t)\, U[x(t), v(t), t]\, dt. \tag{14}$$

Ordinarily, it is assumed that if the chosen policy leads to a constant felicity, (14) will converge. This is equivalent to the condition,

$$\int_0^{+\infty} \alpha(t)\, dt \text{ converges.} \tag{15}$$

If we follow the earlier line of argument, we would be interested in the maximum utility obtainable starting at some time t_0, analogous to (4),

$$V(x, t_0) = \max \int_{t_0}^{+\infty} \alpha(t)\, U[x(t), v(t), t]\, dt, \quad \text{where } x(t_0) = x. \tag{16}$$

However, this means that felicities for times beyond t_0 are being discounted to time 0. It is more natural to discount them to time t_0. Since one unit of felicity at time t_0 is equivalent to $\alpha(t_0)$ units at time 0, it is necessary to divide $V(x, t_0)$ by $\alpha(t_0)$ to obtain the *current-value return function*,

(17) $$W(x, t_0) = V(x, t_0)/\alpha(t_0).$$

Previously we obtained the auxiliary variables, $p(t)$, as the marginal contributions of the state variables to the utility functional, $p_i = \partial V/\partial x_i$. In the present context it seems more reasonable to define

(18) $$p_i = \partial W/\partial x_i = (\partial V/\partial x_i)/\alpha.$$

In applying Proposition 4 (apart from the transversality condition) to the discounted infinite-horizon case, it is then necessary to replace $U(x, v, t)$ by $\alpha(t)\, U(x, v, t)$ and $p(t)$ by $\alpha(t) p(t)$. The Hamiltonian becomes

(19) $$\alpha(t)\, U(x, v, t) + \alpha(t) p(t)\, T(x, v, t) = \alpha(t) H(x, v, p, t),$$

where we now define the *current-value Hamiltonian*

(20) $$H(x, v, p, t) = U(x, v, t) + pT(x, v, t).$$

Then $\alpha(t) H$ must replace H throughout the restatement of Proposition 4.

Since $\alpha(t) > 0$, the choice of instruments to maximize $\alpha(t) H$ is the same as that to maximize H, so that Proposition 4(d) remains unchanged. If we interpret the Lagrange multipliers q and r as referring to the maximization of H as now defined subject to the constraints, then L must be replaced by $\alpha(t) L$. Proposition 4(e) becomes

$$d[\alpha(t) p_i]/dt = -\partial[\alpha(t) L]/dx_i$$

or

$$\dot{\alpha} p_i + \alpha \dot{p}_i = -\alpha(\partial L/\partial x_i).$$

Divide through by α, and define

(19) $$\rho(t) = -\dot{\alpha}(t)/\alpha(t).$$

Then

(20) $$\dot{p}_i = \rho(t) p_i - (\partial L/\partial x_i).$$

In economic terms, $\rho(t)$ is a short-term interest rate, corresponding to the system of discount factors $\alpha(t)$. The definition (19) can be integrated back to yield the familiar form

(21) $$\alpha(t) = \exp\left[-\int_0^t \rho(u)\, du\right]$$

if we adopt the convention that $\alpha(0) = 1$. If (20) is written

$$\dot{p}_i + (\partial L/\partial x_i) = \rho(t) p_i(t),$$

it is the familiar equilibrium relation for investment in capital goods; the sum of capital gains and marginal productivity should equal the interest on the investment.

The infinite-horizon analogue of Proposition 4 (apart from transversality conditions) becomes

PROPOSITION 6. *Let $v^*(t)$ be a choice of instruments* ($t \geqq 0$) *which maximizes $\int_0^{+\infty} \alpha(t) U[x(t), v(t), t] dt$ subject to the conditions* (a), (b), *and* (c) *of Proposition* 4. *If the Constraint Qualification holds, then there exist auxiliary variables $p(t)$ satisfying* (d) *of Proposition* 4;

(e) $\dot{p}_i = \rho p_i - (\partial L/\partial x_i)$, *evaluated at* $x = x(t)$, $v = v^*(t)$, $p = p(t)$, $q = q(t)$, $r = r(t)$,

where $\rho(t) = -\dot{\alpha}(t)/\alpha(t)$, and (f) *and* (g) *of Proposition* 4 *hold.*

The sufficiency theorem, Proposition 5, remains valid if the transversality condition is replaced by (13) where, however, $p(t)$ is replaced by $\alpha(t) p(t)$.

PROPOSITION 7. *In the notation of Propositions* 4 *and* 6, *if*

$$H^0(x, p, t) = \max_v H(x, v, p, t),$$

where the maximization is over the range specified in Proposition 4(d), *is a concave function of x for given p and t, then any policy satisfying the conditions of Proposition* 6 *and the transversality conditions*

$$\lim_{t \to +\infty} \alpha(t) p(t) \geqq 0, \qquad \lim_{t \to +\infty} \alpha(t) p(t) x(t) = 0$$

is optimal.

It is frequently appropriate to make an assumption that the basic conditions of the optimization problem are stationary; the sequence of conditions to be encountered in the future is much the same as today or can be made so after some simple renormalizations. This property will be heavily exploited in our subsequent discussions. The basic stationarity assumptions are that the functions $U(x, v, t)$, $T(x, v, t)$, $F(x, v, t)$, and $\rho(t)$ are all independent of time. With ρ constant, it follows from (21) that

$$\alpha(t) = e^{-\rho t}, \tag{22}$$

and the convergence condition (15) becomes

$$\rho > 0. \tag{23}$$

Under the stationarity assumption, the current-value return function, $W(x, t_0)$, defined by (17), is in fact independent of t_0;

this can be seen by writing, in view of the previous remarks,

$$W(x, t_0) = \frac{1}{\exp[-\rho t_0]} \max \int_{t_0}^{+\infty} e^{-\rho t} U[x(t), v(t)]\,dt$$

$$= \max \int_{t_0}^{+\infty} \exp[-\rho(t - t_0)]\, U[x(t), v(t)].$$

Since the constraints $F(x, v) \geqq 0$ and the transition relations, $\dot{x} = T(x, v)$, do not involve time explicitly, it is clear that replacing t_0 by 0, say, leaves completely unaffected the form of the optimal policy. This is an illustration of Bellman's [**3**] well-known "principle of optimality". But if $W(x, t_0) = W(x)$, independent of t, then, from (18), p is completely determined by the state x in the following sense. Suppose we have two optimization problems of the type dealt with in Proposition 6 (but also satisfying the stationarity assumptions) which are identical in all respects except for initial conditions. Let $x^1(t)$ and $x^2(t)$ be the paths of the state variables along the optimal solutions for the two problems, respectively, and let $p^1(t)$ and $p^2(t)$ be the corresponding paths of the auxiliary variables. Then if $x^1(t) = x^2(t')$, $p^1(t) = p^2(t')$.

Note that since p is determined by x, and U and T do not depend on t along the optimal path, $H(x, v, p, t)$ is a function of x and v alone, and therefore the value of v which maximizes H depends only on x. The optimum policy can be represented as a *strategy* or *feedback control,* with v a function of x.

Also note that, for given x, v, and p, H is independent of t, and therefore Propositions 4 and 6(d), by itself, implies that v^* is determined by x and p; independent of t. The stationarity assumptions then imply that t does not enter explicitly into the system of differential equations defined by (a) and (e). Such a system is termed autonomous.

PROPOSITION 8. *Under the assumptions and in the notation of Proposition* 6, *suppose in addition that*

(a) $U(x, v, t) = U(x, v)$, $T(x, v, t) = T(x, v)$, $F(x, v, t) = F(x, v)$, *and* $\rho(t) = \rho$, *all independent of* t.

Then

(b) *the optimal policy,* $v^* = v^*(x)$, *and the values of the auxiliary variables,* p, *along the optimal path, are functions of* x *alone, independent of* t *for given* x;

(c) *the system of differential equations defined by* (a), (d), *and* (e) *is autonomous.*

For an autonomous system, considerable interest usually relates to its stationary point or *equilibrium,* where all motion ceases, i.e. the values of x and p for which $\dot{x} = 0$ and $\dot{p} = 0$. This notion in economics is that of long-run stationary equilibrium (as opposed to temporary or short-run equilibrium in which capital stocks are given). In the present system an equilibrium is defined by x^*, p^*, v^* satisfying the conditions:

$T(x^*, v^*) = 0.$

$\rho p_i^* = L_{xi}^*.$

v^* maximizes $H(x^*, v, p^*)$ under the constraints $F(x^*, v) \geqq 0$, $T_i(x^*, v) \geqq 0$ if $x_i^* = 0$.

If the initial state of the system is x^*, then all the conditions of Proposition 6 can be satisfied by setting $x(t) = x^*$, $v(t) = v^*$, $p(t) = p^*$ for all t. It may be asked under what conditions this solution is optimal. More generally, suppose we can find a path satisfying the conditions of Proposition 1 which converges to the stationary values; when is this optimal?

For simplicity of reference, define a *Pontryagin* path as a system, $x(t)$, $p(t)$, $v^*(t)$, satisfying the conditions of Proposition 6.

PROPOSITION 9. *Let* $x(t)$, $p(t)$, $v^*(t)$ *be a Pontryagin path for the problem of Proposition* 6. *Suppose further that the concavity hypothesis of Proposition* 7 *and the stationarity hypothesis of Proposition* 8, *with* $\rho > 0$, *are satisfied. Then, if* $x(t)$ *and* $p(t)$ *converge to an equilibrium,* x^*, p^*, *where* $p^* \geqq 0$, *they constitute an optimal path.*

PROOF. From Proposition 7 it suffices to note that the transversality condition $\alpha(t) p(t) x(t) \to 0$ is satisfied. But $p(t)$ and $x(t)$ converge to finite limits, and $\alpha(t) = e^{-\rho t}$ approaches zero since $\rho > 0$.

It should be remarked, however, that (a) there may be more than one equilibrium, and (b) there may exist optimal paths which do not converge to any finite equilibrium; for examples, see Kurz [**9**], [**10**, Section B].

3. **Optimal investment planning in a one-commodity model.** In this lecture we review in detail the simplest possible capital accumulation model, first studied by Ramsey [**16**]; for further important contributions, see Mirrlees [**14**] and Koopmans [**7**]. We assume there is only one commodity, which can be either consumer or invested. We take the viewpoint of a government which is in a position to control the economy completely and to plan perfectly so as to optimize with respect to all possible instruments of the economic system—in this case, only investment and consumption (which are subject to the constraint that their sum not exceed total output).

We first assume a constant population and a constant labor force. The felicity at any moment is taken to be a function of consumption, C, only. Then the aim of the economy is to maximize

$$V = \int_0^{+\infty} e^{-\rho t} U[C(t)]\, dt, \tag{24}$$

where ρ is the rate of interest on felicity, $C(t)$ is consumption at time t, and $U(C)$ is the felicity derived from consumption C. It is assumed that

(25) $U(C)$ is strictly concave and increasing.

The output at any moment of time is a function of the stock of capital and of the labor force. Since the latter is assumed constant, we assume simply

$$Y(t) = F[K(t)], \tag{26}$$

where $Y(t)$ is output at time t and K is the stock of capital. With the labor force held constant, increases in capital may be supposed to yield lower and lower returns; also it is assumed that capital is indispensable to production,

(27) F is strictly concave, $F(0) = 0$.

It is not necessarily assumed that $F(K)$ is increasing; for example, if the stock of capital depreciates at a rate proportional to its quantity, then the depreciation ought properly to be subtracted from the gross output to get a true measure of net output available for consumption and net investment (increase of the capital stock). It is possible that if K is very large, the marginal gross output of an additional unit may be less than the depreciation on that unit.

Finally, the accumulation of capital is precisely investment, I,

$$\dot{K} = I, \tag{28}$$

and the conservation of product flow implies that consumption and investment, together, do not exceed output, i.e. $C + I \leqq Y$ or, in view of (26),

$$F(K) - C - I \geqq 0. \tag{29}$$

It also follows from the very definition of capital that it cannot be negative; $K(t) \geqq 0$, all t.

At present it will be assumed that I may be positive, zero, or negative; the latter means that existing capital can be turned into consumption goods. The case where I is necessarily nonnegative will be considered later. It will be assumed that $C \geqq 0$; but to simplify matters it will be also assumed for the moment that

$$U'(0) = +\infty, \tag{30}$$

which, as will be seen, implies that the choice of the instrument C at any moment of time will necessarily be positive, so that the nonnegativity constraint is ineffective.

Propositions 6—9 can be applied to this model. The state of the system is represented by the single variable, K. There are two instruments, C and I. The felicity function depends only on the one instrument C; the transition function (28) depends only on I. The choice of C and I is constrained by (29), which corresponds to Proposition 6(b) or 4(b). There will be one auxiliary variable, p, so that the current-value Hamiltonian is $H = U(C) + pI$, and the Lagrangian, L, is

$$\begin{aligned} L &= U(C) + pI + q[F(K) - C - I] \\ &= [U(C) - qC] + (p - q)I + qF(K). \end{aligned} \tag{31}$$

By Proposition 6(g) or 4(g), C and I must be chosen so that $\partial L/\partial C = 0$ and $\partial L/\partial I = 0$. The latter implies that

$$p = q. \tag{32}$$

The former implies that $U'(C) = q$ and, by (32),

$$U'(C) = p. \tag{33}$$

Because of (30) and the concavity of $U(C)$, it is assured that the solution to (33) will be positive.

The auxiliary equation, Proposition 6(e), becomes

$$\begin{aligned} \dot{p} &= \rho p - q(\partial L/\partial K), \\ \dot{p} &= [\rho - F'(K)]p, \end{aligned} \tag{34}$$

in view of (31) and (32). Since $U' > 0$ by (25), $p > 0$ by (33) and $q > 0$ by (32); the constraint (29) is effective, so that from (28)

$$\dot{K} = F(K) - C(p), \tag{35}$$

where

$$C(p) \text{ is the solution of (33).} \tag{36}$$

Equations (34) and (35) are a pair of autonomous equations. An equilibrium is defined by $\dot{p} = 0$ and $\dot{K} = 0$. But $\dot{p} = 0$ means $p = 0$ or $F'(K) = \rho$. Since $U' > 0$, the first alternative is impossible from (33). Let K^∞, p^∞ be the equilibrium values of K and p, respectively, and let C^∞ and I^∞ be the values of the instruments at the equilibrium. It has just been shown that

$$F'(K^\infty) = \rho. \tag{37}$$

Since F is strictly concave, F' is strictly decreasing. It will be assumed that (37) has a solution with $K^* > 0$. This is equivalent to assuming that $F'(0) > \rho$, $F'(K) < \rho$ for K sufficiently large.

From (28) and the definition of equilibrium,

$$I^\infty = 0. \tag{38}$$

From (35) and (36), with $\dot{K} = 0$,

$$C^\infty = F(K^\infty), \tag{39}$$

$$p^\infty = U'(C^\infty). \tag{40}$$

Consider now all solutions of the differential equations (34) and (35). Their movements may be represented in a phase diagram (Figure 1). Since $F'(K)$ is decreasing, $\rho - F'(K)$ is increasing; then from (34) and (37), $\dot{p} > 0$ if $K > K^\infty$, $\dot{p} < 0$ if $K < K^\infty$. Since U' is a decreasing function, it follows from (36) and (33) that $C(p)$ is a decreasing function of p. The curve for which $\dot{K} = 0$ is, from (35), defined by the equation $F(K) = C(p)$. This equation can be solved uniquely for p in terms of K; call the solution $\pi(K)$. Since $F(K)$ is concave, and $F'(K^\infty) = \rho > 0$, $F(K)$ is either always increasing or increasing up to a value $\overline{K} > K^\infty$; therefore, $\pi(K)$ is decreasing for all K or else decreasing to $\overline{K}$ and increasing thereafter. Further, for fixed K, $\dot{K} = F(K) - C(p)$ is an increasing function of p so that $\dot{K} > 0$ above the curve and $\dot{K} < 0$ below. The components of the directions of movement in the four quadrants into which the diagram is divided by the curve $\dot{K} = 0$ and the vertical line $\dot{p} = 0$ are indicated in Figure 1 by arrows; note that $\dot{K} > 0$ is a movement to the right, and $\dot{p} > 0$ an upward movement.

Regions II and IV are traps in the sense that a Pontryagin path which enters either of these regions never leaves it. Further, any path which comes to a boundary of either region must enter that region and then remain in it permanently. It will now be shown that a path which enters either region cannot be optimal.

Consider first region IV. Without loss of generality, suppose that the path is in region IV at time 0. Then $\dot{p} > 0$, $\dot{K} > 0$, all t. Since $K(0) > K^\infty$, $K(t) \geqq K(0) > K^\infty$, all t. Since F' is decreasing, $F'[K(t)] \leqq F'[K(0)] < F'(K^\infty) = \rho$, so that, from (34), $\dot{p}/p = \rho - F'[K(t)] \geqq \epsilon = \rho - F'[K(0)] > 0$; $p(t) \geqq p(0)e^{\epsilon t}$ by integration, so

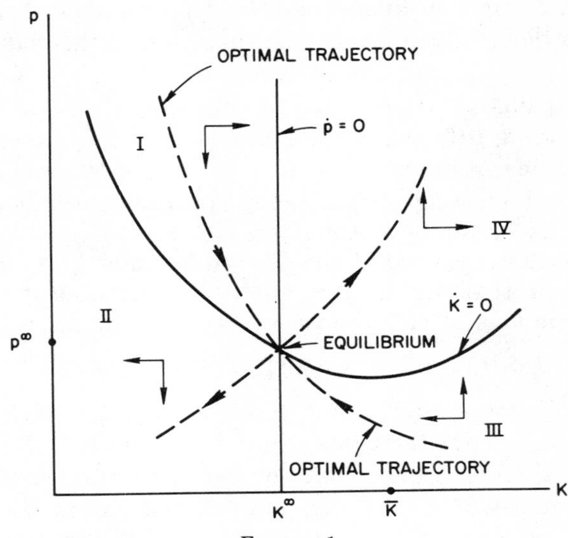

FIGURE 1.

that, certainly,

$$p(t) > p^{\infty} \quad \text{for all } t \geqq t_0,$$

for some t_0. Since $C(t) = C[p(t)]$ is a decreasing function of p,

$$C(t) < C(p^{\infty}) = C^{\infty} \quad \text{for } t \geqq t_0.$$

Since $K(t_0) > K^{\infty}$, it follows that we can always improve on the given path by consuming the capital stock (disinvesting) in some interval beginning at t_0 until K diminishes to K^{∞}, after which the equilibrium policy, $C = C^{\infty} = F(K^{\infty})$, $K = K^{\infty}$, is maintained perpetually.

Now consider any trajectory in region II. By the same reasoning $\dot{p}/p \leqq \epsilon$, where now $\epsilon = \rho - F'[K(0)] < 0$. But then $p(t) \to 0$, which implies $C[p(t)] \to +\infty$. Since $F(K)$ is uniformly bounded on the closed interval $[0, K^{\infty}]$, $\dot{K} = F[K(t)] - C[p(t)] \leqq \delta < 0$ from some time on. Then $K(t)$ must become zero at some finite time. Since $C > 0$, then, $I = F(0) - C < 0$, and K will become negative, violating the nonnegativity of K.

Consider now a path starting in quadrant I. If it stays in quadrant I forever, then both p and K are bounded from below. Since both are decreasing, they approach limits which, by a general theorem on differential equations, can only be the equilibrium values. By Proposition 9 such a path is necessarily optimal. If the path did not remain in quadrant I for all t, then it reaches either the boundary with quadrant IV or that with quadrant II; then, as already noted, the path cannot be optimal.

Similarly, any path in quadrant III which remains there forever approaches the equilibrium and is optimal; any other path is non-optimal.

It only remains to argue that, for any initial $K = K(0)$, there is a corresponding $p(0)$, with the point $(K, p(0))$ in quadrant I or quadrant III according as $K < K^\infty$ or $K > K^\infty$, such that the Pontryagin path starting at that point approaches the equilibrium. Such a path is certainly optimal.

The approach makes use of the fact that, under the stationarity assumptions of this problem, p and the optimal instruments C and I are functions of K. Divide (34) by (35) to see that

$$dp/dK = p[\rho - F'(K)]/[F(K) - C(p)]. \tag{41}$$

Consider, for values of $K \leqq K^\infty$, solutions of (41) which intersect the line $K = K^\infty$ above the equilibrium, i.e. for $p(K^\infty)$ a prescribed value greater than p^∞. Such a solution can be continued for smaller and smaller values of K. We first note that it can never cross the curve $\dot{K} = 0$, which has been written $p = \pi(K)$. Let $p(K)$ be the solution of (41), and suppose it intersected the curve $p = \pi(K)$ at $\widetilde{K} < K^\infty$. Then $p'(K) \to -\infty$ as $K \to \widetilde{K} + 0$. But $p(K) > \pi(K)$ for K in a right-hand neighborhood of $\widetilde{K}$, and therefore $p'(\widetilde{K}) \geqq \pi'(\widetilde{K})$, a contradiction, since $\pi'(\widetilde{K})$ is certainly finite.

Since the denominator of (41) is finite and the numerator is bounded from above, it is clear that the solution of (41) can be continued for all positive values of $K \leqq K^\infty$. There is one such solution for each value of $p(K^\infty) > p^\infty$. These solutions never cross because of the uniqueness of solutions of this differential equation away from the equilibrium point. Hence, for any given $K < K^\infty$, there is a lower bound, $\underline{p}(K)$, on the values of p for which there exists a solution of (41) passing through (K, p) and for which $p(K^\infty) > p^\infty$. It is obvious and can easily be demonstrated that $\underline{p}(K)$ also satisfies (41), and that $\underline{p}(K^\infty) = p^\infty$. This path in (p, K)-space defines the optimal trajectory. If $K(0) < K^\infty$, choose $p(0) = \underline{p}[K(0)]$. Then the points of the time solution, $p(t)$, $K(t)$, for (34) and (35) move along the trajectory $\underline{p}(K)$ and converge to the equilibrium.

The solution in this form is very convenient, for the choice of the instruments, C and I, is determined as a function of K by (33) and (29) (with equality), through the dependence of ρ on K.

The analysis in quadrant III is the same, except that we find for each $K > K^\infty$ the upper bound of p-values for which the solution of (41) passes below the equilibrium.

It should, however, be noted that we could apply the same procedure in quadrants II and IV; but then the limiting solution would be the divergent dashed curves marked in Figure 1.

The optimal solution, then, is defined by a solution of (41) which passes through the equilibrium; but there are two such solutions. The equilibrium is a singular point of (41), so the solution through that point need not be unique. In this case it is clear that the optimal solution is identified as the one with the negative slope at the equilibrium.

We will analyze the nonuniqueness at equilibrium a little more closely. The right-hand side of the differential equation (41) is, strictly speaking, not defined at $K = K^\infty$, $p = p^\infty$, since both numerator and denominator vanish. Since p is to be a function of K, both numerator and denominator are functions of K, directly and through p. Let

$$\phi(K) = p[\rho - F'(K)], \qquad \psi(K) = F(K) - C(p), \tag{42}$$

$$p'(K) = \phi(K)/\psi(K). \tag{43}$$

Since both ϕ and ψ vanish at $K = K^\infty$, we can define $p'(K) (= dp/dK)$ there by L'Hôpital's rule:

$$p'(K^\infty) = \phi'(K^\infty)/\psi'(K^\infty), \tag{44}$$

and it remains to evaluate these derivatives.

$$\phi'(K) = p[-F''(K)] + [\rho - F'(K)]p'(K).$$

From (37),

$$\phi'(K^\infty) = -p^\infty F''(K^\infty).$$

$$\psi'(K) = F'(K) - C'(p)\,p'(K).$$

Again make use of (37); then

$$\psi'(K^\infty) = \rho - C'(p^\infty)\,p'(K^\infty).$$

Substitute into (44); then

$$p'(K^\infty) = \frac{-p^\infty F''(K^\infty)}{\rho - C'(p^\infty)p'(K^\infty)},$$

or, clearing fractions,

$$-C'(p^\infty)[p'(K^\infty)]^2 + \rho p'(K^\infty) + p^\infty F''(K^\infty) = 0, \tag{45}$$

a quadratic equation in the slope of the solution to (41) which passes through the equilibrium. Since $C(p)$ is decreasing, the coefficient of the quadratic term is positive. Since $F'' < 0$, the constant term is negative. Thus, the product of the roots is negative, which implies that both are real, with one positive and one negative. As already noted, the negative root is the appropriate one.

Since $C(p)$ is defined by (36) and (33), we must have $U''[C(p)]C'(p) = 1$, so that $C'(p^\infty) = 1/U''(C^\infty)$.

Proposition 10. *Suppose the aim of the economic system is to maximize* $\int_0^{+\infty} e^{-\rho t} U[C(t)]dt$, *where* $\rho > 0$, *subject to the conditions* $\dot{K} = I$, $C + I \leqq F(K)$, $K \geqq 0$, *where* $U(C)$ *is a strictly concave increasing function and* $F(K)$ *is a strictly concave function with* $F(0) = 0$, *and also assume* $K(0)$ *given. Define* K^∞, C^∞, p^∞ *by the relations* $F'(K^\infty) = \rho$, $C^\infty = F(K^\infty)$, $p^\infty = U'(C^\infty)$.

Then the optimal strategy can be characterized by finding that solution $p(K)$ *of the differential equation* (41) *for which* $p(K^\infty) = p^\infty$ *and for which* $p'(K^\infty)$ *is the negative root of the quadratic equation* $[-1/U''(C^\infty)][p'(K^\infty)]^2 + \rho p'(K^\infty) + p^\infty F''(K^\infty) = 0$. *Then, for any* K, C *is so chosen that* $U'(C) = p(K)$, *and* $I = F(K) - C$.

Proposition 10 has been stated without the hypothesis, $U'(0) = +\infty$, which was used in the proof. It will be an interesting exercise in the use of Proposition 6 to consider the case where $U'(0)$ is finite. In this case, the constraints $C \geqq 0$ and $K \geqq 0$ may become effective. Consider the first for regions in which $K > 0$. Let w be the multiplier associated with the constraint, $C \geqq 0$. Then (31) is modified to read

$$(46)\qquad \begin{aligned} L &= U(C) + pI + q[F(K) - C - I] + wC \\ &= [U(C) - (q - w)C] + (p - q)I + qF(K), \end{aligned}$$

where

$$(47)\qquad w \geqq 0, \qquad wC = 0.$$

If $C = 0$, then the condition $\partial L/\partial C = 0$ becomes

$$U'(0) = U'(C) = q - w \leqq q.$$

We still have the condition $p = q$, so that $C = 0$ if $p \geqq U'(0)$. The system (34) and (35) is still valid, but the definition of $C(p)$ is slightly modified;

$C(p)$ is the solution of the equation, $U'(C) = p$ if $p \leqq U'(0)$, $C(p) = 0$ if $p > U'(0)$.

The previous analysis is completely unchanged; in Figure 1, the curve $\dot{K} = 0$ intersects the p-axis at $p = U'(0)$ instead of being asymptotic to it. Since the optimal trajectory lies above the curve $\dot{K} = 0$, there will be a $\underline{k} > 0$ for which $p(\underline{k}) = U'(0)$. For $k \leqq \underline{k}$, C will then be zero.

Now consider the possibility that the constraint $K = 0$ becomes effective. As has already been seen, this question arises only for paths which start in or have entered region II. It will be shown, even with $U'(0)$ finite, such paths are nonoptimal.

Recall the basic definition (18) of p as dW/dK, where W is maximum value of the utility functional if the initial state is K. Clearly, in this model an increase in K is always beneficial; given an increase in K, one can always consume a somewhat higher amount for some period until the value of $K(t)$ falls to that on the original path, and then follow the latter thereafter. Hence, p must be positive.

Now consider a path that has reached the p-axis at time t_0 from region II. Since the initial value of p was finite, the time to reach the p-axis was finite, and the right-hand side of (34) is bounded over this path, $p(t_0)$ is finite. Now the constraint $K \geqq 0$ becomes effective and therefore the constraint $I \geqq 0$ is imposed. The constraints,

$$C \geqq 0, \qquad I \geqq 0, \qquad C + I \leqq F(K) = F(0) = 0,$$

imply that C and I are 0; from the latter, it follows that $K(t) = 0$ for all $t \geqq t_0$. The Lagrangian (46) is modified by the addition of a term corresponding to the constraint $I \geqq 0$, with multiplier r,

$$\begin{aligned} L &= U(C) + pI + q[F(K) - C - I] + wC + rI \\ &= [U(C) - (q - w)C] + (p + r - q)I + qF(K), \end{aligned}$$

where $r \geqq 0$ when $I = 0$. No longer does the equality $p = q$ hold; instead,

$$p = q - r \leqq q.$$

Equation (34) is modified to

$$\dot{p} = \rho p - qF'(K) = \rho p - qF'(0) \quad \text{for } t \geqq t_0.$$

Recall that $F'(0) > \rho$. Use in turn the inequalities $p \leqq q$, $U'(0) \leqq q$,

$$\dot{p} = \rho p - qF'(0) \leqq q[\rho - F'(0)] \leqq U'(0)\,[\rho - F'(0)].$$

The last term is a negative constant. Hence, p must become negative in finite time, which contradicts the assumption that the path being studied is optimal.

It may be asked what happens if the initial stock of capital is 0. The only feasible path is that of zero investment and consumption. The argument just given would show that for any finite $p(0)$,

$p(t)$ would become negative eventually. The answer evidently is that $p(0)$ must be chosen $+\infty$ initially, and then $p(t)$ would remain $+\infty$.

4. Further aspects of the Ramsey problem: Irreversibility; growth of population and labor force and technical change. It is sometimes reasonable to argue that investments, once made in physical form, cannot be converted into consumer goods. Hence, investment should be irreversible, i.e. subject to the constraint $I \geqq 0$. On the other hand, there is in real life a somewhat more subtle way in which capital can, within limits, be run down to permit more consumption; namely, capital goods depreciate and failure to replace them constitutes a way of increasing consumption at the expense of capital.

A reasonable assumption about depreciation is that a fixed fraction of the existing capital becomes useless in each time period. Thus, the net rate of increase of capital is the amount of (gross) investment, i.e. new output devoted to capital uses, less the amount of depreciation. This amounts to replacing (28) by

$$\dot{K} = I - \delta K, \tag{48}$$

for some $\delta > 0$. We also assume that investment is nonnegative,

$$I \geqq 0. \tag{49}$$

Otherwise the model is identical with that of the last section, including (24) and (29), with the assumptions (25), (27), and (30) (though the last is dispensable). Let p be the auxiliary variable corresponding to (48), q the multiplier corresponding to (29), and s that corresponding to (49). The Lagrangian becomes

$$\begin{aligned} L &= U(C) + p(I - \delta K) + q[F(K) - C - I] + sI \\ &= [U(C) - qC] + (p + s - q)I + qF(K) - p\delta K, \end{aligned} \tag{50}$$

where

$$s \geqq 0, \qquad sI = 0. \tag{51}$$

Equating derivatives with respect to C and K to zero yields

$$U'(C) = q, \qquad q = p + s, \tag{52}$$

which can be combined with (51) to yield

$$p \leqq q; \qquad \text{if } p < q, \text{ then } I = 0. \tag{53}$$

The auxiliary equation is

$$\dot{p} = (\rho + \delta) p - qF'(K). \tag{54}$$

Since the problem is stationary, we know that the instruments and the auxiliary variable are functions of K. From (53), for any given K, there are two possibilities: either $p = q$, or $p < q$ with $I = 0$. Therefore, the K-axis is divided, in general, into alternating *blocked* and *free* intervals:

$$I = 0, \quad C = F(K), \quad q = U'[F(K)] > p \text{ on a blocked interval.} \tag{55}$$

$$I \geqq 0, \quad C \leqq F(K), \quad q = p \geqq U'[F(K)] \quad \text{on a free interval.} \tag{56}$$

From (48), $\dot{K} = -\delta K < 0$ if $I = 0$. Hence, the system cannot have an equilibrium in a blocked interval. Then $p = q$ at an equilibrium and, from (54), (48), and (29), the following relations hold at equilibrium:

$$\begin{aligned} F'(K^\infty) - \delta &= \rho, & I^\infty &= \delta K^\infty, \\ C^\infty &= F(K^\infty) - \delta K^\infty, & p^\infty &= U'(C^\infty). \end{aligned} \tag{57}$$

It will be observed that the equilibrium is the same as that for the reversible Ramsey problem, where $F(K)$ is replaced by $F(K) - \delta K$. Indeed, if we define $I_N = I - \delta K$, then, if the constraint (49) is ineffective, the problem is identical with the reversible Ramsey problem, with I replaced by I_N. For the reversible Ramsey problem the optimal policy would have $I_N \geqq 0$, for $K \leqq K^\infty$, and therefore $I = I_N + \delta K > 0$. Hence, starting with any such K, if the optimal policy for the reversible Ramsey problem is followed, it will always satisfy constraint (49) and therefore remain feasible under irreversibility. It will therefore remain optimal. Indeed, the same must be true for K in some right-hand neighborhood of K^∞, for while $I_N = 0$, $I_N + \delta K > 0$ for $K = K^\infty$ and, by continuity, $I \geqq 0$ for $K^\infty \leqq K \leqq \underline{K}$ for some $\underline{K}$ (which might even be $+\infty$). On the optimal path K decreases in this interval, and therefore K never goes outside the interval, so that the optimal path for the reversible Ramsey path is still feasible and therefore optimal for $K \leqq \underline{K}$.

The general method for finding the optimal strategy can now be sketched. As before, we are interested in the differential equation defining $p'(K) = dp/dK$. From (48) and (54),

$$p'(K) = [(\rho + \delta)p - qF'(K)]/(I - \delta K). \tag{58}$$

Here q and I can be determined as functions of K and p from (55) and (56). In the neighborhood of the equilibrium, the solution, as noted, is the same as for the reversible case. From (55) and (56), (58) specializes in the two kinds of intervals as follows:

(59) $$p'(K) = p[\rho + \delta - F'(K)]/[F(K) - \delta K - C(p)]$$ in a free interval.

(60) $$p'(K) = [(\rho + \delta)p - U'[F(K)]F'(K)]/(-\delta K)$$ in a blocked interval.

Let

(61) $$r = U'[F(K)]/p.$$

From (55) and (56),

(62) $r \leqq 1$ on a free interval, $r > 1$ on a blocked interval.

We know there is a free interval, $\langle 0, \underline{K} \rangle$, with $\underline{K} > K^{\infty}$. We therefore solve (59) around the equilibrium and continue the solution first for all smaller values of K. Then continue it for larger values of K until r reaches 1. The K-value where this occurs is $\underline{K}$, and there is a calculated value of p, $p(\underline{K})$. We then solve (60) with this starting point until r comes down to 1 from above. At this point we start a new free interval, and solve (59), but with the starting point being that achieved at the end of the previous blocked interval. This process can be continued indefinitely.

Thus, the problem is capable of numerically meaningful solution. Analytically sharper characterization cannot be obtained in general, though more specific hypotheses imply some limits on the numbers of blocked and free intervals. In particular, though, it can be shown that it is possible to have a denumerable or arbitrarily large finite number of alternations between free and blocked intervals. For these and other results, see Arrow and Kurz [**2**].

Another modification of the Ramsey model consists of allowing for growth in population and labor force and for technological change. Under certain simple but by no means absurd assumptions, these factors can be introduced into the Ramsey model by a simple reinterpretation of variables.

Population by itself affects the utility functional. Let $N(t)$ be the number of individuals at time t. Assume for simplicity that the aggregate amount of consumption at any time t is divided equally among the existing population. Assume also that each individual has the same felicity function. Then the felicity of any individual at time t is $U[C(t)/N(t)]$. Since there are $N(t)$ individuals, it is reasonable to conclude that the total felicity of society at time t is $N(t)\,U[C(t)/N(t)]$, and the utility functional then is

$$\int_0^{+\infty} e^{-\rho t} N(t)\, U\left[\frac{C(t)}{N(t)}\right] dt. \tag{63}$$

The production possibilities of society are of course influenced by the size of the labor force. This effect has been ignored until now because the labor force has been assumed constant. The growth of the labor force is roughly proportional to that of population, but it will be convenient to ignore this relation for the moment. We assume in any case that the size of the labor force is a known function of time, independent of the instruments or the state variables. Let $L(t)$ be the number of workers at time t. For any given supplies of capital and labor, output is determined by the production function

$$Y = F(K, L), \tag{64}$$

where it is assumed that

(65) F is concave and homogeneous of degree 1, and $F(0, L) = 0$.

The property of homogeneity of degree 1 is known to economists as constant returns to scale; if labor and capital are varied in the same proportion, then the same productive *methods* can be employed, with only their scale changed, and therefore output can be changed in the same proportion. This assumption is not fully true but may be accepted as an approximation. The assumption $F(0, L) = 0$ amounts to saying that capital is indispensable in production.

The transition equation of the system is still

$$\dot{K} = I - \delta K. \tag{48}$$

The constraint on the instruments, C and I, is now

$$F(K, L) \geqq C + I. \tag{66}$$

Technological progress can be stated formally as

$$Y = F(K, L, t), \tag{67}$$

that is, the output obtainable from fixed amounts of capital and labor varies with time, presumably increasing. A particular hypothesis about technological progress for which there is some evidence is that it is *labor-augmenting*, which has the more specific form:

$$Y = F[K, A(t)L],$$

that is, each worker at time t can do, in every way, exactly what $A(t)$ workers could do at time 0. In this form, however, we can see that we may as well retain (64) where, however, it is understood that L now represents not the number of workers in the usual sense but the number of efficiency-equivalent workers. Thus, in the new definition, L can and usually will be increasing more rapidly than N.

The Lagrangian is

$$H = N(t)\,U[C(t)/N(t)] + p(I - \delta K) + q[F(K,L) - C - I]. \tag{68}$$

The necessary conditions, with I unrestricted as to sign, become

$$p = q, \qquad U'[C(t)/N(t)] = p, \tag{69}$$

$$\dot{p} = p[\rho + \delta - (\partial F/\partial K)]. \tag{70}$$

The system looks much as it did before, but it is not autonomous since time enters explicitly through $N(t)$ and through $L(t)$ in $\partial F/\partial K$. It is possible to use a nonautonomous system, but autonomous systems are much more convenient; with appropriate changes of variables, together with additional assumptions, it is possible to state the system in autonomous form.

Since labor is growing, we can hardly expect an equilibrium in terms of the original variables, but it is reasonable to suppose there will be one in terms of ratios to the labor force. Divide all variables by L, and let small letters denote the resulting intensive magnitudes:

$$c = C/L, \qquad k = K/L, \qquad i = I/L. \tag{71}$$

Let

$$f(k) = F(k, 1). \tag{72}$$

Then, from (65),

$$f(k) \text{ is concave}, \quad f(0) = 0, \tag{73}$$

$$F(K,L) = LF(K/L, 1) = Lf(K/L), \tag{74}$$

so that $f(k)$ expresses the output per (effective) worker as a function of the capital per worker. Differentiate (74) partially with respect to K,

$$F_K = L(1/L)\,f'(K/L) = f'(K/L). \tag{75}$$

(70) can then be written

$$\dot{p}/p = \rho + \delta - f'(k). \tag{76}$$

Since $\log k = \log K - \log L$ (natural logarithms),

$$\dot{k}/k = (\dot{K}/K) - (\dot{L}/L).$$

Multiply through by k, note that $k/K = 1/L$, substitute from (48), and use the definitions (71),

$$\dot{k} = i - (\gamma + \delta)k, \tag{77}$$

where

$$\gamma = \dot{L}/L, \tag{78}$$

frequently referred to as the *natural rate of growth* of the economy (remember that L has been so defined as to reflect technical progress as well as labor force growth). Divide through in (66) by L, and use the definition (71) and (72),

$$f(k) \geqq c + i. \tag{79}$$

The equality will certainly always hold in (79). Elimination of i between (77) and (79) yields

$$\dot{k} = f(k) - (\gamma + \delta)k - c. \tag{80}$$

Now define

$$g(k) = f(k) - (\gamma + \delta)k; \tag{81}$$

from (73), $g(k)$ is concave, $g(0) = 0$. Then (76) and (80) can be written

$$\dot{p}/p = \rho - \gamma - g'(k). \tag{82}$$

$$\dot{k} = g(k) - c. \tag{83}$$

Finally, (69) can be written

$$U'[(L/N)c] = p. \tag{84}$$

The system (82-84) would be autonomous if the following two conditions are satisfied:

$$\gamma \text{ constant}, \tag{85}$$

$$L(t)/N(t) \text{ constant}. \tag{86}$$

This is the case of no technological progress and a constant rate of population and labor force growth. Then the equations have exactly the same form as those for the Ramsey model, with K, C, $F(K)$, and ρ replaced by k, c, $g(k)$, and $\rho - \gamma$, respectively. The importance of the last substitution must be stressed. The optimality analysis of the Ramsey case made use of the hypothesis, $\rho > 0$, to show that the transversality conditions were satisfied. This condition seems reasonable. But in the case of growth, the corresponding condition is $\rho > \gamma$; this is somewhat odd because the value of ρ is a value judgment while that of γ is an empirical fact. There seems no intrinsic reason why the inequality should hold in one direction or the other.

It must be remarked, moreover, that the hypothesis cannot be essentially weakened. In the Ramsey model without growth it can be shown that if $\rho < 0$, there is no optimal path in any meaningful sense; for a detailed analysis see Koopmans [**7**, p. 251-252, 279-285]; as just seen, the same result holds if the economy is growing at a constant rate γ and $\rho < \gamma$. The borderline case, $\rho = 0$ in the model without growth or $\rho = \gamma$ in a growing economy, has been studied in considerable detail by Ramsey [**16**], Koopmans [**7**, pp. 239-243 and 269-275], von Weizsäcker [**19**]. Alternative definitions of optimality are possible since the utility functional need not converge, and in general the existence of an optimal program in the borderline case depends on the specific properties of the production function.

To allow for technological progress, we wish to relax (86) and allow the ratio $L(t)/N(t)$ to be increasing. We still wish to arrive at an autonomous system. In the system (82)-(84) it is (84) which will no longer be autonomous. In general there is no transformation of the variables in (82)-(84) which will make the system autonomous, but such a transformation is possible if U' is homogeneous of some degree. Note that U' must be decreasing; therefore it must be homogeneous of some negative degree, say $-\sigma$.

(87) Assume that $U'(c)$ is homogeneous of degree $-\sigma$, $\sigma > 0$.

We also assume, to replace (86),

(88) $L(t)/N(t)$ has a constant rate of growth, τ,

which may be interpreted as the rate of (labor-augmenting) technological progress.

From (87) and (84),

$$[L(t)/N(t)]^{-\sigma}U'(c) = p,$$

or

$$U'(c) = p[L(t)/N(t)]^{\sigma}.$$

In an effort to reach an autonomous system it is then a good idea to define

$$\bar{p} = p[L(t)/N(t)]^{\sigma}, \tag{89}$$

so that

$$U'(c) = \bar{p}, \tag{90}$$

and then seek a differential equation for $\bar{p}$ to replace (82). Take the logarithm of both sides in (89), differentiate with respect to time and substitute from (82) and (88),

$$\dot{\bar{p}}/\bar{p} = \rho + \sigma\tau - \gamma - g'(k). \tag{91}$$

The system of equations (83), (91), and (90) is now again of the same form as the Ramsey model, with K, C, p, $F(K)$, and ρ being replaced by k, c, $\bar{p}$, $g(k)$, and $(\rho + \sigma\tau) - \gamma$, respectively. The last conditions mean that, for optimality, we need

$$\rho + \sigma\tau > \gamma, \tag{92}$$

with some possible cases of optimality when equality holds. It is also worth noting that, from (91), the equilibrium capital-labor ratio, k^{∞}, is defined by

$$g'(k^{\infty}) = \rho + \sigma\tau - \gamma.$$

From (81), this can be written

$$f'(k^{\infty}) - \delta = \rho + \sigma\tau. \tag{93}$$

The left-hand side is thus the equilibrium *net* marginal productivity of capital (net of depreciation, that is) and so, in usual economic terminology, the right-hand side is an equilibrium rate of interest.

REMARK 1. The existence condition (92) amounts to saying that the equilibrium rate of interest exceeds the rate of growth.

REMARK 2. The equilibrium rate of interest is higher the higher the rate of technological progress. Notice also that if $\tau = 0$, then the entire equilibrium does not depend in any way on the felicity function but only on the production function and the utility rate of discount, ρ. With technological progress, on the other hand, this ceases to be true; other things being equal, the marginal productivity of capital is higher (and therefore the capital-labor ratio, k^∞, is smaller) the higher σ, i.e. the more rapidly the individual becomes surfeited with goods.

REMARK 3. Note also that c is consumption per effective worker, not consumption per capita. As the optimal path converges, c converges to a limit; but since L/N increases at the constant rate τ, it follows that, asymptotically, consumption per capita will grow exponentially at the rate τ.

5. **Optimal growth in a dual economy.** It is a common hypothesis among economists that in underdeveloped countries there exist side-by-side two economic systems, one advanced and the other backward. The economic significance of this separation is that workers in the advanced economy receive a wage which may be much higher than anything received in the backward sector. At the same time, it is assumed that these workers save nothing, so that any capital accumulation must come out of the surplus of output over wage payments. For simplicity, assume there is no relevant product at all in the backward sector. It still may not be optimal for the economy to have full employment of the labor force in the advanced section; each additional worker creates more product, on the one hand, and a claim to a fixed portion of that product, on the other. Thus capital accumulation might be lower under full employment than with some unemployment.

For simplicity, it is assumed here that the population and available labor force are constant and that there is no technological progress; generalization in these directions can easily be carried out by the methods of the last section. The following discussion is based on the work of Marglin [12] and Dixit [4]. The Ramsey model is modified by adding one instrument and two constraints. The additional instrument is the amount of labor to be employed, L; the additional constraints are that there is a fixed parameter, w (wage rate in terms of goods), such that,

$$C - wL \geqq 0, \tag{94}$$

and that the amount of labor employed not exceed the fixed amount available,

$$\overline{L} - L \geqq 0. \tag{95}$$

Otherwise, the Ramsey conditions remain:

(24) maximize $\int_0^{+\infty} e^{-\rho t} U[C(t)]\,dt$.

(28) $\dot{K} = I$.

(66) $F(K, L) - C - I \geqq 0$.

(66) is substituted for (29) since the labor force is a variable of the problem; the function F is assumed to satisfy (65).

The Lagrangian can be written

$$U(C) + pI + q_1[F(K,L) - C - I] + q_2(C - wL) + q_3(\overline{L} - L). \tag{96}$$

Equate to zero the derivatives of the Lagrangian with respect to the three instruments, C, I, and L,

$$U'(C) = q_1 - q_2, \quad p = q_1, \quad q_1 F_L(K,L) = q_2 w + q_3,$$

or,

$$U'(C) = p - q_2, \tag{97}$$

$$pF_L = q_2 w + q_3 \tag{98}$$

where $F_L = \partial F / \partial L$. Of course,

$$q_2 \geqq 0, \qquad q_2(C - wL) = 0; \qquad q_3 \geqq 0, \qquad q_3(\overline{L} - L) = 0. \tag{99}$$

Since the constraint (66) is certainly effective, (28) and (66) imply,

$$\dot{K} = F(K,L) - C. \tag{100}$$

The auxiliary equation, as before, is

$$\dot{p}/p = \rho - F_K(K,L). \tag{101}$$

From (100) and (101), at an equilibrium,

$$F_K(K^\infty, L^\infty) = \rho, \qquad C^\infty = F(K^\infty, L^\infty). \tag{102}$$

To be an equilibrium of this system, however, (94) must be satisfied. Since F is homogeneous of degree 1, it is easy to prove that F_K is homogeneous of degree 0; the first equation in (102) can therefore be solved for K^∞ / L^∞. Write the second equation as

$$C^\infty / L^\infty = F(K^\infty / L^\infty, 1),$$

since $F(K, L)$ is homogeneous of degree 1. We will *assume* then that,

$$C^\infty / L^\infty > w. \tag{103}$$

Then the constraint (94) is not binding at equilibrium, and $q_2^\infty = 0$. Then, from (98), $q_3^\infty > 0$, so that (95) is binding, i.e. there is full employment. Thus, for K in the neighborhood of K^∞, the optimal path is identical with that for the Ramsey problem. Since, in the Ramsey problem, p is a decreasing function of K, and therefore C is an increasing function of K, it follows that the constraint (94) is fulfilled and ineffective for $K \geqq K^\infty$. It follows that there is $\overline{K} < K^\infty$ such that the optimal solution for the dual economy coincides with the Ramsey solution in the interval $\langle \overline{K}, +\infty \rangle$, which will be termed interval I.

$\overline{K}$ is defined by the condition that (94) becomes effective there. Since $p(K)$ is the same as for the Ramsey solution for $K \geqq \overline{K}$, it is now known for $K = \overline{K}$. Also, $q_3(\overline{K}) = p(\overline{K}) F_L(\overline{K}, \overline{L}) > 0$. As K decreases below $\overline{K}$, it must be that q_3 remains positive, at least for some interval, while q_2 rises above 0. Then constraints (94) and (95) are both effective in an interval to the left of $\overline{K}$, termed interval II, in which $C = w\overline{L}$, $\dot{K} = F(K, \overline{L}) - w\overline{L}$, so that, from (101),

$$dp/dK = p[\rho - F_K(K, \overline{L})]/[F(K, \overline{L}) - w\overline{L}] \text{ in interval II.} \tag{103}$$

Since $p(\overline{K})$ is known, this equation can be solved rather easily for smaller values of K.

Also in interval II, $q_2 = p - U'(w\overline{L})$, from (97), so that, from (98),

$$q_3 = p[F_L(K, \overline{L}) - w] + wU'(w\overline{L}). \tag{104}$$

Thus the lower end of interval II is defined by the condition $q_3 = 0$.

Since $F(0, L) = 0$ for all L, by (65), there exists K_1 so that $F(K_1, \overline{L}) = w\overline{L}$.

As K approaches $K_1 + 0$, the denominator of (103) is asymptotically equivalent to $F_K[K_1, \overline{L}]\ (K - K_1)$, so that clearly $p(K)$ approaches infinity. Also, from Euler's theorem on homogeneous functions,

$$w\overline{L} = F(K_1, \overline{L}) = F_L(K_1, \overline{L})\, \overline{L} + F_K(K_1, \overline{L})\, K_1 > F_L(K_1, \overline{L})\, \overline{L},$$

so that $F_L(K_1, \overline{L}) < w$. The first term of (104) then approaches $-\infty$, while the second is constant. Hence, $q_3(\underline{K}) = 0$ for some $\underline{K} > 0$.

Interval III is the interval $\langle 0, \underline{K} \rangle$. In this interval, the full employment condition, (95), ceases to be binding, and $q_3 = 0$. From (94), (97), and (98), we deduce

$$U'(wL) = p\{1 - [F_L(K, L)/w]\}, \tag{105}$$

which defines L as a function of K and p. The basic differential equation takes the form in interval III,

$$dp/dK = p[\rho - F_K(K, L)]/[F(K, L) - wL]. \tag{106}$$

It is to be noted that $dL/dK > 0$ in this interval (the more capital, the more labor can be employed). This means that, as we push the solution to lower values of K, the full employment constraint will never become binding again. To see that $dL/dK > 0$ in interval III, first note, from (105), that

$$1 - [F_L(K, L)/w] > 0 \text{ in interval III.}$$

Differentiate (105) totally with respect to K and group terms,

$$\begin{aligned}&[U''(wL)\,w + (p/w)\,F_{LL}(K, L)](dL/dK)\\ &\quad = -\,(p/w)\,F_{LK}(K, L) + \{1 - [F_L(K, L)/w]\}\,(dp/dK).\end{aligned} \tag{107}$$

From the concavity of U and F, it follows that $U'' < 0$, $F_{LL} < 0$. Since F_L is homogeneous of degree 0,

$$F_{LL}L + F_{LK}K = 0$$

by Euler's theorem; but since $F_{LL} < 0$, and $L, K > 0$, $F_{LK} > 0$. It is then easy to calculate, from (107), that

$$\text{if } dp/dK < 0, \text{ then } dL/dK > 0 \text{ in interval III.} \tag{108}$$

We now show that the denominator of (106) is positive and the numerator negative in the open interval $(0, \underline{K})$. If the denominator were not everywhere positive in the interval,

$$F[K_2, L(K_2)] - wL(K_2) = 0, \quad \text{for some } K_2,\ 0 < K_2 < \underline{K};$$

choose the largest such. Since $F(\underline{K}, \overline{L}) = F[\underline{K}, L(\underline{K})] > wL(\underline{K})$,

$$F[K, L(K)] - wL(K) > 0, \quad K_2 < K < \underline{K}, \tag{109}$$

and therefore

$$d\{F[K, L(K)] - wL(K)\}/dK \geqq 0 \quad \text{at } K = K_2. \tag{110}$$

Since F is homogeneous of degree one, strictly increasing in K, and concave,

$$F[K_2/L(K_2), 1] = w < F(K^\infty/L^\infty, 1),$$

which implies $K_2/L(K_2) < K^\infty/L^\infty$, and therefore $F_K[K_2, L(K_2)] > \rho$. From (106) and (109), $\lim_{K \to K_2+0}(dp/dK) = -\infty$ and, by (107), $\lim_{K \to K_2+0}(dL/dK) = +\infty$. Then

$$\begin{aligned}&\lim_{K \to K_2+0} d\{F[K, L(K)] - wL(K)\}/dK\\ &\quad = F_K[K_2, L(K_2)] + \{F_L[K_2, L(K_2)] - w\} \lim_{K \to K_2+0} (dL/dK) = -\infty,\end{aligned}$$

a contradiction to (110).

Now we show that $F_K[K, L(K)] > \rho$ for $0 < K < \underline{K}$. Since $\underline{K} < K^\infty$, $F_K(\underline{K}, \overline{L}) > F_K(K^\infty, L^\infty) = \rho$. Suppose

$$F_K[K^*, L(K^*)] = \rho, \quad \text{for some } K^*,\ 0 < K^* < \underline{K};$$

choose the largest such, so that $F_K(K, L) > \rho$, $K^* < K < \underline{K}$, and

$$K/L < K^\infty/\overline{L} \quad \text{for } K^* < K \leqq \underline{K},$$

while $K^*/L^* = K^\infty/\overline{L}$. Since $F_{LK} > 0$, F_L increases with K for fixed L; but since F_L is a function of K/L, F_L increases with K/L. Hence, $F_L(K^*, L^*) > F_L(K, L)$ for K in a right-hand neighborhood of K^*, where, it will be recalled, L is a function of K defined by (105), and L^* is its value at $K = K^*$. Therefore,

$$dF_L/dK \leqq 0, \quad \text{at } K = K^*.$$

But, $dF_L/dK = F_{LL}(dL/dK) + F_{LK}$. Compute dL/dK from (107), and recall that $dp/dK = 0$ at $K = K^*$. Then,

$$\begin{aligned} &dF_L/dK \\ &\quad = F_{LK}(K^*, L^*)\, U''(wL^*)\, w / [U''(wL^*)\, w + (p/w)\, F_{LL}(K^*, L^*)] \\ &\qquad\qquad > 0 \quad \text{at } K = K^*, \end{aligned}$$

a contradiction. Hence, the numerator of (106) is negative and the denominator positive, so that $p'(K) < 0$ for $0 < K < \underline{K}$; by (108) L is an increasing function of K in interval III (capital permits employment), and consumption is proportional to L.

References

1. K. J. Arrow, L. Hurwicz and H. Uzawa, *Constraint qualifications in nonlinear programming,* Naval Res. Logist. Quart. **8** (1961), 175-191.

2. K. J. Arrow and M. Kurz, *Optimal growth with irreversible investment in a Ramsey model,* Tech. Rep. No. 1 (NSF GS-1440), Inst. Math. Studies Social Sci., Stanford University, Stanford, Calif., 1967 (to be published in Econometrica).

3. R. Bellman, *Dynamic programming,* Princeton Univ. Press, Princeton, N. J., 1957.

4. A. Dixit, *Optimal development in the labor-surplus economy,* Rev. of Economic Studies **35** (1968), 23-34.

5. W. M. Gorman, *Convex indifference curves and diminishing marginal utility,* J. Political Econ. **65** (1957), 40-50.

6. H. Halkin, *On the necessary conditions for optimal control of nonlinear systems,* J. Analyse Math. **12** (1964), 1-82.

7. T. C. Koopmans, "On the concept of optimal economic growth" in Study Week on *The econometric approach to development planning,* North-Holland, Amsterdam, 1965, pp. 225-287.

8. H. W. Kuhn and A. W. Tucker, *Non-linear programming,* Proc. Second Berkeley Sympos. Math. Stat. and Prob., Univ. of California Press, Berkeley, Calif., 1951, pp. 481-492.

9. M. Kurz, *Optimal economic growth and wealth effects*, Tech. Rep. No. 136, Inst. Math. Studies Social Sci., Stanford University, Stanford, Calif., 1965, published in Internat. Economic Rev. **9** (1968), No. 2.

10. ———, *The general instability of a class of competitive growth processes*, Inst. Math. Studies Social Sci., Stanford University, Stanford, Calif., 1967, published in Rev. Economic Studies **35** (1968).

11. O. L. Mangasarian, *Sufficient conditions for the optimal control of nonlinear systems*, SIAM J. Control **4** (1966), 139-152.

12. S. A. Marglin, *Industrial development in the labor-surplus economy: An essay in the theory of optimal growth*, 1966 (unpublished manuscript).

13. P. Massé, *Les réserves et la régulation de l'avenir dans la vie économique*. I, II, Actualités Sci. Indust., nos. 1007, 1008, Hermann, Paris, 1946.

14. J. A. Mirrlees, *Optimum growth when the technology is changing*, Rev. Economic Studies **34** (1967), 95-124.

15. L. S. Pontryagin, V. G. Boltyanskii, R. V. Gamkrelidze and E. F. Mischenko, *The mathematical theory of optimal processes*, Interscience, New York, 1962.

16. F. P. Ramsey, *A mathematical theory of savings*, Economic J. **38** (1928), 543-559.

17. P. A. Samuelson, *The foundations of economic analysis*, Harvard Univ. Press, Cambridge, Mass., 1947.

18. J. Tinbergen, *On the theory of economic policy*, North-Holland, Amsterdam, 1952.

19. C. C. von Weizsäcker, *Existence of optimal programs of accumulation for an infinite time horizon*, Rev. Economic Studies **32** (1965), 85-104.

STANFORD UNIVERSITY
STANFORD, CALIFORNIA

David Gale
W. R. Sutherland

Analysis of a One Good Model of Economic Development

These notes were prepared by the first author above for presentation at the 1967 Summer Seminar of the American Mathematical Society on the Mathematics of the Decision Sciences, at Stanford University. The notes present the more basic results and techniques of the theory in the first three sections. The last section applies these techniques to derive some new results obtained jointly by the two authors during the past year. Acknowledgement is here made to the National Science Foundation which is supporting the Summer Seminar as well as the preparation of theses notes, and also to the Logistics and Mathematical Statistics Branch of the Office of Naval Research for support of some of the research reported on here.

1. **Introduction.** In these notes we are going to analyze an idealized, or better, an imaginary economy in which there is only one good. This good can be used for two purposes; (A) it can be consumed, thus creating satisfaction or *utility* for the people who consume it, or (B) it can be *invested*, in which case it creates additional amounts of itself. We will be concerned with an operation of this economy throughout time and therefore the problem at each instant will be to decide how much to consume and how much to invest in order to maximize utility throughout time in some suitably defined sense.

What is the purpose in considering this sort of imaginary situation which bears little resemblance to any actual economy, living or dead? The answer is that in analyzing this model we shall run into certain mathematical and economic techniques which turn out to be basic not only for the study of this make-believe economy but also for the more realistic (but more complicated) models which may come up in practice. Our aim is thus to isolate this technique in a simple context. The technique we refer to is what economists describe as the use of a *price system* and what mathematicians refer to as the method of *dual variables*. By whatever name one calls it, this subject is the central one both in economic analysis and modern optimization theory. Mathematically, it enables one to answer such questions about optimal development programs as: Do they exist? Are they unique? What are their qualitative properties? Economically it allows one to give a competitive market interpretation to these optimal paths along which, it turns out, producers are maximizing profits and consumers are maximizing utility subject to their budgetary limitations.

The above is a rough preview of what will be found in the rest of these notes. The overture is now ended and the show will begin.

2. **The model; finite time horizon.** The model will involve a single commodity which we will refer to as "goods". It is described by two functions, a *production function* $f_t(x)$ and a *utility function* $u_t(c)$ where $f_t(x)$ is the amount of goods produced at time $t+1$ from an investment of x units of goods at time t, and $u_t(c)$ is the satisfaction gained by consuming c units of goods at time t. The domain of t is the nonnegative integers and that of x and c the nonnegative reals.

DEFINITION 1. A *program* with *initial stocks* s is a sequence of pairs $\langle x_t, c_t \rangle$ finite or infinite such that

$$c_0 = s - x_0, \tag{2.1}$$

$$c_t = f_{t-1}(x_{t-1}) - x_t \quad \text{for } t > 0. \tag{2.2}$$

If $\langle x_t, c_t \rangle$ is a program, the corresponding *utility sequence* is is given by $\langle u_t(c_t) \rangle$.

Clearly, conditions (2.1) and (2.2) state that the sum of consumption and investment in period t is equal to the amount produced in the previous period. If the sequence $\langle x_t, c_t \rangle$ is finite with $t = 1, \cdots, T$ then it is called a *T-period program* and if $f_T(x_T) = s'$ we refer to the program as a T-period program with *initial stocks* s and *final stocks* s' or, more briefly, a *T-period program from s to s'*. The *value* of such a program is $\sum_{t=0}^{T} u_t(c_t)$.

DEFINITION 2. A T-period program for s to s' is called *optimal* if it has maximum value among all such programs.

Although our principal interest will be in infinite rather than finite programs it will be necessary first to develop the basic properties of finite optimal programs.

We now introduce the central concept of these notes.

DEFINITION 3. The program $\langle x_t, c_t \rangle$ is called *competitive* if there exist nonnegative numbers (prices) p_t such that

(A) $u_t(c) - p_t c$ is maximized at c_t for all t,

(B) $p_{t+1} f_t(x) - p_t x$ is maximized at x_t for all t.

These conditions have an important economic interpretation. Regarding p_t as prices we see that $p_t x$ is the cost of investing x units at time t, while $p_{t+1} f_t(x)$ is the return or value of $f_t(x)$ units at time $t+1$. The difference, therefore, represents profit and condition (B) requires that investment be chosen at each time t so so as to maximize profits.

To motivate condition (A) we note from (2.2) that $p_t c_t = p_t(f_{t-1}(x_{t-1}) - x_t)$ and the right-hand side here might be thought of as disposable income since it represents the value of goods just produced minus the cost of goods to be invested. If we then require consumers to spend no more than the amount $p_t c_t$ (budget constraint) condition (A) says that consumers will then consume so as to maximize their utility subject to this constraint.

The following simple result is the starting point for the theory.

THEOREM 1. *If $\langle x_t, c_t \rangle$ is a T-period program from s to s' which is competitive, then it is optimal.*

PROOF. Let (p_t), $t = 1, \cdots, T+1$ be the competitive prices and let $\langle x'_t, c'_t \rangle$ be any other program from s to s'. Then from (A), (2.1) and (2.2),

$$u_0(c'_0) - u(c_0) \leqq p_0(c'_0 - c_0) = p_0(s - x'_0) - p_0(s - x_0)$$
$$= -p_0 x'_0 + p_0 x_0,$$
$$u_t(c'_t) - u(c_t) \leqq p_t(c'_t - c_t)$$
$$= p_t(f_{t-1}(x'_{t-1}) - x'_t) - p_t(f_{t-1}(x_{t-1}) - x_t)$$
$$\text{for } t = 1, \cdots, T$$

and

$$0 = p_{T+1}(s' - s') = p_{T+1}f_T(x_T) - p_{T+1}f_T(x'_T),$$

and summing on t gives

$$\sum_{t=0}^{T}(u_t(c'_t) - u_t(c_t))$$

$$\leqq \sum_{t=0}^{T}\{(p_{t+1}\, f_t(x'_t) - p_t x'_t) - (p_{t+1}f_t(x_t) - p_t x_t)\}$$

where we have collected terms in x_t. But since each term in the sum on the right-hand side above is nonpositive from (B), it follows that

$$\sum u_t(c'_t) - \sum u_t(c_t) \leqq 0,$$

so $\sum u_t(c_t)$ is a maximum as asserted.

What we have here shown is that competitive programs are optimal. We need a converse to this theorem and for this purpose must make some assumptions about the functions f and u. These are

(I) The function f is nonnegative, concave and increasing in x (for each t) and $f_t(0) = 0$.

(II) The function u is concave and increasing in c, but possibly $u_t(0) = -\infty$.

The last condition of (II) is important for we would like to permit functions such as $u(c) = \log c$, $c \geqq 0$. The condition $u(0) = -\infty$ would mean that to consume nothing (starvation) is "infinitely bad". Unless otherwise stated it will be assumed henceforth that conditions (I) and (II) are satisfied.

We now recall the fundamental mathematical result needed for this work (which may well be the fundamental result of all optimization theory), namely the Kuhn-Tucker Theorem. We can get by with the following weak form.

KUHN-TUCKER THEOREM. *Let $u(x)$ and $f_i(x)$, $i = 1, \cdots, m$, be convex functions defined on a convex set X and let $\bar{x}$ minimize $u(x)$ in X subject to*

(2.3) $$f_i(x) \leqq 0, \qquad i = 1, \cdots, m.$$

Then if (2.3) *has a strict solution there exist numbers $p_i \geqq 0$ such that*

(2.4) $$u(x) - \sum p_i f_i(x) \text{ is minimized at } \bar{x}.$$

SUGGESTION FOR A DO-IT-YOURSELF PROOF. Let Y be the set of all $y = (y_1, \cdots, y_m)$ such that the inequalities

$$f_i(x) \leqq y_i$$

have a solution. Show that Y is convex and has 0 as an interior point (here we use the strict solution hypothesis). Now let $\mu(y) = \min_{f_i(x) \leqq y_i} u(x)$ and show that μ is a convex function of y. Then use the fact that a convex function ϕ has a *support* at every interior point $\bar{x}$ of its domain, i.e. there is a linear function $p \cdot x$ such that $p \cdot (x - \bar{x}) \leqq \phi(x) - \phi(\bar{x})$ for all x and X. The support of μ at 0 is the p we are looking for.

We can now get the desired converse for Theorem 1.

THEOREM 2. *Let $\langle \bar{x}_t, \bar{c}_t \rangle$ be an optimal program from s to s' and assume*

$$\bar{c}_t > 0 \text{ for at least one } t. \tag{2.5}$$

Then $\langle \bar{x}_t, \bar{c}_t \rangle$ is competitive.

REMARK. Without (2.5) the theorem would not be true. Suppose $f_t(x) = \rho x$ for some fixed ρ (i.e. f is linear) and suppose $u(c) = \log c$, $s = 1$, $s' = \rho^T$. Then clearly the only program from s to s' is $\langle x_t, c_t \rangle = \langle \rho^t, 0 \rangle$, but condition (A) requires there exist p_t such that

$$\log c - p_t c = \max \text{ at } c = 0$$

and clearly no such p_t exist. The fact that $\log 0 = -\infty$ is not crucial here. The same situation would occur for $u(c) = c^{1/2}$. The difficulty comes from the fact that the *slope* of u is infinite at $c = 0$.

PROOF. Replace conditions (2.1) and (2.2) by

$$\begin{gathered} c_0 + x_0 - s \leqq 0, \\ c_t + x_t - f_{t-1}(x_{t-1}) \leqq 0, \qquad t = 1, \cdots, T, \\ s' - f_T(x_T) \leqq 0. \end{gathered} \tag{2.6}$$

Now clearly $\langle \bar{x}_t, \bar{c}_t \rangle$ satisfies (2.6) and it also maximizes $\sum u_t(c_t)$ subject to (2.6), since each u_t is nondecreasing in c. Hence the Kuhn-Tucker Theorem applies provided we can show that (2.6) has a strict solution. Assuming this for the moment, we obtain numbers $p_t \geqq 0$, $t = 0, \cdots, T+1$ such that

$$(2.7)\qquad \sum_{t=0}^{T} u_t(c_t) - p_0(c_0 + x_0) - \sum_{t=1}^{T} p_t[(c_t + x_t) - f_{t-1}(x_{t-1})] + p_{T+1}f_T(x_T)$$

is maximized at $\langle \bar{x}_t, \bar{c}_t \rangle$. Rearranging (2.7) gives

$$(2.8)\qquad \sum_{t=0}^{T} (u_t(c_t) - p_t c_t) + \sum_{t=0}^{T} (p_{t+1} f_t(x_t) - p_t x_t)$$

is maximized at $\langle \bar{c}_t, \bar{x}_t \rangle$, but note that the terms of (2.8) are *independent*, hence (2.8) is maximized at $\langle \bar{x}_t, \bar{c}_t \rangle$ if and only if $u_t(c_t) - p_t c_t$ is maximized at $\bar{c}_t$ and $p_{t+1} f_t(x_t) - p_t x_t$ is maximized at $\bar{x}_t$ for all t, and these are precisely conditions (A) and (B).

To show that (2.6) has a strict solution we consider the new program $\langle \bar{x}_t, 0 \rangle$ and note that we have

$$(2.9)\qquad \bar{x}_0 - s \leqq 0, \qquad \bar{x}_t - f_{t-1}(\bar{x}_{t-1}) \leqq 0, \qquad s' - f_T(\bar{x}_T) \leqq 0,$$

and at least one of the above inequalities is strict by assumption (2.5). We therefore reduce the problem to the following.

LEMMA 1. *If* (2.9) *has a solution with one strict inequality then it has a strict solution.*

PROOF. Induction on T. If $T = 0$ we have

$$x_0 - s \leqq 0, \quad s' - f(x_0) \leqq 0.$$

If $x_0 - s < 0$, then by slightly increasing x_0 if necessary, we can assure that $s' - f(x_0) < 0$ also, since f is increasing. If $s' - f(x_0) < 0$ then by slightly decreasing x_0 we can assure $x_0 - s < 0$ as well.

Now suppose one of the inequalities (2.9) is strict for some $t_0 > 0$. Then by the induction hypothesis there is a solution x_t' giving strict inequality for all but the first inequality, and to get a strict solution we slightly decrease x_0' if necessary. In the other case we have $x_0 - s < 0$ so inductively there is a solution x_t' satisfying all but the last inequality strictly and this will be satisfied too by a slight increase in x_T'.

The fact that the class of optimal and competitive programs are identical is of economic interest in itself as it shows that if the "prices are right" optimality is attained by allowing producers and consumers to act purely selfishly and maximize profits and utility respectively. We shall now show how the price theorem can be used to gain qualitative information about the nature of optimal programs. For the rest of this section we will assume that the functions f and u are independent of the time.

DEFINITION 4. The function f will be called *productive* if for any $x \geqq 0$, $h > 0$, $f(x+h) > f(x) + h$. In words, increasing the input by some amount will increase the output by more than that amount so production is always better than storage. For f differentiable, this is equivalent to $f'(x) > 1$.

THEOREM 3. *If f is productive, then for any competitive program $\langle x_t, c_t; p_t \rangle$*

(a) *prices p_t are positive and decreasing in t,*

(b) *consumption c_t (and hence utility) is nondecreasing in t,*

(c) *stocks x_t are nondecreasing up to some time t_0 and decreasing thereafter.*

PROOF. (a) We first note from (A) that c_t maximizes $u(c) - p_t c$. This shows that $p_t > 0$ since otherwise u, being increasing, would have no maximum. Next, from (B)

$$p_{t+1} f(x_t) - p_t x_t \geqq p_{t+1} f(x) - p_t x \quad \text{for all } x \geqq 0,$$

or

$$p_t / p_{t+1} \geqq (f(x) - f(x_t)) / (x - x_t) \quad \text{for all } x > x_t;$$

but since f is productive the right-hand side above is greater than 1; hence $p_{t+1} < p_t$.

(b) From (A)

$$u(c_t) - u(c_{t+1}) \geqq p_t (c_t - c_{t+1}),$$

$$u(c_{t+1}) - u(c_t) \geqq p_{t+1} (c_{t+1} - c_t);$$

hence

$$0 \geqq (p_t - p_{t+1})(c_t - c_{t+1}),$$

(this relation is sometimes called La Chatelier's principle, I think). But from (a) $p_t - p_{t+1} > 0$ hence $c_t - c_{t+1} \leqq 0$ as asserted.

(c) It will suffice to show that if $x_t < x_{t-1}$ then $x_{t+1} < x_t$. Now

$$x_{t+1} - x_t = f(x_t) - f(x_{t-1}) - (c_{t+1} - c_t) \leqq f(x_t) - f(x_{t-1}) \text{ from (b).}$$

But since $x_{t-1} < x_t$ and if f is nondecreasing it follows that

$$x_{t+1} - x_t < 0.$$

3. **Infinite programs.** The finite horizon programs are not of great interest in economic development. It is true that if one were devising, say a five year plan and had decided on the final stocks s' then it would be natural to try to solve the problem of the previous section. However, the important decision would in this case already have been made, namely the choice of s'. The main problem in economic planning is to set reasonable goals for capital accumulation and it appears that the only way to attack this is to consider infinite programs. The first thing needed is a notion of optimality.

DEFINITION 5. If $\langle x_t, c_t\rangle$ and $\langle x_t', c_t'\rangle$ are infinite programs we say that $\langle x_t, c_t\rangle$ *overtakes* $\langle x_t', c_t'\rangle$ if there exists a time T such that

$$\sum_{t=0}^{T'} u_t(c_t) > \sum_{t=0}^{T'} u_t(c_t') \quad \text{for all } T' \geqq T.$$

We say that $\langle x_t, c_t\rangle$ *catches up to* $\langle x_t', c_t'\rangle$ (at infinity) if

$$\liminf_{T\to\infty} \sum_{t=0}^{T} (u(c_t) - u(c_t')) \geqq 0.$$

A program will be called *optimal* (*strongly optimal*) if it catches up to (overtakes) every other program.

We remark that if it should happen that the series $\sum_{t=0}^{\infty} u_t(c_t)$ converge for all programs (as may occur, for instance if future utilities are suitably discounted) then Definition 5 corresponds to choosing as the optimal program the one whose utility sum is greatest, just as in the finite case. However, Definition 5 is more general for, as we shall see, optimal programs in this broader sense may exist although all the utility series are divergent. In the next section we shall give a specific rather general existence theorem for the case when u and f are independent of the time. In the present section we shall obtain infinite analogues to Theorems 1 and 2 relating optimal and competitive programs. Note in this connection that the definition of a competitive program requires no modification for the infinite case since conditions (A) and (B) carry over as given.

THEOREM 4. *Any optimal program* $\langle x_t, c_t\rangle$ *is competitive.*

PROOF. We first dispose of a trivial case in which $c_t = 0$ for all but a finite number of times t. This means that all stocks are completely consumed by the end of T time periods for some T, so we are back in the finite case of Theorem 2 where the final stocks s' are zero.

In all other cases $c_t > 0$ for infinitely many t. Note that if we truncate the program at $t = T$ we have an optimal T-period program (with final stocks $f_T(x_T)$), so for each T there exist prices p_t^T which satisfy (A) and (B) for $t \leqq T$. Denoting by Π_t^T the set of all such prices one verifies that Π_t^T is a closed interval, possibly unbounded above, of nonnegative numbers. Also $\Pi_t^{T+1} \subset \Pi_t^T$ since if (A) and (B) are satisfied for $t \leqq T+1$ they are satisfied for $t \leqq T$, so it remains to show $\Pi_t = \bigcap_{T=1}^{\infty} \Pi_t^T$ is nonempty, and this will follow from the nested interval theorem if we can show that for any t there exists T such that Π_t^T is bounded. We first note that Π_T^T is bounded if $c_T > 0$, for from (A) if $p_T^T \in \Pi_T^T$ then

$$u_T(c_T) - p_T^T c_T \geqq u_T(c_T/2) - p_T^T c_T/2$$

or

$$p_T^T \leqq 2(u_T(c_T) - u_T(c_T/2))/c_T.$$

Now for any t, from (B),

$$p_{t+1}^T f_t(x_t) - p_t^T x_t \geqq 0, \qquad (\text{since } f_t(0) = 0),$$

so $p_t^T \leqq (f_t(x_t)/x_t) p_{t+1}^T$. Letting $q_t = f_t(x_t)/x_t$, we have

$$p_t^T \leqq q_t q_{t+1} \cdots q_{T-1} p_T^T$$

and this establishes the desired bound, and shows the existence of the competitive prices.

We would like now to establish some sort of analogue of Theorem 1 asserting that competitive programs are optimal, but since we do not have the concept of final stocks, some additional condition will be required. Before continuing we consider a concrete example.

Example 1. Let $u(c) = -1/c$, $f(x) = \rho x$, $s = 1$, where ρ is some positive constant.

Proposition 1. *The sequence $\langle x_t, c_t \rangle$ is a program if and only if $\sum_{t=0}^{\infty} c_t/\rho^t \leqq 1$.*

Proof. We have

$$c_0 = 1 - x_0, \qquad c_t = \rho x_{t-1} - x_t.$$

Multiplying the tth equation by $1/\rho^t$ and summing gives

$$\sum_{t=0}^{T} c_t/\rho^t = 1 - x_T/\rho^T.$$

Conversely, let $q_T = \sum_{t=0}^{T} c_t/\rho^t$ and let $x_T = \rho^T(1 - q_T)$. Then $x_0 = 1 - c_0$ and $x_T - \rho x_{T-1} = \rho^T(1 - q_T) - \rho\rho^{T-1}(1 - q_{T-1}) = \rho^T(q_T - q_{T-1}) = c_T$, so $\langle x_T, c_T \rangle$ is a program.

PROPOSITION 2. *Competitive programs exist if and only if* $\rho > 1$.

PROOF. Let p_t be competitive prices. Then from (A) $u(c) - p_t c = -1/c + p_t c$ is maximized at c_t; thence $u'(c_t) = p_t = 1/c_t^2$ or

$$c_t = 1/p_t^{1/2} \tag{3.1}$$

and hence c_t and x_t are positive for all t. From (B), $(p_{t+1}\rho - p_t)\,x$ is maximized at x_t, so $p_{t+1} = p_t/\rho$; hence

$$p_t = p_0/\rho^t. \tag{3.2}$$

Letting $\sigma = \rho^{1/2}$ we have from (3.1) $c_t = \sigma^t/p_0^{1/2}$ and $c_t/\rho^t = 1/(p_0^{1/2}\sigma^t)$. The series $\sum_{t=0}^{\infty} c_t/\rho^t$ will converge if and only if $\rho > 1$, in which case

$$\sum_{t=0}^{\infty} c_t/\rho^t = \sigma/p_0^{1/2}(\sigma - 1); \tag{3.3}$$

so $\langle x_t, c_t \rangle$ is competitive if and only if $c_t = 1/p_0^{1/2}\sigma^t$, where $p_0 \geqq (\sigma/\sigma - 1)^2$.

PROPOSITION 3. *The optimal program is* $\langle \bar{x}_t, \bar{c}_t \rangle$, *where* $\bar{c}_t = \sigma^t - \sigma^{t-1}$.

If there is any optimal program it must be competitive by Theorem 4 and clearly the best of the competitive programs (3.1) is the one for which $p_0 = (\sigma/\sigma - 1)^2$. However, we can prove directly that this program is optimal. Let $\langle c_t, x_t \rangle$ be any other program. From (A) and (3.2) we have

$$u(c_t) - (p_0/\rho^t)\,c_t \leqq u(\bar{c}_t) - (p_0/\rho^t)\,\bar{c}_t$$

or

$$u(c_t) - u(\bar{c}_t) \leqq p_0(c_t/\rho^t - \bar{c}_t/\rho^t). \tag{3.4}$$

Further, if for some t, $\bar{c}_t \neq c_t$, then (3.4) is strict. Summing on t gives

$$\sum_{t=1}^{T} u(c_t) - u(\bar{c}_t) < p_0 \sum_{t=1}^{T} (c_t/\rho^t - \bar{c}_t/\rho^t), \tag{3.5}$$

but from Proposition 1 and the fact that $\sum_{t=1}^{\infty} \bar{c}_t/\rho^t = 1$, it follows that the right-hand side of (3.5) converges to some nonpositive number and hence the left-hand side must eventually become and remain negative, proving the asserted optimality.

We now prove a converse of Theorem 4.

THEOREM 5. *If* $\langle \bar{x}_t, \bar{c}_t; p_t \rangle$ *is competitive and*

$$\lim_{t \to \infty} p_t \bar{x}_t = 0, \tag{3.6}$$

then $\langle \bar{x}_t, \bar{c}_t \rangle$ *is optimal.*

PROOF. Let $\langle x_t, c_t \rangle$ be any program from s and let π_t denote the profit from this program at prices p_t in the period t; that is,

$$\pi_t = p_t f(x_{t-1}) - p_{t-1} x_{t-1}, \qquad \bar{\pi}_t = p_t f(\bar{x}_{t-1}) - p_{t-1} \bar{x}_{t-1}.$$

From (A) we have

$$\begin{aligned} u(c_0) - u(\bar{c}_0) &\leqq p_0(c_0 - \bar{c}_0) = p_0(s - x_0) - p_0(s - \bar{x}_0) \\ &= -p_0(x_0 - \bar{x}_0) \\ u(c_t) - u(\bar{c}_t) &\leqq p_t(c_t - \bar{c}_t) \\ &= p_t(f(x_{t-1}) - x_t) - p_t(f(\bar{x}_{t-1}) - \bar{x}_t), \quad t \geqq 1, \end{aligned} \tag{3.7}$$

so

$$\sum_{t=0}^{T} u(c_t) - \sum_{t=0}^{T} u(\bar{c}_t) \leqq \sum_{t=1}^{T} (\pi_t - \bar{\pi}_t) + p_T \bar{x}_T - p_T x_T. \tag{3.8}$$

From (B) $\pi_t \leqq \bar{\pi}_t$ so the sum on the right is nonpositive and since $p_T \bar{x}_T \to 0$, it follows that the entire right-hand side becomes less than any preassigned positive number for T sufficiently large, which is the definition of optimality.

COROLLARY. *If* u *is strictly concave, then* $\langle \bar{x}_t, \bar{c}_t \rangle$ *is strongly optimal.*

PROOF. In this case, if $c_t \neq \bar{c}_t$ for some t, then the corresponding inequality of (3.7) becomes strict and the argument above shows that (3.8) becomes negative.

We now give an important equivalent interpretation to condition (3.6).

We define $p_0 s$ to be the *initial wealth* of the economy. We define

$$W_T = p_0 s + \sum_{t=1}^{T} \pi_t, \text{ the } \textit{accumulated wealth} \text{ up to period } T,$$

$$E_T = \sum_{t=0}^{T} p_t c_t, \text{ the } \textit{expenditure on consumption} \text{ up to period } T,$$

$$p_T x_T = \textit{value of stocks} \text{ in period } T.$$

Then we have the following obvious identity:

$$p_T x_T = W_T - E_T,$$

which is obtained by multiplying the tth equation of (2.1), (2.2) by p_t and adding.

In particular, we have

$$E_T \leqq W_T, \tag{3.9}$$

which is an obvious budget inequality, stating that expenditure on consumption cannot exceed accumulated wealth. Condition (3.6) now becomes

$$\lim_{T \to \infty} (W_T - E_T) \equiv 0 \tag{3.10}$$

so that "at infinity" all wealth has been used up in consumption.

The condition seems like a reasonable one for optimality. However, it is not a necessary condition. One can show that for cases in which the function f is not productive, so that eventually $f(x) < x$, then $p_T x_T$ converges to some positive value rather than to zero.

We call a program *efficient* if it satisfies (3.10).

4. **The time independent case.** In this section we confine ourselves to the case where f and u are independent of time. We need one more assumption which is a strengthening of Definition 4.

DEFINITION 5. The function f is *strongly productive* if there is a constant $\rho > 1$ such that

$$f(x+h) > f(x) + \rho h.$$

For f differentiable this is equivalent to $f'(x) > \rho$.

EXISTENCE THEOREM. *If f is strongly productive, there exists an optimal program if and only if u is bounded.*

This theorem was originally proved by D. McFadden for the case of f a linear function. We first prove the necessity of the boundedness condition.

LEMMA 2. *If $\langle x_t, c_t; p_t \rangle$ is competitive, then $p_{t+1}/p_t \leqq 1/\rho$ and $p_t \leqq p_0/\rho^t$.*

PROOF. From (B),

$$p_{t+1} f(x_t) - p_t x_t \geqq p_{t+1} f(x) - p_t x \quad \text{for all} \quad x \geqq 0$$

or

$$p_{t+1}(f(x_t) - f(x)) \geqq p_t(x_t - x)$$

or

$$p_{t+1}/p_t \leqq (x_t - x)/(f(x_t) - f(x)) \leqq 1/\rho \quad \text{for } x > x_t$$

from which the result follows.

Lemma 3. *If $\langle x_t, c_t; p_t \rangle$ is competitive then p_t approaches 0 and x_t and c_t approach ∞ monotonically.*

Proof. The first assertion follows from the previous lemma. Suppose (c_t) were bounded. Then there would exist $\bar{c}$ such that $c_t < \bar{c}$ and $u(\bar{c}) - u(c_t) \geqq \delta > 0$ for all t. But from (A),

$$p_t(\bar{c} - c_t) \geqq u(\bar{c}) - u(c_t) > \delta \quad \text{for all } t,$$

and we have seen that the left-hand side above approaches zero, giving a contradiction. Since c_t becomes infinite, so does x_t and monotoneity follows from Theorem 3.

Theorem 6. *If there is an optimal program, then u must be bounded.*

Proof. Let $\langle x_t, c_t \rangle$ be an optimal, hence competitive, program. From (A),

$$u(c_1) - u(c_0) \leqq p_0(c_1 - c_0) = p_0(f(x_0) - x_1) - p_0(s - x_0),$$

$$u(c_{t+1}) - u(c_t) \leqq p_t(c_{t+1} - c_t)$$

$$= p_t(f(x_t) - x_{t+1}) - p_t(f(x_{t-1}) - x_t).$$

Summing from $t = 1$ to T,

$$u(c_{T+1}) - u(c_0) \leqq p_T(x_T - x_{T+1})$$

$$+ \sum_{t=1}^{T} [(p_t f(x_t) - p_{t-1}x_t) - (p_t f(x_{t-1}) - p_{t-1}x_{t-1})] + p_0(f(x_0) - s),$$

but the first term above is nonpositive by Lemma 3 and the terms in the summation are nonpositive from (B). Hence

$$u(c_{T+1}) \leqq u(c_0) + p_0(f(x_0) - s),$$

so $u(c_t)$ is bounded for all t, but since $c_t \to \infty$ this means that u is bounded.

If u is bounded, we establish the existence of an optimal program by taking the limit as $T \to \infty$ of T-period programs as follows.

Let $P^T = \langle x_t^T, c_t^T; p_t^T \rangle$ be a T-period program which maximizes $\sum_{t=0}^T u(c_t)$ (the final stocks in this case are zero). Now for a fixed t, the sets $\{x_t^T\}$ and $\{c_t^T\}$ are bounded for all T. If in addition we knew that $\{p_t^T\}$ was bounded, then a standard "diagonal process" argument would establish the existence of a competitive program $\bar{P}$, a pointwise limit of the programs P^T. Our procedure will be first to prove the boundedness of $\{p_t^T\}$ and then to show that $\bar{P}$ is efficient and hence optimal, by Theorem 5.

We first need a fundamental inequality.

Lemma 4. *If $\langle x_t, c_t; p_t \rangle$ is competitive, then*

$$(*) \qquad \sum_{t=T_1}^{T} p_t c_t \leqq (\rho/\rho - 1)\,[p_{T_1} c_{T_1} + u(c_{T+1}) - u(c_{T_1})].$$

Proof. From (A),

$$u(c_{T+1}) - u(c_{T_1}) = \sum_{T_1+1}^{T+1} [u(c_t) - u(c_{t-1})] \geqq \sum_{T_1+1}^{T+1} p_t(c_t - c_{t-1})$$

$$= p_{T+1}c_{T+1} + \sum_{T_1}^{T} (p_t - p_{t+1})\, c_t - p_{T_1} c_{T_1}$$

$$\geqq (1 - 1/\rho) \sum p_t c_t - p_{T_1} c_{T_1} \quad \text{from Lemma 2,}$$

and rearranging gives (*).

An immediate economic application of this lemma is

Theorem 7. *If u is bounded (above and below) then there is a number M such that for any competitive program $\langle x_t, c_t; p_t \rangle$ the quantity $E_T = \sum_{t=0}^T p_t c_t \leqq M$ for all T.*

Proof. Applying (*) with $T_1 = 0$ gives

$$E_T \leqq (\rho/\rho - 1)\,[u(c_{T+1}) - (u(c_0) - p_0 c_0)],$$

but from (A), $u(c_0) - pc_0 \geqq u(0)$, so if $u(c) \leqq \mu$ for all c then choose $M = \rho/\rho - 1\,[\mu - u(0)]$.

Corollary. If $\langle x_t, c_t; p_t \rangle$ *is an infinite competitive program, then $\sum_{t=0}^{\infty} p_t c_t$ converges.*

Lemma 5. *For the programs P^T the prices p_0^T satisfy*

$$p_0^T \leqq M/s.$$

Proof. Since the final stocks $x_T = 0$ in P^T, inequality (3.9) becomes

$$p_0^T s + \sum_{t=1}^{T} \pi_t^T = E_T^T \leqq M \text{ by Theorem 7,}$$

and since $\pi_t^T \geqq 0$ the result follows.

COROLLARY. *The prices* p_t^T *satisfy* $p_t^T \leqq M/s\rho^t$.

PROOF. Lemma 2.

THEOREM 8. *There exists an infinite competitive program.*

PROOF. Take the pointwise limit of the programs P^T and call this limit $\overline{P}$. It is a standard exercise to verify that $\overline{P}$ is a competitive program.

To complete the existence theorem we must prove that $\overline{P}$ is efficient. Let $\overline{E} = \sum_{t=0}^{\infty} \bar{p}_t \bar{c}_t$ which exists by the Corollary to Theorem 7. Let $\overline{W} = \bar{p}_0 s + \sum_{t=1}^{\infty} \bar{\pi}_\epsilon$. We must show that this expression converges and that $\overline{E} = \overline{W}$.

LEMMA 6. *If* $\langle x, c_t; p_t \rangle$ *is competitive and* u *is bounded above, then* $\lim_{t\to\infty} p_t c_t = 0$.

PROOF. Let $\mu = \sup_{c \geqq 0} u(c)$. For $\epsilon > 0$ choose $\bar{c}$ so that $\mu - u(\bar{c}) \leqq \epsilon/2$ and choose T so that for all $t > T$, $c_t \geqq 2\bar{c}$. Then from (A),

$$u(c_t) - p_t c_t \geqq u(\bar{c}) - p_t \bar{c}$$

so

$$p_t c_t - p_t \bar{c} \leqq u(c_t) - u(\bar{c}) \leqq \mu - u(\bar{c}) \leqq \epsilon/2,$$

hence

$$p_t c_t/2 \leqq \epsilon/2 \text{ and } p_t c_t \leqq \epsilon.$$

It will be convenient to consider the program P^T to be infinite with the convention that for $t > T$ $x_t = c_t = 0$.

LEMMA 7. *For any* $\epsilon > 0$ *there exist* t_ϵ *such that*

$$\sum_{t=t_\epsilon}^{\infty} p_t^T c_t^T < \epsilon \text{ and } \sum_{t=t_\epsilon}^{\infty} \pi_t^T < \epsilon \text{ for all } T.$$

PROOF. From Lemma 5 (Corollary) we can choose t_1 so that $p_{t_1}^T$ is arbitrarily small, but as $p_t \to 0$, we have $c_t \to \infty$ (Lemma 3) hence $u(c_t) \to \mu = \sup_{c \geqq 0} u(c)$ and $p_t c_t \to 0$ since p_t is a support of u at c_t (Lemma 5) then we can choose t_ϵ so that $(\rho/\rho - 1) p_{t_\epsilon}^T c_{t_\epsilon}^T \leqq \epsilon/2$ and $(\rho/\rho - 1)(\mu - u(c_{t_\epsilon}^T)) \leqq \epsilon/2$ for all T. Now apply (*) and we have

$$\sum_{t=t_\epsilon}^{T} p_t^T c_t^T \leqq (\rho/\rho - 1)\left[p_{t_\epsilon}^T c_{t_\epsilon}^T + (u(c_T) - u(c_{t_\epsilon}))\right] \leqq \epsilon \text{ for all } T.$$

Finally, by (3.9), $\sum_{t=0}^{t_\epsilon - 1} p_t^T c_t^T \leqq p_0^T s + \sum_{t=1}^{t_\epsilon - 1} \pi_t^T$ so

$$\sum_{t=t_\epsilon}^{\infty} \pi_t^T = \sum_{t=t_\epsilon}^{T} \pi_t^T \leqq \sum_{t=t_\epsilon}^{T} p_t^T c_t^T = \sum_{t=t_\epsilon}^{\infty} p_t^T c_t^T \leqq \epsilon.$$

THEOREM 9. *The program* $\bar{P}$ *is efficient.*

PROOF. Let $E^T = \sum_{t=0}^{\infty} p_t^T c_t^T$ and let $W^T = p_0^T s + \sum_{t=1}^{\infty} \pi_t^T$ and let $\bar{E} = \sum_{t=0}^{\infty} \bar{p}_t \bar{c}_t$. Now for $T', T > t_\epsilon$ it follows from Lemma 7 that $E^{T'} - E^T \leqq \epsilon$ and $W^{T'} - W^T < \epsilon$, so (E^T) and (W^T) are Cauchy sequences and converge to their pointwise limits $\bar{E}$ and $\bar{W}$. But $E^T = W^T$ for all T; hence $\bar{E} = \bar{W}$, completing the proof.

APPENDIX. THE CASE OF MORE THAN ONE GOOD.

In these notes the entire analysis of optimal programs has been based on the use of competitive prices, and the existence of these prices therefore played a key role. To establish their existence we were at some pains in the proofs of Theorems 4 and 8 to obtain an a priori bound on the values of prices for finite horizon programs. This boundedness requirement is no mere mathematical technicality but is quite essential to the understanding of the models. We will here illustrate this further by considering a very simple two good model in which there is an obvious optimal program which, however, is not competitive.

The model involves both a *production good* P and a *consumption good* Q, and there is a single joint process for producing both. Namely, from x units P invested in period t one obtains ρx units of P and x units of Q in period $t+1$. Assuming initial stock of P is 1 a program $\langle x_t, c_t \rangle$ must satisfy

$$x_0 \leqq 1, \quad x_t \leqq \rho x_{t-1} \text{ and } c_t \leqq x_t \text{ for all } t.$$

The inequalities here simply have the meaning that one can throw away either production or consumption goods.

Now, it is perfectly clear that by any reasonable definition of optimality the only optimal program is $x_t = c_t = \rho^t$ since any other program involves needless throwing away. It also follows from the Kuhn-Tucker Theorem that every T-period optimal program is competitive for any utility function u. However,

THEOREM 10. *If* $\rho > 1$ *and the utility function* u *is unbounded, then the optimal program is not competitive.*

PROOF. We must first write down the competitive conditions. Let p_t and q_t be the prices of P and Q in period t. Condition (A) then remains

(A′) $u(c) - q_t c$ is maximized at c_t,

and the profit condition (B) at time t is clearly

(B′) $q_t x + p_{t+1} \rho x - p_t x$ is maximized at x_t.

Suppose now that $\langle \rho^t, \rho^t \rangle$ is optimal. Then from (B′) we must have

$$q_t = (p_t - \rho p_{t+1}) \tag{1}$$

and from (A′)

$$u(\rho^t) - u(\rho^{t+1}) \geqq q_t(\rho^t - \rho^{t+1}) = (1 - \rho)\, \rho^t (p_t - \rho p_{t+1}).$$

Summing from $t = 0$ to $T - 1$ gives

$$u(1) - u(\rho^T) \geqq (1 - \rho)(p_0 - \rho^T p_T)$$

or

$$(\rho - 1)\, p_0 \geqq u(\rho^T) - u(1) + (\rho - 1)\, \rho^T p_T \geqq u(\rho^T) - u(1) \quad \text{for all } T,$$

but if u is unbounded this is impossible, since p_0 would have to be infinite.

UNIVERSITY OF CALIFORNIA, BERKELEY
UNIVERSITY OF TORONTO

VIII
DYNAMIC PROGRAMMING

Cyrus Derman

Markovian Decision Processes— Average Cost Criterion[1]

1. **Introduction.** We are concerned with the optimal control of certain types of dynamic systems. We assume such a system is observed periodically at times $t = 0, 1, \cdots$. After each observation the system is classified into one of a possible number of states. Let I denote the space of possible states. I will be assumed to be either finite or denumerable. After each classification one of a possible number of decisions is implemented. Let K_i denote the number of possible decisions when the system is in state i, $i \in I$. The sequence of implemented decisions interacts with the chance environment to effect the evolution of the system.

More specifically, let $\{Y_t\}$, $t = 0, 1, \cdots$, denote the successive observed states of the system. Let $\{\Delta_t\}$, $t = 0, 1, \cdots$, denote the successive decisions. Assume that when $Y_t = i$ and $\Delta = k$ a known cost w_{ik} is incurred. The numbers $\{w_{ik}\}$ may be expected costs rather than actual costs. In such a case we assume a distribution of costs dependent upon the state i and decision k from which the expected cost can be computed.

A *rule* or *policy* R for controlling the system is a set of functions

[1] This research was supported by the Army, Navy, Air Force and NASA under a contract administered by the Office of Naval Research; Contract Nonr 266(55)-NR-042-099. Reproduction in whole or in part is permitted for any purpose of the United States Government.

$\{D_k(h_{t-1}, Y_t)\}$, $t = 0, 1, \cdots$, where $h_t = \{Y_0, \Delta_0, \cdots, Y_t, \Delta_t\}$, $D_k(\cdot) \geqq 0$, and $\sum_k D_k(\cdot) = 1$; $D_k(h_{t-1}, Y_t)$ is to be interpreted as the probability of implementing decision k at time t given the "history" h_{t-1} and the "present state" Y_t. Thus a rule specifies, at each point in time, a chance mechanism to be used in deciding which action to take. The rule is only permitted to depend on the history of states and decisions.

Given a rule R and a probability distribution over the initial state Y_0, we assume the sequence $\{Y_t, \Delta_t\}$ is a stochastic process defined over the joint space of I and the possible decisions. Throughout we shall assume that $P(Y_0 = i)$, $i \in I$, is known. Moreover, we assume that there are known transition probabilities $\{q_{ij}(k)\}$ such that for every i, j and k

$$q_{ij}(k) = P\{Y_{t+1} = j \mid h_{t-1}, Y_t = i, \Delta_t = k\}$$

independent of t and h_{t-1}. In words, when, at any time t, the state i is observed and decision k is made, then $q_{ij}(k)$ denotes the probability that the system will be observed in state j at time $t + 1$. Under this latter assumption, we refer to the process $\{Y_t, \Delta_t\}$ as a *Markovian Decision Process.* It shall be emphasized, however, that $\{Y_t, \Delta_t\}$ is not necessarily a Markov process. For when R is such that $D_k(h_{t-1}, Y_t)$ is a function of h_{t-1}, the process $\{Y_t, \Delta_t\}$ will not be Markovian. If, however, $D_k(h_{t-1}, Y_t)$ is a function of Y_t and t for every $t = 0, 1, \cdots$, then $\{Y_t\}$ is a Markov process. And, if $D_k(h_{t-1}, Y_t)$ is only a function of Y_t for every t, then $\{Y_t\}$ is a Markov chain with stationary transition probabilities.

Let W_t, $t = 0, 1, \cdots$, be defined as follows:

$$W_t = w_{ik} \text{ if } Y_t = i,\ \Delta_t = k, \qquad k = 1, \cdots, K_i,\ i \in I.$$

Given a policy R and an initial state $Y_0 = i$, then the sequence $\{W_t\}$, $t = 0, 1, \cdots$, is a stochastic process. We can speak of the expected cost at time t as

$$E_R W_t = \sum_j \sum_k w_{jk} P_R\{Y_t = j,\ \Delta_t = k \mid Y_0 = i\}$$

where E_R and P_R denote the expectation and probability under the policy R. We, of course, assume that the costs $\{w_{ik}\}$ are such that $E_R W_t$ exists.

Let

$$\phi_{R,T}(i) = \frac{1}{T+1} \sum_{t=0}^{T} E_R W_t,$$

i.e. $\phi_{R,T}(i)$ is the expected average cost incurred by the system up to time T given $Y_0 = i$ and R is the policy controlling the system.

Let

$$\phi_R(i) = \liminf_{T \to \infty} \phi_{R,T}(i).$$

The problem under consideration in this exposition is that of finding R to minimize $\phi_R(i)$.

For what follows it is convenient to consider three classes of policies. The first is the class of *all* policies of the form described. That is, all policies which, at each point in time, use past states and decisions as a basis for making a decision. We let C denote this class. The second class is the class which uses, at each point in time, the state of the system at that instant as a basis for making a decision. We shall refer to this class as the class of *stationary Markovian* policies and denote this class by C'. The third class is the subclass of C' in which the policies are not of a random character. We denote this class by C''. A policy $R \in C''$ may be thought of as a function defined over the states with range in the set of possible decisions; to each state there corresponds a unique decision. We refer to C'' as the class of *deterministic stationary Markovian* policies.

We shall divide the following discussion into two parts. One for the case where I is finite; the other, for the case where I is denumerable. In going from the finite to the denumerably infinite, mathematical questions arise which are not yet settled.

2. **Finite number of states.** The fundamental fact concerning the problem at hand for this finite state case can be summarized as

THEOREM 1. *If $K_i < \infty$, $i \in I$, and I is finite, then there exists a policy $R \in C''$ which minimizes $\phi_R(i)$, $i \in I$.*

The above theorem has been proved, more or less, by several authors (see [**1**]—[**4**]). Each proof relies on what we refer to as

THEOREM 2. *If $K_i < \infty$, $i \in I$, and I is finite, then there exists* *a policy* $R_\alpha \in C''$ *which minimizes*

$$\psi_R(i,\alpha) = \sum_{t=0}^{\infty} \alpha^t E_R W_t, \qquad i \in I,$$

where α *is a given number between zero and one.*

$\psi_R(i,\alpha)$ is often of economic relevance. It is referred to as the expected discounted (with discount factor α) cost criterion.

Theorem 2 is usually taken as self-evident. However, for proof, see [5] and [6]. On letting $\alpha \to 1$ and using an appropriate Tauberian theorem (e.g. $\lim_{\alpha\to 1}(1-\alpha)\,\psi_R(i,\alpha) = \phi_R(i)$ when R is such that $\phi_R(i) = \lim_{T\to\infty} \phi_{R,T}(i)$), Theorem 1 can be established.

Theorem 1 carries with it two advantages. The class C'' contains only a finite number of policies (even though the number may be astronomically large) and under any $R \in C''$, $\{Y_t\}$ is a Markov chain with stationary transition probabilities. As a result, finite algorithms for minimizing $\phi_R(i)$ over $R \in C''$ are obtainable.

Two methods for obtaining the optimal $R \in C''$ have been advanced. One method involves linear programming (see [7], [3]). The other is a derivative of dynamic programming (see [8], [2]). We indicate the latter method first.

Let R be an arbitrary policy in C''. It can be shown that there exists a unique set of numbers $\{g_i^R, v_i^R\}$, $i \in I$, such that $g_i^R = g_j^R$ if i and j are in the same ergodic class and which satisfies

$$g_i^R + v_i^R = w_{iR} + \sum_j q_{ij}(R)\, v_j^R, \qquad i \in I, \tag{1}$$

and

$$\sum_j \Pi_{ij}^R v_j^R = 0, \qquad i \in I. \tag{2}$$

Here w_{iR} and $q_{ij}(R)$ denote the cost w_{ik} and transition probability $q_{ij}(k)$ involved at state i under policy R; Π_{ij}^R is the limiting (as $T \to \infty$) expected proportion of time that the system is in state j given it is initially in state i. The theory of Markov chains indicates how the limiting values $\{\Pi_{ij}^R\}$ can be obtained.

One can also see that $g_i^R = \sum_{j\in I} \Pi_{ij}^R w_{jR} = \phi_R(i)$.

For each $i \in I$, let E_i denote the set of decisions k for which either

(a) $$\sum_j q_{ij}(k)\, g_j^R < g_i^R$$

or

(b) $$\sum_j q_{ij}(k)\, g_j^R = g_i^R$$

and

$$w_{ik} + \sum_j q_{ij}(k)\, v_j^R < g_i^R + v_i.$$

Define a policy R' as follows. For at least one i such that E_i is nonempty, prescribe a decision in E_i when in state i. For all states i where E_i is empty or where one does not prescribe a decision in E_i, make the decision dictated by policy R. We shall call the mapping from R to R' a *policy iteration*.

It can be shown that $\phi_{R'}(i) \leqq \phi_R(i)$, $i \in I$. The equality may hold because of the possible presence of transient states in the Markov chain associated with policy R'. Thus the *policy improvement procedure* starts with an arbitrary policy R_0 and carries out successive policy iterations until no more can be made. Since there are only a finite number of policies in C'' this stage must be reached. (Actually, this is not obvious because of the possibility that $\phi_{R'}(i) = \phi_R(i)$, $i \in I$. That is, an argument must be made to show that cycles of policies will not occur.) At the termination of the successive policy iterations we must have, if R^* is the terminal policy,

(3) $$g_i^{R^*} + v_i^{R^*} = \min_k \left\{ w_{ik} + \sum_j q_{ij}(k)\, v_j^{R^*} \right\}, \qquad i \in I,$$

and

(4) $$\sum_j q_{ij}(k)\, g_j^{R^*} \geqq g_i^{R^*}, \qquad k = 1, \cdots, k_i, \quad i \in I.$$

From this system it can be shown that R^* is optimal. Thus, the sequence of policy iterations terminates at an optimal policy within a finite number of iterations.

The policy improvement procedure simplifies under the asumption

(A) I contains at most one ergodic class of states for every $R \in C''$.

Under (A), $g_i^R = g^R$ independent of the initial state i, part (a) of the definition of E_i is unnecessary, and (4) of the terminal system will always hold.

Let $\phi_R^{(m)}(i)$, $m = 1, \cdots, M$, be defined in a manner similar to $\phi_R(i)$ except that $\phi_R^{(m)}(i)$ is defined with respect to costs $\{w_{ik}^{(m)}\}$,

$m = 1, \cdots, M$. Although the policy improvement procedure is powerful enough to obtain an optimal solution to the problem under discussion, under general conditions (provided the number of states is not too large), it does not provide an algorithm for solving the more complicated problem.

Minimize $\phi_R(i)$ subject to

$$\phi_R^{(m)}(i) \geqq b_m, \qquad m = 1, \cdots, M,$$

where b_m, $m = 1, \cdots, M$, are given constants.

When (A) holds, the method of linear programming is effective for obtaining an optimal policy. Under (A), $\Pi_{ij}^R = \Pi_j^R$, independent of i, $\phi_R(i) = \sum_j \sum_k \Pi_j^R D_{jk}^R w_{jk}$,

$$\phi_R^{(m)}(i) = \sum \sum \Pi_j^R D_{jk}^R w_{jk}^{(m)}, \qquad m = 1, \cdots, M,$$

where D_{ik}^R denotes the probability of making decision k when in state i; the $\{D_{ik}^R\}$ define a rule $R \in C'$. The $\{\Pi_j^R\}$ must satisfy the steady state equations

$$\Pi_j^R \geqq 0,$$

$$\Pi_j^R - \sum_i \sum_k \Pi_i^R D_{ik}^R q_{ik}(k) = 0,$$

$$\sum_j \Pi_j^R = 1.$$

If one makes the transformation $X_{ik} = \Pi_i^R D_{ik}^R$, one gets the linear programming problem

$$\min \sum_i \sum_k X_{ik} w_{ik}$$

subject to

$$X_{ik} \geqq 0 \qquad k = 1, \cdots, K_i, \qquad i \in I,$$

$$\sum_k X_{jk} - \sum_i \sum_k X_{ik} q_{ij}(k) = 0, \qquad i \in I,$$

$$\sum_j \sum_k X_{jk} = 1,$$

$$\sum \sum X_{ik} w_{ik}^{(m)} \geqq b_m, \qquad m = 1, \cdots, M.$$

If $\{X_{jk}^*\}$ is a solution to the above problem then the optimal policy $R^* \in C'$ is defined by setting

$$D_{ik}^{R^*} = X_{ik}^* / \sum_k X_{ik}^* \text{ if } \sum_k X_{ik}^* > 0,$$

and arbitrary if $\sum_k X_{ik}^* = 0$.

If $M = 0$, i.e. no additional constraints are imposed, the method is an alternative to the policy iteration procedure under condition (A).

The question of how to solve the problem with additional constraints without assuming (A) remains open. In general, an optimal policy need not exist in C'. One can assert, however, (see [**9**]) that under (A) an optimal policy will always exist in C'.

It should be pointed out that the solution to the problem of minimizing $\phi_R(i)$ need not be unique. The question arises as to whether some solutions might not be better than others. That is, other criterion, not explicitly put into the problem, may, in part, be relevant. For example, it has been shown (see [**2**]), that there exists a policy $R^* \in C''$ such that $\psi_R(i, \alpha)$ is minimized for all i and for all α near enough to 1. When this is the case R^* also minimizes $\phi_R(i)$. No computational procedure has yet been given to find such a policy. However, a procedure (see [**10**]) has been given for finding a policy R^{**} having the property that

$$\lim_{\alpha \to 1} [\psi_{R^*}(i, \alpha) - \psi_{R^{**}}(i, \alpha)] = 0.$$

Needless to say, R^{**} also minimizes $\phi_R(i)$.

3. **Denumerable state case.** Let us turn from the finite case to the case where I is denumerably infinite and ask ourselves whether the basic facts as put forth in Theorems 1 and 2 remain the same.

The following modification of Theorem 2 can be shown (see [**5**] and [**6**]).

THEOREM 2′. *If $K_i < \infty$, $i \in I$, and $\{w_{ik}\}$ bounded, then there exists a policy $R_\alpha \in C''$ which minimizes $\psi_R(i, \alpha)$, $i \in I$.*

If $K_i = \infty$ it can be easily shown that Theorem 2′ need not hold. If the $\{w_{ik}\}$ are not bounded it can also be shown that the result does not hold in general (see [**6**]). Thus, we might ask if the conclusions of Theorem 1 hold under the conditions of Theorem 2′.

An example [**11**] has been given showing that the conditions of Theorem 2′ do not guarantee the conclusions of Theorem 1. Moreover, the example shows that an optimal policy R for minimizing $\phi_R(i)$ may not exist. This counterexample also implies that there may not be a policy $R \in C''$ which minimizes $\psi_R(i, \alpha)$ for all α near enough to 1, as in the case when I is finite. For if such were the case, it would be possible to show that $\phi_R(i)$ could always be optimized.

A more surprising counterexample (see [12]) shows that an optimal policy may not exist in C''; but it may in C'. Thus, if restricted to stationary policies, a policy involving randomization may prove to be more effective than one that is deterministic. The counterexample exploits the fact that denumerable state Markov chains may have recurrent *null* states—a property denied finite state chains.

An even more surprising counterexample (see [13]) shows that $\phi_R(i)$ may be minimized by a policy in $C - C'$, whereas no optimal policy exists in C'. Thus, in order to obtain an optimal policy one may find it necessary to go beyond the class of stationary Markovian policies. This fact seems to run counter to one's intuition regarding the problem under discussion. The literature has numerous remarks asserting the reasonableness of assuming that an optimal policy is stationary.

What develops as an interesting mathematical question is that of determining the weakest conditions under which it can be asserted that a policy $R \in C''$ is optimal. In [12] and [14] the following was proved.

Theorem 3. *Suppose* $\{w_{ik}\}$ *are bounded. If there exists bounded numbers* g, $\{v_j\}$, $j \in I$, *satisfying*

$$g + v_i = \min_k \left\{ w_{ik} + \sum_j q_{ij}(k)\, v_j \right\}, \qquad i \in I, \tag{5}$$

then there exists a policy $R^* \in C''$ *which is optimal. The policy* R^* *is: implement decision* $k = k_i$ *which minimizes the right side of* (5) *for each* $i \in I$. *Also,*

$$g = \phi_{R^*}(i), \qquad i \in I.$$

Note that (5) is related to (1) with $g_i = g$. Actually, Theorem 3 can be generalized to the case where the g_i's are not all equal and (1) and (2) replace (5).

Conditions implying the hypothesis of Theorem 3 can be given. The approach is to define a policy improvement procedure similar to the one discussed in the finite state case and show that in the limit one gets a policy $R \in C''$ satisfying (5).

We define a *policy iteration* as follows. Let $R \in C''$ be given. Assume a solution g^R, $\{v_j^R\}$, $j \in I$, to the system of equations

$$g^R + v_i^R = w_{iR} + \sum_j q_{ij}(R)\, v_j^R, \qquad i \in I \tag{6}$$

exists. (The system (6) is treated in [15].) Define R' by choosing, for each $i \in I$, that decision $k = k_i$ which minimizes

$$w_{ik} + \sum_j q_{ij}(k)\, v_j^R.$$

The transformation from R to R' is the policy iteration. Note that the policy iteration, here, is defined more stringently than for the finite state case. By a *policy improvement procedure* we mean starting with an arbitrary policy $R_0 \in C''$, letting R_{n+1} denote the policy obtained by a policy iteration on R_n, $n = 0, 1, \cdots$. If, for any n, $R_{n+1} = R_n$, then equations (5) are satisfied and R_n is optimal. Otherwise, one may or may not obtain an optimal policy as $n \to \infty$. It is of interest to provide conditions under which $\{R_n\}$ converges to an optimal policy.

Let us list the following conditions:

(B) For every $R \in C''$, the associated Markov chain is irreducible and positive recurrent.

(C) For every $R \in C''$ there exists a bounded solution g^R, $\{v_j^R\}$, $j \in I$ to (6). The solutions are uniformly bounded over $R \in C''$.

(D)[2] For every $i \in I$, $\inf_{R \in C''} \Pi_i^R > 0$.

We can assert (see [12])

THEOREM 4. *If $K_i < \infty$, $i \in I$, $\{w_{ik}\}$ are bounded, and* (B), (C), *and* (D) *hold, then the policy improvement procedure converges to an optimal policy $R^* \in C''$.*

The proof of Theorem 4 involves showing that the limiting policy does yield a solution to (5). Under weaker conditions, i.e. (B) and (C), it can be shown that a solution to (5) exists. Therefore, an optimal policy is in C''. However, it is not clear that a policy improvement procedure will converge to an optimal policy.

Conditions are given in [15] guaranteeing (C); slightly weaker conditions are given in [13]. A better approach to the existence question would, in all likelihood, avoid the equations (5) and (6).

References

1. Dean Gillette, *Stochastic games with zero stop probabilities*, Ann. of Math. Studies, no. 39, Princeton Univ. Press, Princeton, N. J., 1957.

2. David Blackwell, *Discrete dynamic programming*, Ann. Math. Statist. **33** (1962), 719-726.

---

[2] A recent result of Lloyd Fisher shows that (B) implies (D).

3. Cyrus Derman, *On sequential decisions and Markov chains,* Management Sci. **9** (1962), 16-24.

4. O. V. Viskov and A. N. Surjaev, *On controls leading to optimal stationary states,* Selected Transl. in Math. Statist. and Prob., Vol. 6, 1964, 71-83.

5. David Blackwell, *Discounted dynamic programming,* Ann. Math. Statist. **36** (1965), 226-235.

6. Cyrus Derman, *Markovian sequential control processes—Denumerable state space,* J. Math. Anal. Appl. **10** (1965), 295-302.

7. Alan S. Manne, *Linear programming and sequential decisions,* Management Sci. (3) **6** (1960), 259-267.

8. Ronald A. Howard, *Dynamic programming and Markov processes,* John Wiley, New York, 1960.

9. Cyrus Derman, *Stable sequential control rules and Markov chains,* J. Math. Anal. Appl. **6** (1963), 257-265.

10. Arthur F. Veinott, Jr., *On finding optimal policies in discrete dynamic programming with no discounting,* Ann. Math. Statist. **37** (1966), 1284-1294.

11. Ashok Maitra, *Dynamic programming for countable state systems,* Ph. D. thesis, University of California, Berkeley, Calif., 1964.

12. Cyrus Derman, *Denumerable state Markovian decision processes—Average cost criterion,* Ann. Math. Statist. **37** (1966), 1545-1553.

13. Sheldon M. Ross, *Non-discounted denumerable Markovian decision models,* Tech. Report No. 94, Dept. of Statistics, Stanford University, Stanford, Calif., 1967.

14. Cyrus Derman and G. J. Lieberman, *A Markovian decision model for a joint replacement and stocking problem,* Management Sci. (9) **13** (1967), 609-617.

15. Cyrus Derman and Arthur F. Veinott Jr., *A solution to a countable system of equalities arising in Markovian decision processes,* Ann. Math. Statist. **38** (1967), 582-584.

Columbia University
New York, New York

Herman Chernoff

Optimal Stochastic Control[1]

1. **Introduction.** Certain techniques which were developed for sequential analysis have been found to apply to a wide variety of problems, including some stochastic control problems. We shall outline these techniques and indicate how they may be applied. The main tool is the representation of incoming information in terms of a continuous time Wiener process which relates these problems to the solution of Free Boundary Problems involving the heat equation.

2. **Some problems.** Consider the following distinct problems.

Problem 1. (*A sequential analysis problem*). The random variable x is normally[2] distributed with unknown mean μ and known variance σ^2. The statistician must decide whether $\mu > 0$ or $\mu < 0$ and the cost of an incorrect decision is $k|\mu|$, $k > 0$. He is permitted to sample sequentially (one observation at a time) at a cost of c per observation. He may stop sampling at any time and make a decision. The total cost will be cn if the decision is correct and $cn + k|\mu|$ if it is wrong where n is the number of observations taken. The cost is a random variable whose distribution depends on the unknown μ and the (sequential) procedure used.

To determine an optimal procedure one must specify some criterion of optimality. It is convenient to treat this problem in

[1] Prepared with the partial support of NSF Grant GP 5705. An amplified version of this paper will be submitted to *Sankhyā*.

[2] We shall use the following notation throughout the paper. Let $\phi(x) = (2\pi)^{-1/2}\exp(-x^2/2)$ and $\Phi(x) = \int_{-\infty}^{x}\phi(y)\,dy$ represent the standard normal density and cumulative distribution functions. The normal distribution with mean μ and variance σ^2 is $\mathscr{N}(\mu, \sigma^2)$ and has density $n(x; \mu, \sigma^2) = \sigma^{-1}\phi[(x-\mu)/\sigma]$ and cdf $\Phi[(x-\mu)/\sigma]$. We denote the probability distribution (law) of a random variable X by $\mathscr{L}(X)$ and its mathematical expectation by $E(X)$. $\mathscr{L}(X|Y)$ and $E(X|Y)$ represent the conditional distribution and expectation of X given Y.

a Bayesian context assuming that the unknown μ is normally distributed with mean μ_0 and variance σ_0^2 (both known). Then a sequential procedure determines an expected cost, and one may seek that procedure which minimizes the expected cost.

Problem 2. (*A stopping problem*). Let $\{X_n, -m \leqq n \leqq 0\}$ be a stochastic process such that $X_{-m} = x$ is specified and $X_{n+1} = X_n + u_n$ where the u_i are independently and normally distributed with mean 0 and variance 1. Thus it is convenient to think of X_n changing as the subscript n increases to 0. For each $n < 0$, an observer can stop the process and collect 0 or he can wait till $n = 0$ at which point he collects 0 if $X_0 \geqq 0$ and X_0^2 if $X_0 < 0$. However, he must pay 1 for each observation. What constitutes an optimal procedure for stopping?

Problem 3. (*Rocket control-infinite fuel*). A rocket is directed toward Mars. At certain time points $\{t_i\}$, instruments make measurements estimating the distance by which it will miss. The miss distance can then be adjusted by an "instantaneous" use of fuel. The cost of missing by an amount y is ky^2. The cost per unit of fuel is c. As much fuel as is desired is available. How should fuel be allocated to minimize expected total cost?

Certain simplifications and specifications are necessary to make this problem meaningful and tractable. The assumption of an infinite supply of available fuel is already such a simplifying assumption. Let us replace the natural two-dimensionsal miss vector by a one-dimensional value which can take on positive and negative values. Let us assume that if an amount of fuel Δ is used at time t the miss distance is changed by $\pm e(t)\Delta$ where $e(t)$ represents the efficiency of fuel at time t. Since fuel is used to change direction, the efficiency $e(t)$ is greatest at the beginning of the flight when the rocket is far away from the target. We shall assume that the amount of fuel used and its effect can be controlled and measured with precision. Let us now assume that the measurements estimating the miss distance are x_i at time t_i where the x_i are independent and normally distributed with mean equal to the miss distance (provided no additional fuel is used) and known variance equal to σ_i^2.

Here, as in the Sequential Analysis Problem, the performance characteristics of a procedure depend on the unknown value of a fundamental parameter. In the Sequential Analysis Problem, that was μ. Here it is the unknown miss distance that would be obtained if no adjustments were made. As in the sequential analysis problem we find it convenient to assume that the unknown miss distance has a normal prior distribution with specified mean μ_0 and variance σ_0^2. Then there is an expected cost for each procedure and the problem of selecting an optimal procedure is meaningful.

The three problems have more flavor if the competing factors are indicated qualitatively. In the sequential problem, after much data has been accumulated one is either reasonably certain of the sign of μ or that $|\mu|$ is so small that the loss of deciding wrong is less than the cost of another observation. Here one expects the proper procedure to be such that one stops and makes a decision when the current estimate of $|\mu|$ is sufficiently large and continues sampling otherwise. What constitutes sufficiently large depends on, and should decrease with, the number of observations or equivalently the precision of the estimate. It can be shown that after a certain sample size it pays to stop no matter what the current estimate of $|\mu|$ is.

In the Stopping Problem, it is clear that if X_n is sufficiently negative (depending on n) one ought to pay the cost of continuing one more step. It is to be expected that there are limits y_n so that for $X_n < y_n$, it pays to continue and for $X_n \geqq y_n$ it pays to stop.

In the rocket problem, when the rocket is close to target, fuel efficiency may be so low that even though the miss distance is practically known one would not use fuel unless the miss distance were very great. When the rocket is far from the target, fuel is effective but the miss distance is not well known and so one is reluctant to make an adjustment for fear of overshooting or adjusting in the wrong direction. Here one should expect the solution to have the following property. There are limits y_i at time t_i such that if the estimated miss distance exceeds y_i in absolute value, enough fuel is used to bring the adjusted estimate to $\pm y_i$. It seems reasonable to expect the y_i to be large at the beginning of flight and at the end of the flight and relatively small in between.

For specified values of the constants, all three of these problems can be solved numerically by the backward induction techniques of dynamic programming. For the sequential analysis problem care must be taken to initiate the backward induction at a sample size n sufficiently large so that no matter what the estimate of μ

is, the optimal procedure will lead to a decision rather than to additional sampling. The technique of the backward induction can be summarized by the equation

$$\rho_n(\xi_n) = \inf_{a_n} E\{\rho_{n+1}[\xi_{n+1}(a_n, \xi_n)]\} \tag{2.1}$$

where $\rho_n(\xi_n)$ is the expected cost of an optimal procedure given the history ξ_n up to stage n, $\xi_{n+1}(a_n, \xi_n)$ describes the history up to stage $n+1$ which may be random, with distribution depending on ξ_n and the action a_n taken at stage n. It is possible to show that in the sequential analysis and rocket control problems ξ_n is adequately summarized by the mean and variance of the *posterior distribution* of μ while in the stopping problem X_n may be used to describe ξ_n. The minimizing a_n which depend on ξ_n determine the optimal procedure.

To illustrate, we note that for the stopping problem, $\rho_0(x) = m$ for $x \geqq 0$ and $\rho_0(x) = m - x^2$ for $x \leqq 0$, since reaching $n = 0$ implies a payment of m for continuing m steps. At $n = -1$, the choice of stopping at $X_{-1} = x$ leads to a cost of $(m-1)$ while the choice of continuing leads to $X_0 = x + u_0$ and an expected cost of $\int_{-\infty}^{\infty} \rho_0(x+u)\,\phi(u)\,du$.
Thus,

$$\rho_{-1}(x) = \min\left[(m-1),\quad m - \int_{-\infty}^{-x} (x+u)^2 \phi(u)\,du\right]$$

and the best action for $X_{-1} = x$ is to stop or continue depending on which of the two terms in the brackets is smaller. Having evaluated $\rho_{-1}(x)$, one can in principle proceed in the same way to obtain ρ_{-2} and an optimal decision for $n = -2$ as a function of X_{-2}, etc.

The rocket and sequential analysis problem seem more complex in that they involve the posterior distributions, but the calculus of posterior distributions (to be discussed in §4) when dealing with normal random variables and normal priors permits these problems to be treated with equal facility.

In a sense then, these problems are trivial. If, however, it is desired to derive some overall view of how the solutions depend on the various parameters, the simple though extensive numerical calculations of the backward induction are not adequate.

An approach which seems to have particular relevance to large sample theory is that of replacing the discrete time random variables by analogous continuous time stochastic processes. The use of the Wiener process seems especially relevant and serves to convert the problem to one in which the analytic methods of partial differential equations can serve fruitfully.

3. **Wiener process.** Suppose $x_1, x_2, \cdots, x_n$, are independently, identically, and normally distributed with mean μ and variance σ^2. Let $X_n = x_1 + x_2 + \cdots + x_n$. Then for $n \geqq m \geqq 0$, $X_n - X_m$ is independent of $X_1, X_2, \cdots, X_m$, and normally distributed with mean $(n - m)\mu$ and variance $(n - m)\sigma^2$. When n is large, a graph of the discrete X_n process resembles the analogous, continuous time process $\{X(t): 0 \leqq t\}$ which has the following properties. The function $X(t)$ is continuous, $X(0) = 0$, and for $0 \leqq t_1 \leqq t_2$, $X(t_2) - X(t_1)$ is independent of $\{X(t): 0 \leqq t \leqq t_1\}$ and is normally distributed with mean $\mu(t_2 - t_1)$ and variance $\sigma^2(t_2 - t_1)$. This may be referred to as a Gaussian process with independent increments or as a Wiener process with drift μ and variance σ^2 per unit time. Typically the drift is not referred to when we consider the case $\mu = 0$. There is a trivial but occasionally useful variation where $X(t)$ is initiated at the point $X(t_0) = x_0$ rather than at $X(0) = 0$. Notationally, it is convenient to refer to the process by the equations

$$E\{dX(t)\} = \mu dt, \quad \mathrm{Var}\{dX(t)\} = \sigma^2 dt. \tag{3.1}$$

This is especially convenient for variations of the Wiener process which are derived by changing the scales. Observe that if $E\{dX(t)\} = 0$ and $\mathrm{Var}\{dX(t)\} = dt$, the transformation $t^* = a^2 t$, $X^*(t^*) = aX(t)$ yields $E\{dX^*(t^*)\} = 0$ and $\mathrm{Var}\{dX^*(t^*)\} = dt^*$.

Although the stopping problem (Problem 2, §2) does not involve an unknown parameter μ, an analogue of this problem can be posed in terms of a continuous time Wiener process without drift originating from a given point (x, t), $t \leqq 0$.

4. **Posterior distributions.** Inasmuch as the Sequential Analysis and Rocket Control Problems involve unknown parameters which may be estimated by incoming data, both have a statistical component. For statistical problems in a Bayesian context, the posterior distribution of the unknown parameter is crucial.

In the discrete case suppose that μ has prior distribution $\mathscr{N}(\mu_0, \sigma_0^2)$ (normal with mean μ_0 and variance σ_0^2), and for given μ, the data $x_1, x_2, \cdots, x_n$ are independent with distribution laws $\mathscr{L}(x_i)$

$= \mathcal{N}(\mu, \sigma_i^2)$, and σ_i^2 known. Then the posterior distribution of μ given the data is (see [18])

$$\mathcal{L}(\mu | x_1, \cdots, x_n) = \mathcal{N}(Y_n, s_n) \tag{4.1}$$

where

$$Y_n = (\mu_0\sigma_0^{-2} + x_1\sigma_1^{-2} + \cdots + x_n\sigma_n^{-2})/(\sigma_0^{-2} + \sigma_1^{-2} + \cdots + \sigma_n^{-2}), \quad n \geqq 0 \tag{4.2}$$

and

$$s_n^{-1} = \sigma_0^{-2} + \sigma_1^{-2} + \cdots + \sigma_n^{-2}, \qquad n \geqq 0. \tag{4.3}$$

Here Y_n, the mean of the posterior distribution, may be called the (posterior) Bayes Estimate of μ. It is a weighted average of the individual estimates x_i weighted by the *precisions* (σ_i^{-2}) where the prior distribution is treated as an estimate with mean μ_0 and precision σ_0^{-2}. Similarly Y_n may be regarded as a summary of the previous information and as an estimate of μ with precision s_n^{-1}.

Since Y_n is the Bayes estimate of μ, one should expect that for $n > m$, $E(Y_n | Y_m) = Y_m$. It is somewhat more surprising to find by routine but tedious calculations that

$$\mathcal{L}(Y_n - Y_m | Y_m) = \mathcal{N}(0, s_m - s_n), \qquad n \geqq m \geqq 0. \tag{4.4}$$

5. **Continuous time problems.** The results of §4 are of consequence in the continuous time analogue of the sequential analysis problem which we now state.

Problem 1*. (*Sequential analysis, continuous time*). Find an optimal procedure for testing $H_1: \mu \geqq 0$ vs. $H_2: \mu < 0$ when the cost of an incorrect decision is $k|\mu|$, the cost of sampling is c per unit time, and the data consists of a Wiener process $X(t)$ with unknown drift μ and known variance σ^2 per unit time. The unknown value of μ has prior distribution $\mathcal{L}(\mu) = \mathcal{N}(\mu_0, \sigma_0^2)$.

As a consequence of the results of §4 we have for problem 1*, the posterior distribution of μ is given by

$$\mathcal{L}(\mu | X(t'), \; 0 \leqq t' \leqq t) = \mathcal{N}(Y(s), s) \tag{5.1}$$

where

$$Y(s) = [\mu_0\sigma_0^{-2} + X(t)\,\sigma^{-2}]/(\sigma_0^{-2} + t\sigma^{-2}), \tag{5.2}$$

$$s^{-1} = \sigma_0^{-2} + t\sigma^{-2}, \tag{5.3}$$

and $Y(s)$ is a Wiener process in the $-s$ scale, originating at $(y_0, s_0) = (\mu_0, \sigma_0^{-2})$, i.e.

$$E\{dY(s)\} = 0, \quad \mathrm{Var}\{dY(s)\} = -ds. \tag{5.4}$$

Note that s decreases from $s_0 = \sigma_0^2$ as information accumulates.

Since the X process can be recovered from the Y process it suffices to deal with the latter which measures the current estimate of μ and which is easier to analyze.

In the sequential problem, when the statistician stops sampling he must decide between $H_1\colon \mu \geqq 0$ and $H_2\colon \mu < 0$. The posterior expected cost associated with deciding in favor of H_1 at time t (when $Y(s) = y$) is then

$$\int_{-\infty}^{0} k|\mu|\, n(\mu; y, s)\, d\mu.$$

This quantity is readily computed and found to be $ks^{1/2}\psi^+(y/s^{1/2})$ where $\psi^+(u) = \phi(u) - u[1 - \Phi(u)]$. Similarly the posterior expected cost associated with deciding $\mu < 0$ is $ks^{1/2}\psi^-(y/s^{1/2})$ where $\psi^-(u) = \phi(u) + u\Phi(u)$. It is easy to see that if sampling is stopped at $Y(s) = y$ the decision should be made on the basis of the sign of y and the expected cost of deciding plus the cost of sampling is given by

$$d(y, s) = c\sigma^2 s^{-1} + ks^{1/2}\psi(y/s^{1/2}) - c\sigma^2/\sigma_0^2 \tag{5.5}$$

where

$$\begin{aligned} \psi(u) &= \phi(u) - u[1 - \Phi(u)], \qquad u \geqq 0, \\ &= \phi(u) + u\Phi(u), \qquad u < 0. \end{aligned} \tag{5.6}$$

Thus the continuous time sequential analysis problem may be regarded simply as the following stopping problem. The Wiener process $Y(s)$ is observed. The statistician may stop at any value of $s > 0$ and pay $d(Y(s), s)$. Find the stopping procedure which minimizes the expected cost. In this version of the problem using the posterior Bayes Estimate, the statistical aspects involving the unknown parameter μ have been abstracted.

The original discrete time sequential problem can also be described in terms of this stopping problem by adding the proviso that allowable stopping values of s are restricted to $s_0, s_1, s_2, \cdots$, where $s_n = (\sigma_0^{-2} + n\sigma^{-2})^{-1}$. At this point it should be reasonably straightforward for the reader to see that the discrete version can be treated numerically by backward induction in terms of the $Y(s)$ process starting from $s_n \leqq c^2/k^2\psi^2(0) = 2\pi c^2/k^2$.

We now present the continuous time version of the Stopping Problem 2 of §2.

Problem 2*. (*Stopping problem*). Let $Y(s)$ be a Wiener process in the $-s$ scale with $E\{dY(s)\} = 0$, $\mathrm{Var}\{dY(s)\} = -ds$, originating from (y_0, s_0). Let

$$d(y,s) = -y^2, \quad \text{if} \quad y < 0 \quad \text{and} \quad s = 0, \tag{5.7}$$
$$= -s, \quad \text{otherwise} \quad (s \geqq 0).$$

Find the stopping time to minimize $E\{d(Y(s),s)\}$.

A more literal translation of Problem 2 of §2 would yield a stopping cost $d^*(y,s) = d(y,s) + s_0$. Since the difference is constant, it does not affect the solution.

While the rules of computing posterior distributions extend to the rocket control problem, that problem is not trivially reduced to a stopping problem. However, we shall see that the continuous version of the infinite fuel problem posed in §2 is also equivalent to a related stopping problem.

6. **Continuous time stopping problems: relevance of stopping sets.** A general class of *stopping problems* may be described as follows. Let $Y(s)$ be a Wiener process in the $-s$ scale originating from (y_0, s_0) with $E\{dY(s)\} = 0$ and $\mathrm{Var}\{dY(s)\} = -ds$. Let $d(y,s)$ be a specified stopping cost. Select a *stopping procedure* S to minimize the risk

$$b(y_0, s_0) = E\{d(Y(S),S)\}. \tag{6.1}$$

A stopping procedure S is a measurable rule which determines the stopping time S in terms of the "past history" of $Y(s)$. Technically, in measure theoretic terms this may be translated to mean

$$\{S \geqq s_1\} \in \mathscr{B}\{Y(s); s_0 \geqq s \geqq s_1\} \tag{6.2}$$

where the right-hand side is the Borel Field generated by the process from s_0 to s_1. Stopping procedures may be subjected to restrictions which are either of the form that stopping is not allowed on certain sets of points (y,s) or that stopping is automatic on other sets. For example, in the continuous time versions of both Problems 1 and 2 of §2, stopping must take place if $s = 0$.

In a trivial sense the discrete time version of the sequential analysis problem may be regarded as a continuous time problem where stopping is not permitted except at a certain set of values of s.

While the discrete time problems of §2 are theoretically trivial insofar as the solutions can be computed by backward induction this is not the case for the continuous time problems. Even discrete time problems with an infinite sequence of possible decision times lead to difficulties. The problems of the existence and characterization of solutions are deep and much remains to be done to obtain precise rigorous results for the continuous time problem. We shall proceed in a heuristic fashion conveniently ignoring some of the more delicate questions which have to be faced ultimately.

Let

$$\rho(y_0, s_0) = \inf b(y_0, s_0) \tag{6.3}$$

among all procedures S. Note that $\rho(y_0, s_0) \leqq d(y_0, s_0)$. Since Y is a process of independent increments, it follows that $\rho(y, s)$ also represents the best that can be expected once $Y(s) = y$ is reached, irrespective of how it was reached. Then, a characterization of an optimal procedure (under regularity conditions) is described by

(I) $\qquad S_0$: "Stop as soon as $\rho(Y(s), s) = d(Y(s), s)$."

Since the optimal procedure S_0 is characterized by the continuation set

$$\mathscr{C}_0 = \{(y, s) : \rho(y, s) < d(y, s)\} \tag{6.4}$$

and the stopping set

$$\mathscr{S}_0 = \{(y, s) : \rho(y, s) = d(y, s)\} \tag{6.5}$$

we shall restrict our attention to procedures which can be represented by a continuation set $\mathscr{C}$ *or its complement the stopping set* $\mathscr{S}$.

It is interesting to note that the characterization (I) does not depend on the initial point (y_0, s_0) and thus it yields the solution for all initial points simultaneously, minimizing $b(y, s)$ uniformly for all (y, s).

Under suitable regularity conditions on $d(y, s)$, the solution of continuous time stopping problems may be approximated by discrete time versions corresponding to a finite sequence of permitted stopping times $(s_1, s_2, \cdots, s_n)$. Since a discrete version permits less choice, the corresponding optimal risk ρ^* is larger and the corresponding optimal continuation set $\mathscr{C}_0^*$ intersects $s = s_1$ on a smaller set than does $\mathscr{C}_0$. As more elements are adjoined to the set of permitted stopping times, ρ^* decreases and the set where $\mathscr{C}_0^*$ intersects $s = s_1$ increases. In this way ρ and $\mathscr{C}_0$ may be derived as limits of monotone sequences.

7. Stopping problems. Relevance of heat equation. The Wiener process is intimately related to the heat equation. Suppose, for example that $b(y,s)$ is the expected cost corresponding to an open continuation set $\mathscr{C}$ and stopping cost $d(y,s)$. Then we shall demonstrate that

$$\tfrac{1}{2} b_{yy}(y,s) = b_s(y,s), \qquad (y,s) \in \mathscr{C} \tag{7.1}$$

while

$$b(y,s) = d(y,s), \qquad (y,s) \in \mathscr{S}. \tag{7.2}$$

Suppose $(y,s) \in \mathscr{C}$. Then, the probability of stopping between $s+\delta$ and s is $o(\delta)$ and Y changes from $Y(s+\delta)$ to $Y(s)$. Consequently

$$\begin{aligned} b(y, s+\delta) &= E\{b(Y(s), s) \mid Y(s+\delta) = y\} + o(\delta) \\ &= E\{b(y + w\delta^{1/2}, s)\} + o(\delta) \end{aligned} \tag{7.3}$$

where we use w as a generic $\mathscr{N}(0,1)$ random variable

$$\begin{aligned} b(y, s+\delta) &= E\{b(y,s) + w\delta^{1/2} b_y(y,s) + \tfrac{1}{2} w^2(\delta)\, b_{yy}(y,s) + \cdots\} + o(\delta) \\ &= b(y,s) + \tfrac{1}{2} b_{yy}(y,s)(\delta) + o(\delta) \end{aligned}$$

and

$$b_s = \tfrac{1}{2} b_{yy}.$$

Doob [15] has elaborated on the relationship between the Wiener process and the heat equation indicating that it represents the natural way in which to study the heat equation. To digress briefly and omitting regularity conditions, a *subparabolic function* u on an open set D is such that for a Wiener process $Y(s)$ originating at (y,s)

$$u(y,s) \leqq E\{u(Y(S), S)\} \tag{7.4}$$

where S is the time when $Y(s)$ first hits the boundary of an open set $\mathscr{C} \subset D$ of which (y, s) is an interior point. A *parabolic function* is one for which the inequality is replaced by equality. If the second derivatives are continuous then

$$\tag{7.5} \tfrac{1}{2} u_{yy} \geqq u_s$$

for subparabolic functions with equality for parabolic functions. Thus solutions of the heat equation are identified with parabolic functions. To solve the Dirichlet Problem (solution of $\frac{1}{2} u_{yy} = u_s$ in $\mathscr{C}$ subject to $u = f$ on the boundary of $\mathscr{C}$) Doob takes $u(y, s) = E\{f(Y(S), S)\}$.

The concept of subparabolic functions provides another characterization of the optimal risk. We observe that

(II) *$\rho(y, s)$ is the maximal subparabolic function which is less than or equal to $d(y, s)$.*

To see that ρ is subparabolic, take an arbitrary set $\mathscr{C}$ of which (y, s) is an interior point. Then $E\{\rho(Y(S), S)\}$ represents the risk associated with the suboptimal procedure which does not stop as long as $(Y(s), s) \in \mathscr{C}$ but which proceeds optimally thereafter. Thus

$$\tag{7.6} \rho(y, s) \leqq E\{\rho(Y(S), S)\}$$

and ρ is subparabolic. Let ρ_1 be any subparabolic function such that $\rho_1 \leqq d$. Using the optimal continuation set $\mathscr{C}_0$ for $\mathscr{C}$, we have, for $(y, s) \in \mathscr{C}_0$,

$$\rho(y, s) = E\{d(Y(S_0), S_0)\} \geqq E\{\rho_1(Y(S_0), S_0)\} \geqq \rho_1(y, s).$$

If $(y, s) \notin \mathscr{C}_0$, $\rho(y, s) = d(y, s) \geqq \rho_1(y, s)$ which completes the proof.

Given the function $u(y, s)$ and a continuation set $\mathscr{C}$, can we determine whether $(u, \mathscr{C}) = (\rho, \mathscr{C}_0)$, i.e. whether $(u, \mathscr{C})$ solve the optimization problem associated with the stopping problem? A sufficient condition is the following.

(III) *If $u \leqq d$ is a subparabolic function which is parabolic on the open continuation set $\mathscr{C}$ and $u = d$ elsewhere, then $(u, \mathscr{C}) = (\rho, \mathscr{C}_0)$, the solution of the optimization problem.*

To show this, note that since $u \leqq d$ is subparabolic, $u \leqq \rho$. But u is the risk corresponding to the continuation set $\mathscr{C}$. Hence $u \geqq \rho$.

8. **Stopping problems—free boundary problem.** Associated with a procedure described by a continuation set $\mathscr{C}$ we have a risk function $b(y,s)$ which satisfies the heat equation in $\mathscr{C}$ subject to the boundary condition $b = d$. The solution of the optimal stopping problem minimizes b everywhere. Now we present the property

$$\text{(IV)} \qquad \rho_y(y,s) = d_y(y,s) \text{ on the boundary of } \mathscr{C}_0.$$

While the *Dirichlet problem* of finding b which satisfies the heat equation in a given $\mathscr{C}$ subject to $b = d$ on the boundary is referred to as a boundary value problem, that of finding b and $\mathscr{C}$ so that $b_y = d_y$ on the boundary also, is referred to as a *Stefan* or *free boundary problem* (f.b.p.). *Property* (IV) *states that the solution of the optimization problem is the solution of the* (*f.b.p.*).

To demonstrate (IV), let us assume that (y_0, s_0) is a point on a portion of the boundary above which are stopping points and below which are continuation points and that d_y exists at (y_0, s_0). Then since $\rho(y, s_0) = d(y, s_0)$ for $y > y_0$ the right-hand derivative $\rho_y^+(y_0, s_0) = d_y(y_0, s_0)$. For $y < y_0$, $\rho(y_0, s_0) \leqq d(y_0, s_0)$ and hence $\rho_y^-(y_0, s_0) \geqq d_y(y_0, s_0)$. Now we note that

$$\rho(y_0, s_0 + \delta) \leqq E\{\rho(y_0 + w\delta^{1/2}, s_0)\}. \tag{8.1}$$

since the right-hand side corresponds to the risk of the suboptimal procedure where one insists on sampling from $s_0 + \delta$ to s_0 and proceeding optimally thereafter. But

$$\begin{aligned} \rho(y_0 + w\delta^{1/2}, s_0) &= \rho(y_0, s_0) + w\delta^{1/2}\rho_y^+(y_0, s_0) + o(\sqrt{\delta}), \qquad w > 0, \\ &= \rho(y_0, s_0) + w\delta^{1/2}\rho_y^-(y_0, s_0) + o(\sqrt{\delta}), \qquad w < 0, \end{aligned} \tag{8.2}$$

$$\begin{aligned} &E\{\rho(y_0 + w\delta^{1/2}, s_0)\} \\ &\quad = \rho(y_0, s_0) + \delta^{1/2}\left\{d_y \int_0^\infty w\phi(w)\,dw + \rho_y^- \int_{-\infty}^0 w\phi(w)\,dw\right\} + o(\delta^{1/2}) \\ &\quad = \rho(y_0, s_0) + (\delta/2\pi)^{1/2}\{d_y - \rho_y^-\} + o(\delta^{1/2}). \end{aligned}$$

Thus

$$\rho(y_0, s_0 + \delta) - \rho(y_0, s_0) \leqq (\delta/2\pi)^{1/2}\{d_y - \rho_y^-\} + o(\delta^{1/2}).$$

Assuming that the difference quotient $[\rho(y_0, s_0 + \delta) - \rho(y_0, s_0)](\delta)^{-1}$ is bounded it follows that $d_y - \rho_y^- \geqq 0$ which combined with the preceding results gives $\rho_y = d_y$ on the boundary which establishes (IV).

Returning to the free boundary problem (f.b.p.) the following question arises. Is a solution of the f.b.p. necessarily a solution of the optimization problem? The answer is *yes* provided certain additional conditions are satisfied. That additional conditions are required is clear from the following considerations. Suppose that $(u, \mathscr{C})$ is a solution of both the f.b.p. ($\frac{1}{2}u_{yy} = u_s$ on $\mathscr{C}$, $u = d$ and $u_y = d_y$ on the boundary) and the optimization problem ($u = \rho, \mathscr{C} = \mathscr{C}_0$). If the problem is modified by sharply decreasing d below u on part of $\mathscr{C}_0$, then $(u, \mathscr{C})$ remains a solution of the free boundary problem but the solution of the optimization problem changes. If d is sharply decreased on a small part of the stopping set near the boundary of $\mathscr{C}$, the optimal continuation region should be enlarged but here again $(u, \mathscr{C})$ remains a solution of the free boundary problem.

These examples lead to sufficient conditions which are related to (III). One of these, (V) may be paraphrased to state that if one cannot trivially improve on u (as was possible in the above counterexamples) then $u = \rho$. Let

(8.3) $$h(y, s; s') = E\{h(y + w(s - s')^{1/2}, s')\}, \qquad s \geqq s'.$$

(V) *If u is the risk corresponding to the continuation set $\mathscr{C}$ and*

(i) $u(y, s) \leqq d(y, s)$ *and*

(ii) $u(y, s; s') \geqq d(y, s)$ *for* $(y, s) \in \mathscr{S}$ *and* $s \geqq s'$,

then $(u, \mathscr{C})$ solves the optimization problem provided that the optimal risk ρ can be approximated by the risk of a procedure where stopping is restricted to a finite number of discrete times.

While (V) does not invoke the f.b.p. condition, that condition can be used to prove condition (ii) of (V). This yields

(VI) *If $(u, \mathscr{C})$ is a solution of the f.b.p. where $\mathscr{C}$ is a continuation set and u and d have bounded derivatives up to third order and*

(i) $u(y, s) \leqq d(y, s)$ *and*

(ii) $\frac{1}{2}d_{yy} > d_s$ *on* $\mathscr{S}$

then $(u, \mathscr{C})$ is a solution of the optimization problem under the proviso of (V).

In some applications (VI) is not enough because some of the conditions break down as s approaches its lower limit s_0 (possibly $-\infty$). In that case it suffices to invoke some supplementary condition which implies

(8.4) $$\sup_{s \to s_0} |u(y, s) - \rho(y, s)| \to 0.$$

9. **Solutions, bounds and expansions for stopping problems.** In the continuous version of Problem 2, we have a stopping problem which may be represented by

$$
\begin{aligned}
d(y,s) &= -s \quad \text{for} \quad s>0 \quad \text{and for} \quad y \geqq 0, s=0, \\
d(y,s) &= -y^2 - s \quad \text{for} \quad y \leqq 0, s=0.
\end{aligned} \tag{9.1}
$$

This problem has the trivial solution where $\mathscr{C}_0 = \{(y,s): y<0, s>0\}$ and

$$
\begin{aligned}
\rho(y,s) &= -s \quad \text{for} \quad y \geqq 0, s \geqq 0, \\
\rho(y,s) &= -y^2 - s \quad \text{for} \quad y<0, s \geqq 0.
\end{aligned} \tag{9.2}
$$

The pair $(\rho, \mathscr{C}_0)$ is a solution of the (f.b.p.) since $\rho = d$ and $\rho_y = d_y$ for $y=0$. Property (V), §8 applies as does a modified version of Property (VI), §8 (a modification is required because ρ and d are not bounded).

Generally, stopping problems are not so easily solved. It is useful to derive bounds on ρ and $\mathscr{C}_0$. To illustrate, let us introduce a new stopping problem, the importance of which will be discussed later.

Problem 4. A stopping problem involving $Y(s)$, $E\{dY(s)\} = 0$, $\mathrm{Var}\{dY(s)\} = -ds$,

$$
\begin{aligned}
d(y,s) &= \quad y, \quad \text{for} \quad s=0, y \geqq 0, \\
&= \quad s^{-1}, \quad \text{for} \quad s>0, y>0, \\
&= \quad 0, \quad \text{for} \quad s \geqq 0, y \leqq 0,
\end{aligned} \tag{9.3}
$$

and stopping is enforced when $Y(s) \leqq 0$ or $s \leqq 0$.

Note that if s is large, the chances of obtaining $Y(s) = 0$ (and zero cost) before $s=0$ is large and so one is encouraged to continue unless Y is large. If s is small, the cost of stopping, (s^{-1}) is large compared to the cost of waiting till $s=0$ (approximately Y) unless Y is large and one is encouraged to continue unless Y is large. Thus one expects $\mathscr{C}_0$ to have a boundary which is high for s large and s small.

Let $u(y,s)$ be an arbitrary solution of the heat equation. Let $\mathscr{B}$ be the set on which $u(y,s) = d(y,s)$. If $\mathscr{B}$ is the boundary of a continuation set $\mathscr{C}$ the risk for the procedure defined by the continuation set $\mathscr{C}$ is $b(y,s) = u(y,s)$ on $\mathscr{C}$ and $b(y,s) = d(y,s)$ on $\mathscr{S}$. But then $\rho(y,s) \leqq b(y,s)$. Thus if (y_0, s_0) is a point of $\mathscr{C}$

where $u < d$, then $\rho(y_0, s_0) < d(y_0, s_0)$ and (y_0, s_0) is a continuation point for the optimal procedure.

For Problem 4, take $u_1(y, s) = y$ which is a solution of the heat equation. $\mathscr{C} = \{(y, s): 0 < y < s^{-1}, s > 0\}$. Since $u_1(y, s) < s^{-1}$ at every point of $\mathscr{C}$, $\mathscr{C} \subset \mathscr{C}_0$ and the boundary $y_1(s) = s^{-1}$ is a *lower bound for the optimal boundary* $\tilde{y}(s)$, i.e.

$$\tilde{y}(s) \geqq y_1(s) = s^{-1}. \tag{9.4}$$

We now describe a method of finding upper bounds for the optimal boundary. In a more general context these represent outer bounds for $\mathscr{C}_0$. Let $u(y, s)$ be a solution of the heat equation. Let $\mathscr{B}$ be the set on which $u_y(y, s) = d_y(y, s)$. Let $\mathscr{C}$ be the continuation set for which $\mathscr{B}$ is the boundary. If $u \neq d$ on $\mathscr{B}$ let $h(s) = u(y, s) - d(y, s)$ along the boundary $\mathscr{B}$ and let $d^*(y, s) = d(y, s) + h(s)$. Then $(u, \mathscr{C})$ is a solution of the f.b.p. for $d^*(y, s)$. Suppose that $(u, \mathscr{C})$ is also a solution of the optimality problem for d^* and $h(s) \leqq 0$ for $s < s_2$ and $h(s_2) = 0$. Then the modified problem is a more "advantageous" problem than the original for $s = s_2$ and $\rho(y, s_2) \geqq u(y, s_2)$. If (y_2, s_2) is a stopping point for the modified problem

$$\rho(y_2, s_2) \geqq u(y_2, s_2) = d^*(y_2, s_2) = d(y_2, s_2) \tag{9.5}$$

and (y_2, s_2) is a stopping point for the original problem.

In review we obtain outer bounds on the continuation set by finding arbitrary solutions of the heat equation which are suitable (i.e. $u - d \leqq 0$ along the boundary where $u_y = d_y$). In principle this method is as elementary as the other method but in application it is usually more delicate.

To illustrate,

$$u_2(y, s) = y - B \exp\left(\tfrac{1}{2} a^2 s\right) \sinh ay$$

is a solution of the heat equation for which $u_{2y} = d_y = 0$ when $y = y_2(s)$ which is determined by

$$1 - Ba \exp\left(\tfrac{1}{2} a^2 s\right) \cosh ay = 0.$$

For $s = 0$, $u_2 - d \leqq 0$. Along the boundary $y = y_2(s)$, $u_2 - d$ takes on negative values for small positive s. The smallest positive value s_2 of s, (if any), where $u_2 - d$ vanishes is described by

$$y - B \exp\left(\tfrac{1}{2} a^2 s\right) \sinh ay = s^{-1}.$$

Any pair of parameters (a, B) which yields such a pair (y_2, s_2) may be used and the corresponding point (y_2, s_2) is a point of $\mathscr{S}_0$. To find the best such point for a given s_2, we select a and B to minimize y_2. If for fixed (a, B), $\partial(u_2 - d)/\partial s > 0$ at (y_2, s_2) it would be possible to adjust a and B to decrease y_2. Thus we impose the third condition $\partial(u_2 - d)/\partial s = 0$,

$$(Ba^2/2) \exp(\tfrac{1}{2} a^2 s) \cosh ay = s^{-2}.$$

Together the three conditions lead to the representation for an outer bound $\tilde{y}_2(s)$ for the boundary, i.e. $\tilde{y}_2(s) \geqq \tilde{y}(s)$ where $y_2(s)$ satisfies

$$\frac{2^{1/2}}{s} (\tilde{y}_2 - s^{-1})^{1/2} = \tanh \left\{ \frac{2^{1/2} \tilde{y}_2}{s} (\tilde{y}_2 - s^{-1})^{-1/2} \right\}. \tag{9.6}$$

It is of interest that $s^{-1} < y_2(s) \leqq s^{-1} + \frac{1}{2} s^2$ which indicates that $\tilde{y}$ is well approximated by s^{-1} for small s. For large s, better approximations are obtained by using similar arguments with solutions of the heat equation of the form

$$u = A\phi \left(\frac{y}{(s+h)^{1/2}} \right) \frac{y}{(s+h)^{3/2}}, \qquad h \geqq 0$$

which yield the lower bound

$$\tilde{y}_3(s) = s^{1/2} \leqq \tilde{y}(s) \tag{9.7}$$

and the upper bound

$$\tilde{y}_4(s) = (s+h)^{1/2} \geqq y(s) \tag{9.8}$$

where $h > e^{1/3}$ satisfies $s = e^{1/2} h (h^{3/2} - e^{1/2})^{-1}$. Together these show that $\tilde{y}(s) = s^{1/2}\{1 + O(s^{-1})\}$ for large s.

Another but related approach to approximating the optimal boundary consists of finding asymptotic expansions for the risk and boundary near distinguished points of s; these distinguished points are typically the end points of the range of interest. For example $s = 0$ and ∞ are important in Examples 1, 2, and 4. The proofs that the formal expansions derived by methods to be briefly described do indeed represent approximations to the desired solution depend on arguments of the type described above.

One important class of solutions of the heat equation used in generating expansions is that generated by "sources of heat" along a vertical ($s =$ constant) line. Thus

(9.9) $$u_0(y,s) = s^{-1/2}\phi(\alpha), \qquad \alpha = y/\sqrt{s},$$

represents a point source of heat at $(y,s) = (0,0)$ and yields a solution of the heat equation for $s > 0$. Similarly, functions of the form

$$u(y,s) = \int \frac{1}{s^{1/2}}\,\phi\left(\frac{y-y'}{s^{1/2}}\right) h(y')\,dy' = \int h(y + w s^{1/2})\,\phi(w)\,dw$$

satisfy the heat equation.

The technique of substituting solutions of these forms into the two boundary conditions leads to asymptotic expansions for the optimal solution of the sequential analysis problem (Problem 1* of §5) of the form

$$\tilde{y}(s) = y(s)\,s^{-1/2} \sim \{\log a^2 s^3 - \log 8\pi - 6(\log a^2 s^3)^{-1} + \cdots\} \qquad \text{as } s \to \infty,$$

$$\tilde{y}(s) = \tilde{y}(s)\,s^{-1/2} \sim \tfrac{1}{4}\,a s^{3/2}\left\{1 - \frac{a^2 s^3}{12} + \frac{7}{15\cdot 16}\,a^4 s^6 - \cdots\right\} \qquad \text{as } s \to 0,$$

where $a = k/c$.

10. **Control problem.** Let us return to the rocket control problem (Problem 3, §2). For reasons to be discussed later an important case can be described in its continuous time version as follows.

Problem 3*. One observes $Y(s)$, a Wiener process in the $-s$ scale originating at (y_0, s_0) with $E\{dY(s)\} = 0$ and $\operatorname{Var}\{dY(s)\} = -ds$. As s decreases to 0, $Y(s)$ may be adjusted instantaneously at any $s > 0$ by an amount Δ at a cost of $|\Delta|\,d(s)$ where $d(s) = s^{-1}$. In addition to the accumulated cost due to the adjustments of $Y(s)$, there is a cost of $\frac{1}{2}Y^2(0)$. Find the rule for adjusting $Y(s)$ which minimizes the expected cost.

The discrete time version of this problem corresponds to the specification of a finite set of s_i, $s_0 \geqq s_1 \geqq s_2 \geqq \cdots \geqq s_n = 0$ at which changes (corresponding to the use of fuel) are permitted. Let $\rho^*(y, s_i)$ represent the expected additional cost associated with the optimal procedure of the discrete time version given $Y(s_i) = y$. Since it is possible to change y instantaneously at a cost of $d(s_i)$ per unit y

$$\rho^*(y, s_i) \leqq \rho^*(y', s_i) + d(s_i)|y' - y|$$

from which it follows that

$$|\partial \rho^*(y, s_i)/\partial y| \leqq d(s_i). \tag{10.1}$$

With a slight variation of this approach let $\rho_0^*(y, s_i)$ represent the optimal risk at $s = s_i$ subject to the restriction that fuel is not used at $s = s_i$. Here

$$\rho^*(y, s_i) = \inf_{y'} \{\rho_0^*(y', s_i) + d(s_i)|y' - y|\}. \tag{10.2}$$

Regarded as a function of y, ρ^* has straight line sections with slope $\partial \rho^*/\partial y = \pm d(s_i)$, where it pays to use fuel. Elsewhere, it does not pay to use fuel and $|\partial \rho^*/\partial y| \leqq d(s_i)$. Thus the optimal policy is described by an *action set* and a *continuation or no-action set*. If (y, s_i) is on the action set one moves to a point $(\tilde{y}^*, s_i)$ on the boundary of the continuation set by applying fuel. Otherwise no fuel is used at this stage. It is possible to show the ρ^* is symmetric and decreasing in $|y|$. Hence the optimal continuation set is described by $-\tilde{y}^*(s) < y < \tilde{y}^*(s)$.

Let us proceed to the continuous time version of the problem for which the above characterization still applies. The boundary of the optimal no-action set is given by $\pm \tilde{y}(s)$. We restrict our attention to procedures which may be described by a symmetric no-action set with boundary $\pm y_1(s)$ and let $b(y, s)$ be the additional expected cost associated with such a procedure given $Y(s) = y$. We shall show that $b(y, s)$ satisfies the following conditions:

$$\tfrac{1}{2} b_{yy}(y, s) = b_s(y, s) \quad \text{on the no-action set,} \tag{10.3}$$

$$\tfrac{1}{2} b_{yyy}(y, s) = b_{ys}(y, s) \quad \text{on the no-action set,} \tag{10.4}$$

$$b_y(y, s) = d(s) = s^{-1} \quad \text{on that part of the boundary and action set for which } y > 0, s > 0, \tag{10.5}$$

$$b_y(0, s) = 0 \quad \text{for} \quad s > 0, \tag{10.6}$$

$$b_y(y, 0) = y \quad \text{for} \quad y > 0. \tag{10.7}$$

However, in Problem 4 of §9, we described a stopping problem whose solution uniformly minimizes b_y subject to the restrictions (10.4—10.7). Consequently the optimal expected cost for our control problem can be obtained by integrating the solution of the stopping problem, Problem 4. The optimal no-action set is the optimal continuation set of the stopping problem.

In this particular case we have been fortunate and profited from the symmetry. Otherwise we would have, for arbitrary procedures described by no-action sets, that $b(y,s)$ satisfies (10.3), (10.4), (10.7), and (10.5) replaced by

(10.5′) $\quad b_y(y,s) = \pm\, d(s)$ on the boundary and action set.

The optimality condition required to determine the free boundary would be

(10.8) $\quad \rho_{yy}(y,s) = 0$ on the boundary.

This corresponds to $(\rho_y)_y = d_y$, and thus *the derivative of the expected cost of the control problem satisfies the same (f.b.p.) as do the optimal risks for the stopping problems.*

In these discussions several claims were made which require some support. First let us deal with the formulation of Problem 3*. Suppose that incoming data used to estimate the miss distance have standard deviation proportional to the distance to target. Then the reasoning of §4 applied to a continuous time version indicates that at any given time the posterior distribution of the miss distance is $\mathcal{N}(Y(s),s)$ where $Y(s)$, the current estimate of the miss distance, is a Wiener process in the $-s$ scale and s, measuring the total cumulated precision, is given by

$$s^{-1} = \sigma_0^{-2} + \int_0^t \sigma^{-2}(t_0 - t)^{-2}\,dt = \sigma_0^{-2} - \sigma^{-2}t_0^{-1} + \sigma^{-2}(t_0 - t)^{-1}$$

where t_0 is the total required time of flight. For simplicity, let us assume that the two quantities, σ_0^{-2} and $\sigma^{-2}t_0^{-1}$, both of which are ordinarily small, cancel giving

$$s = \sigma^2(t_0 - t).$$

Let us also assume that the amount of fuel required to change the miss distance by an amount Δ is inversely proportional to the distance to the target. Then the cost of fuel required per unit change of y is

$$d(s) = c\,[e(t)]^{-1} = ck'(t_0 - t)^{-1}$$

and hence

(10.9) $$d(s) = as^{-1}.$$

Since $s = 0$ corresponds to infinite precision and time of arrival, the cost of missing is $kY^2(0)$. Now let $s^* = h^2 s$ and $Y^*(s^*) = hY(s)$.

Then Y^* is a Wiener process in the $-s^*$ scale and, in terms of s^* and Y^*

$$d(s) = d^*(s^*) = ah^2 s^{*-1}, \quad kY^2(s) = kh^{-2}[Y^*(s^*)]^2.$$

Selecting $kh^{-2} = \frac{1}{2}ah^2$ gives us a starred problem where the costs are proportional to those of Problem 3*. If any of the above assumptions concerning the rate of incoming information or the efficiency of fuel were changed, it would result only in modifying the stopping cost of the related stopping problem.

The fact that b satisfies the heat equation in the interior of the no-action set follows by the typical argument. This in turn implies that b_y satisfies the heat equation. Equations (10.6) and (10.7) follow from the symmetry and terminal cost.

To justify (10.5) one must consider behavior near the boundary. Here, one puzzling aspect of the continuous time version of our policy which was deliberately evaded must now be faced. Suppose that the boundary $y_1(s)$ is a well-behaved function of s. Then if (y, s) is on the action set, the policy calls for bringing $Y(s)$ to $y_1(s)$. Later the unadjusted $Y(s)$ is a rather complicated function and is bound to leave the no-action region "immediately" after it is brought to the boundary. How does one compute the amount of fuel that is used in the many infinitesimal departures and returns? Fortunately, this can be conveniently expressed in terms of

$$M(s) = \max\left\{0, \sup_{s_0 \geqq s' \geqq s} [Y_1(s') - \tilde{y}_1(s')]\right\} \tag{10.10}$$

where $Y_1(s')$ is the original (unadjusted) Wiener process and in the neighborhood of an upper boundary point $(y_1(s_0), s_0)$ of the no-action set, the adjusted process behaves like

$$Y(s) = Y_1(s) - M(s). \tag{10.11}$$

If $Y(s_0) = y_1(s_0)$ and $y_1(s)$ has finite slope then for small δ, $\mathscr{L}(M(s_0 - \delta)) \approx \mathscr{L}(\delta^{1/2} M^*)$ where $M^* = \sup_{0 \leqq t \leqq 1} W(t)$ and $W(t)$ is a standard Wiener process. Moreover, the use of a reflection principle yields $P\{M^* > a\} = 2P\{W(1) > a\}$ for $a > 0$ and hence

$$\mathscr{L}\{M(s_0 - \delta)\} \approx \mathscr{L}(\delta^{1/2}|w|) \quad \text{as} \quad \delta \to 0, \tag{10.12}$$

where $\mathscr{L}(w) = \mathscr{N}(0, 1)$.

Now, to demonstrate (10.5) at an upper boundary point (y_0, s_0) it is easy to see that $b_y^+(y_0, s_0) = d(s_0)$. We shall now show that

$b_y^-(y_0, s_0) = d(s_0)$ assuming that $b(y,s) - b(y, s-\delta) = O(\delta)$. Between time s_0 and $s_0 - \delta$, the process originating from (y_0, s_0) is adjusted by a total amount $M(s_0 - \delta)$ and at time $s_0 - \delta$ is at $Y(s_0 - \delta) = Y_1(s_0 - \delta) - M(s_0 - \delta)$.

$$\begin{aligned} b(y_0, s_0) &= E\{d(s_0) M(s_0 - \delta) + b(Y(s_0 - \delta), s_0 - \delta)\} + O(\delta) \\ &= E\{b(y_0, s_0) + b_y^- [Y_1(s_0 - \delta) - y_0 - M(s_0 - \delta)] \\ &\qquad + d(s_0) M(s_0 - \delta) + O(\delta)\} \\ &= b(y_0, s_0) - [b_y^- - d(s_0)] E\{M(s_0 - \delta)\} + o(\delta^{1/2}). \end{aligned}$$

Since $E\{M(s_0 - \delta)\}$ is approximately $(2\delta/\pi)^{1/2}$, the desired result follows.

Finally, we demonstrate that optimality implies that $\rho_{yy} = 0$ on the boundary. First, since ρ_y is constant above (y_0, s_0), $\rho_{yy}^+(y_0, s_0) = 0$. Second, since $|\rho_y| \leqq d(s_0)$ in the no-action set, $\rho_{yy}^-(y_0, s_0) \geqq 0$. The suboptimal procedure in which no action is taken from $s_0 + \delta$ to s_0, and an optimal policy is followed thereafter has risk b where

$$\rho(y_0, s_0 + \delta) \leqq b = E\{\rho(y_0 + w\sqrt{\delta}, s_0)\},$$

$$\rho(y_0, s_0 + \delta) \leqq \rho(y_0, s_0) + Ew\rho_y\delta + \frac{1}{2}\left[(\rho_{yy}^+\delta) \int_0^\infty w^2\phi(w)\,dw \right.$$

$$\left. + (\rho_{yy}^-\delta) \int_{-\infty}^0 w^2\phi(w)\,dw\right] + o(\delta),$$

$$\rho_s\delta = \tfrac{1}{2}\rho_{yy}^-\delta \leqq \tfrac{1}{4}(\delta)[\rho_{yy}^-] + o(\delta),$$

$$\rho_{yy}^- \leqq 0,$$

which implies $\rho_{yy} = 0$.

Thus we see that ρ_y and $\mathscr{C}$, the partial derivative of the optimal risk and the optimal no-action set for the control problem, correspond to a solution of the free boundary problem determined by $d(s)$. This is the case even without the benefit of the symmetry which is used. Furthermore in the more general case where the cost of fuel per unit change of y is represented by $d(y,s)$, the boundary conditions for ρ_y would be $\rho_y = d$ and $\rho_{yy} = d_y$.

11. **Summary and remarks.** Approximating discrete time problems by continuous time problems invoking the Wiener process makes it possible to apply the analytic methods of partial differential equations to problems of sequential analysis, which are

basically special examples of stopping problems, and to certain stochastic control problems. It was seen that the solution of stopping problems reduce to the solution of free boundary problems involving the heat equation. Almost arbitrary solutions of the heat equation could be used to provide bounds on the solution of stopping problems. Asymptotic expansions for the solution are obtainable by use of relatively simple classes of solutions of the heat equation.

It is difficult to state and prove rigorously "nice" theorems of the kind "the solution of the optimization problem is a solution of the f.b.p.". It would be desirable to have such theorems which invoke only conditions on the elements in the statement of the problem such as the function $d(y, s)$. Most proofs seem to involve conditions on the nature of the unknown solutions. This problem seems hard to avoid because the solutions of certain problems of interest have singular points where the f.b.p. condition breaks down. On the other hand the sufficiency theorems which permit one to recognize when a given candidate is a solution of the optimization problem, are much more amenable to useful statements which can be reasonably applied. Fortunately, these sufficiency results are the more important ones because they are the ones invoked in applying the methods of bounding solutions of stated problems in terms of arbitrary solutions of the heat equation and solution of related optimum problems.

The rocket control problem has a continuous time formulation similar in certain aspects to that of the stopping problem and here the derivative of the optimal expected cost also is a solution of the f.b.p.

A rocket control problem where fuel is free but only a finite amount is available is more difficult to treat. A role analagous to that of ρ_y in the infinite fuel case is played by $V = \rho_u + e(s)\,\rho_y$ where u is the amount of fuel available and $e(s)$ is the change in y obtainable from a unit of fuel. Since V measures the rate of gain derived from using fuel, $V \leqq 0$ on the no-action set and $V = 0$ on the action set. Bounds and expansions have been derived for the solution of this problem subject to the following conjecture. Let a taxed version of the control problem be such that at $s = s_0$ fuel is free, but later $(s < s_0)$, fuel must be paid for and fuel remaining at $s = 0$ must be taxed. The conjecture states that for a situation where the original untaxed problem calls for the use of fuel at $s = s_0$, the taxed version also calls for the use of fuel then.

An approximation has been derived which relates the continuous time solution to discrete time solutions of stopping problems with finely spaced intervals between the permitted stopping times.

Work has been carried out on continuous time stopping problems which do not involve the heat equation. These include that of Mikhalevich [**19**] where the Poisson process leads to a difference differential equation and a pair of diffusion equations of Shiryaev [**22**] where the number of possible values of μ are finite and past history is summarized by a few posterior probabilities rather that $(Y(s), s)$. Bather [**4**], [**5**] has considered certain problems which involve ordinary differential equations because of the stationarity of these problems.

The history of these ideas and methods is long and complicated and the following is a bare outline of related references.

Backward induction—Dynamic programming—[**2**], [**8**].
Stopping problems (discrete time)—[**14**], [**17**], [**24**], [**25**].
Stopping problems (continuous time)—[**1**], [**3**], [**4**], [**5**], [**9**], [**10**], [**11**], [**12**], [**13**], [**16**], [**19**], [**20**], [**22**].
Rocket control—[**6**], [**7**], [**21**], [**26**].
Heat Equation—[**15**].
Expository article—[**23**].

References

1. F. J. Anscombe, *Sequential medical trials,* J. Amer. Statist. Assoc. **58** (1963), 365-383.

2. K. J. Arrow, D. Blackwell and M. A. Girschick, *Bayes and minimax solutions of sequential decision problems,* Econometrica **17** (1949), 213-244.

3. J. A. Bather, *Bayes procedures for deciding the sign of a normal mean,* Proc. Cambridge Philos. Soc. **58** (1962), 599-620.

4. ———, *Control charts and the minimization of costs (with discussion),* J. Roy. Statist. Soc. Ser. B. **25** (1963), 49-80.

5. ———, *A continuous time inventory model,* J. Appl. Probability **3** (1966), 538-549.

6. J. A. Bather and H. Chernoff, *Sequential decisions in the control of a space-ship,* Proc. Fifth Berkeley Sympos. Math. Statist. and Prob., Vol. 3, pp. 181-207, University of California Press, Berkeley, California, 1967.

7. ———, *Sequential decisions in the control of a space-ship (finite-fuel),* J. Appl. Probability **4** (1967).

8. R. Bellman, *Dynamic programming,* Princeton Univ. Press, Princeton, N. J., 1956.

9. J. Breakwell and H. Chernoff, *Sequential tests for the mean of a normal distribution.* II. (*Large t*), Ann. Math. Statist. **35** (1964), 162-173.

10. H. Chernoff, *Sequential tests for the mean of a normal distribution,* Proc. Fourth Berkeley Sympos. Math. Statist. and Prob., Vol. I, pp. 79-91, Univ. of California Press, Berkeley, Calif., 1961.

11. ———, *Sequential tests for the mean of a normal distribution*. III. (*Small t*), Ann. Math. Statist. **36** (1965), 28-54.

12. ——— *Sequential tests for the mean of a normal distribution*. IV. (*Discrete case*), Ann. Math. Statist. **36** (1965), 55-68.

13. H. Chernoff and S. N. Ray, *A Bayes sequential sampling inspection plan*, Ann. Math. Statist. **36**(1965), 1387-1407.

14. Y. S. Chow and H. E. Robbins, *On values associated with a stochastic sequence*, Proc. Fifth Berkeley Sympos. Math. Statist. and Prob., Vol. 1, pp. 419-426, University of California Press, Berkeley, California, 1967.

15. J. L. Doob, *A probability approach to the heat equation*, Trans. Amer. Math. Soc. **80** (1955), 216-280.

16. B. I. Grigelionis and A. N. Shiryaev, *On Stefan's problem and optimal stopping rules for Markov processes*, English transl., Theor. Probabilility Appl. (11) **4** (1966), 612-631.

17. G. W. Haggstrom, *Optimal stopping and experimental design*, Ann. Math. Statist. **37** (1966), 7-29.

18. D. V. Lindley, *Introduction to probability and statistics from a Bayesian point of view*. Part II: *Inference*, Cambridge Univ. Press, Cambridge, 1965.

19. V. S. Mikhalevich, *Sequential Bayes and optimal methods of statistical acceptance control*, Theory of Probability Appl. (1) **4** (1956), 395-420, English transl.

20. S. Moriguti and H. E. Robbins, *A Bayes test of "$p \leqq 1/2$" versus "$p > 1/2$"*, Rep. Statist. Appl. Res. Un. Japan. Sci. Engrs. **9** (1961), 39-60.

21. R. J. Orford, *Optimal stochastic control systems*, J. Math. Anal. Appl. **6** (1963), 419-429.

22. A. N. Shiryaev, *On the theory of decision functions and control by a process of observation on incomplete data*, Trans. III Prague Conference on Information Theory 1964, pp. 657-681.

23. ———, *Sequential analysis and controlled random processes* (*Discrete time*), Kibernetika **3** (1965), 1-24.

24. D. O. Siegmund, *Some problems in the theory of optimal stopping rules*, Ann. Math. Statist. **38** (1967), 1627-1640.

25. J. L. Snell, *Applications of martingale system theorems*, Trans. Amer. Math. Soc. **73** (1952), 293-312.

26. F. Tung and C. J. Striebel, *A stochastic control problem and its applications*, J. Math. Anal. Appl. **12** (1965), 350-359.

STANFORD UNIVERSITY
STANFORD, CALIFORNIA

Arthur F. Veinott, Jr. [1]

On the Optimality of (s, S) Inventory Policies: New Conditions and a New Proof [2]

Abstract. Scarf [**6**] has shown that the (s, S) policy is optimal for a class of discrete review dynamic nonstationary inventory models. In this paper a new proof of this result is found under new conditions which do not imply and are not implied by Scarf's hypotheses. We replace Scarf's hypothesis that the one period expected costs are convex by the weaker assumption that the negatives of these expected costs are unimodal. On the other hand we impose the additional assumption not made by Scarf that the absolute minima of the one period expected costs are (nearly) rising over time. For the infinite period stationary model, this last hypothesis is automatically satisfied. Thus in this case our hypotheses are weaker than Scarf's. The bounds on the optimal parameter values given by Veinott and Wagner [**12**] are established for the present case. The bounds in a period are easily computed, and depend only upon the expected costs for that period. Moreover, simple conditions are given which ensure that the optimal parameter values in a given period equal their lower bounds. When there is no fixed charge for ordering, this reduces to earlier results of Karlin

---

[1] Department of Operations Research, Stanford University, Stanford, California. This research was supported by the Office of Naval Research under Contract Nonr-225(77) and by a grant from the Western Management Science Institute.

[1] Reprinted with permission from J. SIAM Appl. Math. **14** (1966), pp. 1067-1083.

[5] and Veinott [9], [10], [11] for the nonstationary case. The above result is exploited to extend the planning horizon theorem of Veinott [9] to the case where there is a fixed charge for ordering.

1. **Model formulation.** We consider a single product dynamic inventory model in which the demands $D_1, D_2, \cdots$, for a single product in periods $1, 2, \cdots$, are independent random variables with distributions $\Phi_1, \Phi_2, \cdots$. Assume $\{\eta_i\}$ are given constants such that[3] $D_i \geqq \eta_i$ for all i. At the beginning of each period the system is reviewed. An order may be placed for any nonnegative quantity of stock. An order placed at the beginning of period i is delivered at the beginning of period $i + \lambda$, where λ is a known nonnegative integer.

Let x_i denote the stock on hand and on order prior to placing any order in period i. Let y_i denote the stock on hand and on order after ordering in period i. It is possible for x_i and y_i to be negative indicating the existence of a backlog. We assume that the amount of stock on hand and on order at the end of period i is a specified Borel function $v_i(y_i, D_i)$ of y_i and D_i. Thus $x_{i+1} = v_i(y_i, D_i)$. If $\lambda > 0$ we assume that all unsatisfied demand is backlogged so $v_i(y_i, D_i) = y_i - D_i$.

When $\lambda = 0$ our formulation provides for the possibilities of deterioration of stock in storage (perishable goods) and partial backlogging of unsatisfied demand [11, p. 766]. For example suppose that whenever $y_i < D_i$, then a fraction b $(0 \leqq b \leqq 1)$ of the unsatisfied demand is backlogged and the remainder leaves immediately. If instead $y_i > D_i$, then a fraction $1 - a$ $(0 \leqq a \leqq 1)$ of the inventory on hand spoils and is not available for future use. These assumptions imply that $v_i(\cdot, \cdot)$ takes the form

$$v_i(y_i, D_i) = \begin{cases} a \cdot (y_i - D_i) & \text{if } y_i \geqq D_i, \\ b \cdot (y_i - D_i) & \text{if } y_i \leqq D_i. \end{cases}$$

Note that if $a = 0$ we have the case of perishable goods while if

---

[3] Actually the main results of the paper given in §2 also hold in the more general case where D_i is a random vector. All that is required is to let $\mathscr{D}_i$ be the Borel set of possible values of D_i and replace the interval $[\eta_i, \infty)$ of possible values of D_i everywhere by $\mathscr{D}_i$. This more general formulation allows for consideration of several classes of demands, random deterioration rates, random departures of backlogged demand, and random prices, for example, by suitable interpretation of the components of D_i.

$a = 1$ we have the case of nonperishable goods. If $b = 0$ we have the lost sales case while if $b = 1$ we have the backlog case. In the literature these last two cases are usually discussed only where $a = 1$.

At the beginning of period i, the inventory manager is assumed to have observed the vector

$$H_i = (x_1, \cdots, x_i, y_1, \cdots, y_{i-1}, D_1, \cdots, D_{i-1}),$$

representing the history of the process up to the beginning of period i. He bases his ordering decision in period i upon H_i.

An ordering policy for period i is a real valued Borel function $\overline{Y}_i(\cdot)$ to be used as follows. At the beginning of period i, after having observed the past history H_i, the manager orders $\overline{Y}_i(H_i) - x_i$ which is assumed to be nonnegative of course. Also let $\widetilde{Y}_i = (\overline{Y}_i, \cdots, \overline{Y}_n)$ denote a sequence of ordering policies for periods $i, \cdots, n$.

Three types of costs are considered: ordering, holding, and shortage. Assume that the cost of ordering z units in period i is $K_i\delta(z) + c_i z$, where $K_i \geqq 0$, $\delta(0) = 0$, and $\delta(z) = 1$ for $z > 0$. The cost is incurred at the time of delivery of the order. Let $g_i(y, D_{i+\lambda})$ denote the holding and shortage cost in period $i + \lambda$ when y is the amount of stock actually on hand after receipt of orders to be delivered before the end of period $i + \lambda$. We assume that $g_i(\cdot, \cdot)$ is a real valued Borel function.

Let $\alpha_i(\geqq 0)$ be the discount factor for period $i + \lambda$. That is, α_i is the value at the beginning of period $i + \lambda$ of one cost unit at the beginning of period $i + \lambda + 1$. Let $\beta_1 = 1$ and $\beta_i = \prod_{j=1}^{i-1} \alpha_j$ for $i > 1$.

For the case $\lambda = 0$ let

$$W_i(y, t) = c_i y + g_i(y, t) - \alpha_i c_{i+1} v_i(y, t).$$

For the case $\lambda > 0$ let

$$W_i(y, t) = c_i y + \int_{-\infty}^{\infty} g_i(y - z, t)\, d\Phi_i^{\lambda}(z) - \alpha_i c_{i+1}(y - E(D_i)),$$

where $\Phi_i^{\lambda}(\cdot)$ is the distribution of $D_i + \cdots + D_{i+\lambda-1}$.

Now for $\lambda \geqq 0$ let

$$G_i(y) = \int_{-\infty}^{\infty} W_i(y, t)\, d\Phi_{i+\lambda}(t).$$

We assume that all integrals given above exist and are finite.

We suppose that each unit of stock left over after $\lambda + n$ periods can be discarded with a return of c_{n+1}. Similarly, each unit of backlogged demand remaining after $\lambda + n$ periods is satisfied at a cost c_{n+1}. In the literature it has often been assumed that $c_{n+1} = 0$.

Thus, the expected discounted cost incurred in periods $\lambda + 1, \cdots, \lambda + n$ when following the policy $\widetilde{Y}_1$ in periods $1, \cdots, n$ is

$$E\left\{\sum_{i=1}^{n} \beta_i \left[K_i\delta(y_i - x_i) + c_i(y_i - x_i) + g_i\left(y_i - \sum_{j=i}^{i+\lambda-1} D_j, D_{i+\lambda}\right)\right] - \beta_{n+1}c_{n+1}\left(x_{n+1} - \sum_{i=n+1}^{n+\lambda} D_i\right)\right\}.$$

By substituting $x_i = v_{i-1}(y_{i-1}, D_{i-1})$ into the above formula we get as in [**10**], [**12**],

$$\sum_{i=1}^{n} \beta_i E[K_i\delta(y_i - x_i) + G_i(y_i)] - \left[c_1x_1 - \beta_{n+1}c_{n+1} \sum_{i=n+1}^{n+\lambda} E(D_i)\right].$$

Since the second bracketed term is not affected by the choice of $\widetilde{Y}_1$, it is convenient to omit it from the analysis. Thus we may define the conditional expected discounted cost incurred in periods $\lambda + i, \cdots, \lambda + n$ when following $\widetilde{Y}_i$ in periods $i, \cdots, n$ given the observed history H_i as

$$f_i(\widetilde{Y}_i | H_i) = \sum_{j=i}^{n} \beta_j E_{H_i}[K_j\delta(y_j - x_j) + G_j(y_j)]. \tag{1}$$

We seek a policy $\widetilde{Y}_1^* = (\overline{Y}_1^*, \cdots, \overline{Y}_n^*)$, called optimal, which satisfies

$$f_i(\widetilde{Y}_i^* | H_i) \leqq f_i(\widetilde{Y}_i | H_i), \qquad i = 1, \cdots, n, \tag{2}$$

for all H_i and $\widetilde{Y}_i$, where of course $\widetilde{Y}_i^* = (\overline{Y}_i^*, \cdots, \overline{Y}_n^*)$. It is easy to show by induction on i (starting with $i = n$) that if there is an optimal policy, then $f_i(\widetilde{Y}_i^* | H_i)$ depends upon H_i only through x_i, so we may write

$$f_i(\widetilde{Y}_i^* | H_i) = f_i(x_i), \qquad i = 1, \cdots, n, \tag{3}$$

where the f_i satisfy $(f_{n+1}(x) \equiv 0)$

$$f_i(x) = \inf_{y \geqq x} \{ K_i\delta(y - x) + G_i(y) + \alpha_i Ef_{i+1}(v_i(y, D_i)) \} \tag{4}$$

for $i = 1, \cdots, n$ and all x with the infimum being attained for each x. Conversely if there is a sequence of functions $\{f_i\}$ which satisfy

(4), with the infimum being attained for each x and being a Borel function of x, then there exists an optimal policy $\widetilde{Y}_1^*$. Moreover, $\overline{Y}_i^*(H_i)$ minimizes the expression in braces on the right side of (4) subject to $y \geqq x$ where $x = x_i$. Since the minimizing value of y is a function only of x_i, it follows that $\overline{Y}_i^*(H_i) \equiv \overline{Y}_i^*(x_i)$ depends upon H_i only through x_i. For notational convenience, we define

$$J_i(y) = G_i(y) + \alpha_i E f_{i+1}(v_i(y, D_i)). \tag{5}$$

In what follows we shall have occasion to impose one or more of the following assumptions for each i $(K_{n+1} \equiv 0)$:

(i) $G_i(y)$ and $v_i(y, t)$ are continuous in y for each $t \geqq \eta_i$;

(ii) $\underline{\lim}_{y \to \infty} G_i(y) > \inf_y G_i(y) + \alpha_i K_{i+1}$;

(iii) $\underline{\lim}_{y \to -\infty} G_i(y) > \inf_y G_i(y) + K_i$;

(iv) $-G_i(y)$ is unimodal in y;

(v) $v_i(y, t)$ is nondecreasing in y for each $t \geqq \eta_i$; moreover, $v_i(y, t)$ is bounded above in t on $[\eta_i, \infty)$ for each fixed y;

(vi) $K_i \geqq \alpha_i K_{i+1}$.

If (i)—(iii) hold, there are a number $\underline{S}_i$ which minimizes $G_i(y)$ on $(-\infty, \infty)$ and numbers $\underline{s}_i$ $(\leqq \underline{S}_i)$ and $\overline{S}_i$ $(\geqq \underline{S}_i)$ such that

$$G_i(\overline{S}_i) = G_i(\underline{S}_i) + \alpha_i K_{i+1}$$

and

$$G_i(\underline{s}_i) = G_i(\underline{S}_i) + K_i.$$

If in addition (vi) holds, there is a number $\overline{s}_i$, $\underline{s}_i \leqq \overline{s}_i \leqq \underline{S}_i$, such that

$$G_i(\overline{s}_i) = G_i(\underline{S}_i) + (K_i - \alpha_i K_{i+1}).$$

2. **The optimality of the (s, S) policy.** In this section we shall show that if (i)—(vi) hold and if

$$v_i(\underline{S}_i, t) \leqq \underline{S}_{i+1} \quad \text{for } t \geqq \eta_i \quad \text{and } i = 1, 2, \cdots, n-1, \tag{vii}$$

then there is an optimal policy which is an (s, S) policy. By this we mean that there is a sequence $\{(s_i, S_i)\}$ of pairs of numbers such that $(s_i \leqq S_i)$,

$$\overline{Y}_i^*(H_i) = \begin{cases} S_i & \text{if } x_i < s_i, \\ x_i & \text{if } x_i \geqq s_i, \end{cases}$$

for all i and H_i. Moreover the numbers satisfy

$$\underline{s}_i \leqq s_i \leqq \overline{s}_i \leqq \underline{S}_i \leqq S_i \leqq \overline{S}_i \tag{6}$$

and

$$G_i(s_i) \geqq G_i(S_i) + (K_i - \alpha_i K_{i+1}) \tag{7}$$

for all i. The bounds in (6) and (7) are depicted in Figure 1. The inequality (7) has the interpretation that if it is optimal to order in period i with the initial inventory level x, then the reduction in the immediate expected costs due to $G_i(\cdot)$ must be at least $K_i - \alpha_i K_{i+1}$. In the special case where $\eta_i = 0$ and $v_i(y, t) = y - t$ for all i, then (vii) reduces to the simpler form: $\underline{S}_i \leqq \underline{S}_{i+1}$, $i = 1, \cdots, n-1$.

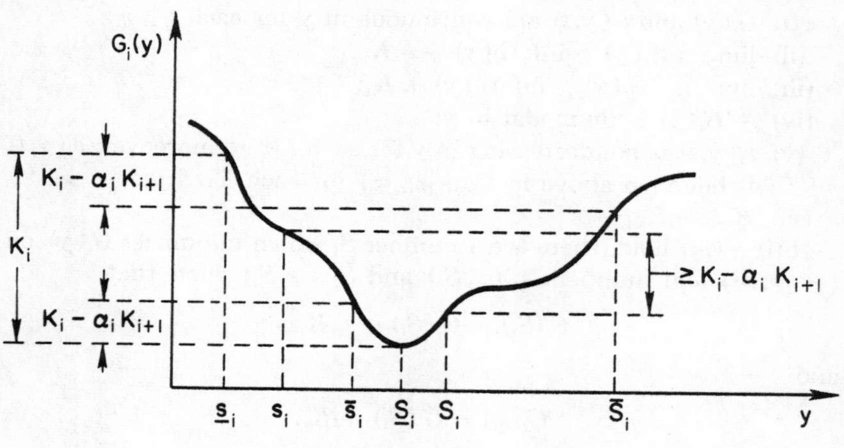

FIGURE 1

The first proof that the (s, S) policy is optimal under reasonably general conditions is due to Scarf [**6**]. (See also [**13**].) Scarf assumes that $G_i(y)$ is convex in y with $G_i(y) \to \infty$ as $|y| \to \infty$, that $v_i(y, t) = y - t$, and that (vi) holds.[4] These assumptions imply (i)—(vi) and are in fact quite a bit stronger. To elaborate on this point we remark that if $W_i(y, t)$ is convex in y, then $G_i(y)$ is convex in y for any distribution $\Phi_{i+\lambda}$. However, if $\Phi_{i+\lambda}$ has a density $\phi_{i+\lambda}$ with $\phi_{i+\lambda}(t - \theta)$ having a monotone likelihood ratio with respect to θ, then $-G_i(y)$ will be unimodal under conditions where $W_i(y, t)$ is not convex in y. Specifically, $-G_i(y)$ is unimodal if $W_i(y, t) = W_i^1(y - t) + W_i^2(t)$ for some functions W_i^1 and W_i^2 where $-W_i^1(z)$ is unimodal in z, [**3**]. See [**4**] for a discussion of the utility of this assumption.

Although Scarf imposes stronger assumptions than (i)—(vi), he

[4] Scarf also assumed that the cost functions and demand distributions do not change over time although that is not essential to his proof [8, p. 200].

does not require that (vii) hold. Thus his results are not implied by ours nor conversely.

As an example to illuminate the significance of (vii) suppose $\lambda = 0$; $D_1, \cdots, D_n$ are identically distributed; and $c_i = c$, $\alpha_i = \alpha$, $g_i(y,t) = g(y,t)$, $v_i(y,t) = y - t$, and $\eta_i = 0$ for $i = 1, 2, \cdots, n$. Then $G_i(y) = G(y)$ so $\underline{S}_i = \underline{S}$ for $i = 1, 2, \cdots, n-1$. If in addition $c_{n+1} = c$, then $G_n(y) = G(y)$ also, so $\underline{S}_n = \underline{S}$ and (vii) holds. On the other hand if $c_{n+1} = 0$ (as is assumed, for example, by Scarf [**6**] and others), then $G_n(y) = G(y) + \alpha c[y - E(D_n)]$. Thus if $\alpha c > 0$, we should ordinarily expect $\underline{S} > \underline{S}_n$ in which case (viii) fails to hold for $i = n - 1$. Both of the above definitions of c_{n+1} are reasonable formulations of a "stationary" version of our model. It is of some interest then that the first assumption assures that (vii) holds while the second does not. Of course in infinite horizon models [**1**], [**2**], the difference between the two stationary formulations vanishes. Thus our hypotheses are actually weaker than Scarf's for the infinite horizon stationary models.

Bounds on s_i and S_i were first established by Iglehart in [**1**], [**2**] under Scarf's hypotheses, the assumption that the cost functions and demand distributions do not change over time, and $c_{n+1} \equiv 0$. Under the above assumptions except $c_{n+1} = c$, Veinott and Wagner [**12**] have established bounds of the form (6). In their analysis $G_i(y) = G(y)$, so the bounds in (6) are independent of i even though this is not true of s_i and S_i. Our present analysis shows that the bounds remain valid under the weaker hypotheses imposed here.

The principal tool of Scarf's proof is the fact that if $J_i(y)$ is K_i-convex (see [**6**] for a definition), then so is $f_i(x)$. This method of proof fails under our hypotheses because $J_i(y)$ need not be K_i-convex.[5] Our proof is based instead upon the following two lemmas which establish properties of functions satisfying (4), (5).

LEMMA 1.

$$f_i(x) \leqq f_i(x') + K_i, \qquad x \leqq x',\ i = 1, \cdots, n. \tag{8}$$

Moreover, if (v) *holds, then*

$$J_i(y') - J_i(y) \geqq G_i(y') - G_i(y) - \alpha_i K_{i+1}, \qquad y \leqq y',\ i = 1, \cdots, n. \tag{9}$$

[5] As an illustration, if $K_i = K < 1$, $G_i(y) = G(y) = \min(1, |y|)$, $\eta_i \geqq 0$, $v_i(y,t) = y - t$, and $\alpha_i = \alpha \leqq 1$, then (i)—(vii) hold. However, $J_n(y) = G(y)$ which is not K_n-convex.

Proof. From (4) and (5) we have for $x \leqq x'$ that

$$f_i(x) \leqq K_i + \inf_{y \geqq x} J_i(y) \leqq K_i + \inf_{y \geqq x'} J_i(y) \leqq K_i + f_i(x'),$$

which establishes (8).

For $y \leqq y'$, we have from (v) that $v_i(y, D_i) \leqq v_i(y', D_i)$. Thus from (4), (5), and (8) we get

$$\begin{aligned} J_i(y') &- J_i(y) \\ &= G_i(y') - G_i(y) + \alpha_i E[f_{i+1}(v_i(y', D_i)) - f_{i+1}(v_i(y, D_i))] \\ &\geqq G_i(y') - G_i(y) - \alpha_i K_{i+1}, \end{aligned}$$

which completes the proof.

The proofs of (8) and (9) are purely analytic. An alternative proof of (8) may be constructed by using the following argument which may be made rigorous. If the initial inventory on hand and on order at the beginning of period i is x but one orders in each period j ($\geqq i$) so as to bring the inventory level after ordering to the level which would be optimal if the initial inventory level in period i were instead x' ($\geqq x$), then the associated expected discounted cost would not exceed $f_i(x') + K_i$. But since the policy just described cannot be better than the optimal policy, (8) must hold. A similar kind of argument may be used to establish (9).

Lemma 2. *If* (v) *holds, if* $\{a_j\}$ *is a sequence of numbers for which* $v_j(a_j, t) \leqq a_{j+1}$ *for* $t \geqq \eta_j$ *and* $j \geqq i$, *and if* $G_j(y)$ *is nonincreasing in* y *on* $(-\infty, a_j]$ *for* $j \geqq i$, *then*

$$J_j(y') - J_j(y) \leqq G_j(y') - G_j(y) \leqq 0, \qquad y \leqq y' \leqq a_j, \tag{10}$$

and

$$f_j(x') - f_j(x) \leqq 0, \qquad x \leqq x' \leqq a_j, \tag{11}$$

for $j \geqq i$.

Proof. The proof is by induction on j. Suppose (10), (11) hold for $j+1$ ($> i$). By (v), $v_j(y, D_j) \leqq v_j(y', D_j) \leqq v_j(a_j, D_j) \leqq a_{j+1}$. Hence using (11) for $j+1$ we get

$$\begin{aligned} J_j(y') &- J_j(y) \\ &= G_j(y') - G_j(y) + \alpha_j E[f_{j+1}(v_j(y', D_j)) - f_{j+1}(v_j(y, D_j))] \\ &\leqq G_j(y') - G_j(y) \leqq 0, \end{aligned}$$

which proves (10) for j.

It follows from (10) for the integer j that

$$f_j(x) = \min\{ J_j(x), K_j + \inf_{y>x} J_j(y) \}$$
$$\geqq \min\{ J_j(x'), K_j + \inf_{y>x'} J_j(y)\} = f_j(x'),$$

which proves (11) for the integer j. The same arguments suffice to establish (10), (11) for $j = n$ which starts the induction and completes the proof.

The proofs of (10) and (11) are purely analytic. An alternative proof of (11) may be devised by using the following argument which can be made rigorous. Suppose the initial inventory level in period j is x'. Suppose also that one orders so as to bring the initial inventory level after ordering in each period k ($\geqq j$) as close as possible to the level which would be optimal if the initial inventory level in period j were x ($\leqq x'$). This policy incurs expected discounted costs which are no greater than $f_j(x)$. But the policy also must incur expected discounted costs at least as large as $f_j(x')$. Combining these remarks proves (11). The inequality (10) can be justified in a similar way.

THEOREM 1. If (i)—(vii) *hold, there exists an optimal policy which is an* (s, S) *policy. Moreover, the parameters of that policy satisfy* (6), (7).

PROOF. The proof is constructive and proceeds in several steps. To begin with suppose $J_i(y)$ is continuous in y.

(a) $J_i(y)$ is nonincreasing on $(-\infty, \underline{S}_i]$.

To see this recall from (iv) that $G_j(y)$ is nonincreasing in y on $(-\infty, \underline{S}_j]$ for $j \geqq i$. Thus by (vii) and Lemma 2, (a) holds.

Since $J_i(y)$ is continuous, there is an S_i which minimizes $J_i(y)$ on $[\underline{S}_i, \overline{S}_i]$. Thus S_i satisfies (6). Moreover,

(b) $\min_y J_i(y) = J_i(S_i)$.

To see this observe from (a) that S_i minimizes $J_i(y)$ on $(-\infty, \overline{S}_i]$. Also by Lemma 1, (iv), and the definition of $\overline{S}_i$ we have for $y > \overline{S}_i$ that

$$J_i(y) - J_i(\underline{S}_i) \geqq G_i(y) - G_i(\underline{S}_i) - \alpha_i K_{i+1}$$
$$\geqq G_i(\overline{S}_i) - G_i(\underline{S}_i) - \alpha_i K_{i+1} = 0.$$

Thus (b) holds.

(c) There exists a number s_i satisfying (6), (7) and

$$J_i(S_i) + K_i - J_i(s_i) = 0. \tag{12}$$

In order to prove this assertion we observe from Lemma 2, (b), and the definitions of $\underline{S}_i$ and $\underline{s}_i$ that

$$\begin{aligned} J_i(S_i) + K_i - J_i(\underline{s}_i) &\leqq J_i(\underline{S}_i) + K_i - J_i(\underline{s}_i) \\ &\leqq G_i(\underline{S}_i) + K_i - G_i(\underline{s}_i) = 0. \end{aligned} \tag{13}$$

On the other hand by Lemma 1 and the definitions of $\underline{S}_i$ and $\bar{s}_i$ we have

$$\begin{aligned} J_i(S_i) + K_i - J_i(\bar{s}_i) &\geqq G_i(S_i) - G_i(\bar{s}_i) + K_i - \alpha_i D K_{i+1} \\ &\geqq G_i(\underline{S}_i) - G_i(\bar{s}_i) + K_i - \alpha_i K_{i+1} = 0. \end{aligned} \tag{14}$$

From (13), (14), and the continuity of $J_i(y)$, it follows that there is an s_i satisfying (6) and (12). Moreover (7) holds also since by Lemma 1 and (12) we have

$$0 = J_i(S_i) + K_i - J_i(s_i) \geqq G_i(S_i) - G_i(s_i) + K_i - \alpha_i K_{i+1}$$

which completes the proof of (c).

(d) The value of y which minimizes the right side of (4) is determined by

$$\underline{y} = \begin{cases} S_i & \text{if } x < s_i, \\ x & \text{if } x \geqq s_i. \end{cases}$$

To prove (d), observe from (a), (b), and (c) that for $x < s_i$,

$$J_i(x) \geqq J_i(s_i) = J_i(S_i) + K_i = \min_y J_i(y) + K_i,$$

so $y = S_i$ minimizes the right side of (4). Now for $s_i \leqq x \leqq S_i$, the same arguments give

$$J_i(x) \leqq J_i(s_i) = \min_y J_i(y) + K_i,$$

so $y = x$ minimizes the right side of (4). Finally for $\underline{S}_i < x < y$, we have from Lemma 1, (iv), and (vi) that

$$J_i(y) + K_i - J_i(x) \geqq G_i(y) - G_i(x) + K_i - \alpha_i K_{i+1} \geqq 0,$$

so $y = x$ minimizes the right side of (4). This completes the proof of (d).

It remains only to verify our assumption that

(e) $J_i(y)$ is continuous in y.

We prove (e) by induction on i. The assertion is trivial for $i = n$ since $J_n(y) = G_n(y)$. Suppose now (e) holds for the integer $i + 1$. Then by (d),

$$f_{i+1}(x) = \begin{cases} K_{i+1} + J_{i+1}(S_{i+1}) & \text{if } x < s_{i+1}, \\ J_{i+1}(x) & \text{if } x \geqq s_{i+1}. \end{cases} \tag{15}$$

Since $J_{i+1}(y)$ is continuous and (12) holds for $i + 1$, $f_{i+1}(x)$ is evidently continuous. Since $G_i(y)$ is continuous by (i), $J_i(y)$ will be continuous if

$$Ef_{i+1}(v_i(y, D_i)) \equiv q(y)$$

is continuous in y on any arbitrary interval, $[a, b]$, say. Since by (i) and the continuity of f_{i+1}, the composite function $f_{i+1}(v_i(y, t))$ is continuous in y, $q(y)$ will be continuous on $[a, b]$ if the composite function is uniformly bounded for $y \in [a, b]$ and all $t\ (\geqq \eta_i)$ by virtue of the dominated convergence theorem. We now show that there is a number B such that

$$J_{i+1}(S_{i+1}) \leqq f_{i+1}(v_i(y, t)) \leqq f_{i+1}(B) + K_{i+1} \tag{16}$$

for all $t\ (\geqq \eta_i)$ and $y \in [a, b]$, which gives the desired bounds. The left-hand inequality follows from (15) and (b). Since by (v), $v_i(y, t) \leqq v_i(b, t) \leqq B$ for some B, the right-hand inequality follows from Lemma 1.

The proof is now complete since we have constructed a solution to (4) with the infimum being attained for each x.

As we have remarked before, if Scarf's hypotheses ($G_i(y)$ convex, $G_i(y) \to \infty$ as $|y| \to \infty$, $v_i(y, t) = y - t$, and (vi), are substituted for (i)—(vii), then there exists an optimal policy which is an (s, S) policy. However, the lower bounds in (6) for s_i and S_i are no longer valid when (vii) fails to hold. The reason for this is clear upon reflection. For example, suppose $D_{n-1} \geqq 0$ and $P(D_{n-1} < \underline{S}_{n-1} - \underline{S}_n) > 0$ so (vii) does not hold. In this event it is apparent from (1) that one would not want to order up to $\underline{S}_{n-1}$ (or more) in period $n - 1$ if $G_n(y)$ increased sufficiently rapidly on the interval $[\underline{S}_n, \underline{S}_{n-1}]$. The reason for this is, of course, that the relatively low expected costs in period $n - 1$ would be more than offset by the extremely high expected costs in period n. For a concrete illustration see footnote 6 below.

Let

$$\underset{\sim}{S}_i = \underline{S}_n \qquad \text{if } i = n,$$
$$= \min(\underline{S}_i, \underset{\sim}{S}_{i+1} + \eta_i) \qquad \text{if } i = 1, 2, \cdots, n-1.$$

Let $\underset{\sim}{s}_i$ $(\leqq \min(\underline{s}_i, \underset{\sim}{S}_i))$ be chosen so that

$$G_i(\underset{\sim}{s}_i) = G_i(\underset{\sim}{S}_i) + K_i.$$

THEOREM 2. *Under Scarf's hypotheses, there is an optimal policy* $\{(s_i, S_i)\}$ *which satisfies* (7) *and*

(6′) $\qquad \underset{\sim}{s}_i \leqq s_i \leqq \bar{s}_i$ *and* $\underset{\sim}{S}_i \leqq S_i \leqq \bar{S}_i, \qquad i = 1, 2, \cdots, n.$

PROOF. We only sketch the proof, leaving the details to the reader. The upper bounds on s_i and S_i and the inequality (7) are established by applying Lemma 1 in exactly the same way as in Theorem 1. The lower bounds on s_i and S_i may be established by applying (10) with $a_j = \underset{\sim}{S}_j$ for all j in a manner similar to that employed in proving Theorem 1.

We remark that if (vii) holds then $\underset{\sim}{S}_i = \underline{S}_i$ and $\underset{\sim}{s}_i = \underline{s}_i$ for all i so that (6′) reduces to (6).

3. Planning horizons and special cases. The next result tells us that if S_k is sufficiently small in comparison with s_{k+1}, then $(\underline{s}_k, \underline{S}_k)$ is optimal for period k. Observe that this is the policy that is optimal for period k when considered by itself or as the final period of a k-period model. Moreover, if $S_i, \cdots, S_k$ are sufficiently small in comparison with s_{k+1}, an optimal policy for periods $i, \cdots, k$ may be determined without evaluating $f_{k+1}(x)$ for any x. In this sense, period k is a planning horizon. The actual calculations are carried out using (4) recursively where $f_{k+1}(x) \equiv 0$ for all x. The theorem generalizes some results in [**9**] to the case where there is a setup cost for placing orders.

THEOREM 3. *Suppose* (i)—(vii) *or Scarf's hypotheses hold, and that* $\{(s_j, S_j)\}$ *is an optimal policy.*

(a) *If* $\{a_j\}$ *is a collection of numbers for which*

(17)[6] $$\underline{S}_k \leqq a_k,$$

[6] The hypothesis (17) is easily seen to be satisfied if (18)—(20) hold for $j = k$ and either (1) (i)—(vii) hold or (2) Scarf's assumptions are fulfilled and $K_{k+1} > 0$ or $S_k < a_k$. The following example shows that (17) cannot be dispensed with under Scarf's hypotheses. Assume $n = k + 1 = 2$, $G_1(y) = |y - 2|$, $G_2(y) = 2|y - 1|$, $K_1 = K_2 = 0$, $P(D_1 = 0) = P(D_2 = 0) = 1$, and $\eta_1 = \eta_2 = 0$. Then $s_1 = S_1 = s_2 = S_2 = 1$ and $\underline{s}_1 = \underline{S}_1 = 2$. Now let $a_1 = a_2 = 1$. Then (18)—(20) hold, but $(\underline{s}_1, \underline{S}_1)$ is not optimal for period 1.

$$a_{k+1} \leqq s_{k+1}, \tag{18}$$

$$v_j(a_j, t) \leqq a_{j+1} \quad \textit{for } t \geqq \eta_j, \tag{19}$$

and

$$S_j \leqq a_j \tag{20}$$

for $j = k$, then $(\underline{s}_k, \underline{S}_k)$ is optimal for period k.

(b) *If (18)—(20) hold for $j = i, i+1, \cdots, k$, then one optimal policy for periods $i, i+1, \cdots, k$, is independent of $f_{k+1}(\cdot)$.*

PROOF. We begin by proving part (a). From Theorem 1, Scarf's theorem, and (18),

$$f_{k+1}(x) = K_{k+1} + J_{k+1}(S_{k+1}) \equiv Q, \qquad x \leqq a_{k+1}. \tag{21}$$

It follows from (v) and (19) that for $t \geqq \eta_k$ and $y \leqq a_k$, $v_k(y, t) \leqq v_k(a_k, t) \leqq a_{k+1}$. Combining this remark and (21) we get

$$J_k(y) = G_k(y) + \alpha_k E f_{k+1}(v_k(y, D_k)) = G_k(y) + \alpha_k Q, \ y \leqq a_k. \tag{22}$$

Now by (20), $J_k(y)$ achieves its minimum on $(-\infty, \infty)$ in $(-\infty, a_k]$. Since this is so it follows from (22) and (17) that $\underline{S}_k$ minimizes $J_k(y)$ on $(-\infty, \infty)$. Moreover, again by (22) we have

$$J_k(\underline{s}_k) = G_k(\underline{s}_k) + \alpha_k Q = K_k + G_k(\underline{S}_k) + \alpha_k Q = K_k + J_k(\underline{S}_k).$$

Hence, by Theorem 1 and Scarf's theorem, $(\underline{s}_k, \underline{S}_k)$ is optimal for period k, which establishes part (a).

In order to prove part (b) we observe from Theorem 1, Scarf's theorem, and (20) that the optimal policy in period j, $i \leqq j \leqq k$, can be determined provided only that we can evaluate $J_j(y)$ for $y \leqq a_j$. Now from (v), (19), and (20) it follows easily by induction on j that $J_j(y)$ may be evaluated for $y \leqq a_j$ without evaluating $f_{k+1}(x)$ for $x > a_{k+1}$, i.e., for $y \leqq a_j$, $J_j(y)$ depends upon f_{k+1} only through the constant Q. (We have already shown this for $j = k$ which starts the induction.)

We may exhibit the dependence of $J_j(y)$ upon Q by writing $J_j^Q(y)$. It is easy to show by induction on j that

$$J_j^Q(y) = J_j^0(y) + Q \prod_{t=j}^{k} \alpha_t, \qquad j = i, i+1, \cdots, k. \tag{23}$$

(Again we have already done this for $j = k$ which starts the induction.) Thus if a policy is optimal for period $j, i \leqq j \leqq k$, for some Q, that same policy is optimal for all Q. Hence if we assume $Q = 0$

and determine the optimal policy for periods $i, i+1, \cdots, k$ in the usual way (taking account of (20)), that policy is optimal for the original problem where (21) holds. We have thus shown that the optimal policy in periods $i, i+1, \cdots, k$ is independent of $f_{k+1}(\cdot)$ as required.

As an illustration of the application of Theorem 3(a), we have the following result.

COROLLARY 1. *If* (i)—(vii) *or Scarf's hypotheses hold, and if*

$$v_k(\overline{S}_k, t) \leqq \underset{\sim}{s}_{k+1} \quad \text{for } t \geqq \eta_k, \tag{24}$$

then $(\underline{s}_k, \underline{S}_k)$ *is optimal for period* k.

PROOF. Let $a_k = \overline{S}_k$ and $a_{k+1} = \underset{\sim}{s}_{k+1}$. Then apply Theorem 3(a).

We remark that if (24) holds, then by (v) and the definitions of $\underline{S}_k$, $\overline{S}_k$, $\underset{\sim}{s}_{k+1}$, $\underline{S}_{k+1}$ (recall $\underset{\sim}{s}_{k+1} = \underline{s}_{k+1}$ if (i)—(vii) hold),

$$v_k(\underline{S}_k, t) \leqq v_k(\overline{S}_k, t) \leqq \underline{s}_{k+1} \leqq \underline{S}_{k+1},$$

so (vii) holds for the integer k. It follows therefore that if (24) holds for all k, then (vii) necessarily holds also. In this event we can replace $\underset{\sim}{s}_{k+1}$ in (24), by $\underline{s}_{k+1}$.

EXAMPLE 1. Suppose unsatisfied demand is backlogged in period k so that $v_k(y, t) = y - t$. Then (24) reduces to $\overline{S}_k - \underset{\sim}{s}_{k+1} \leqq \eta_k$. Thus if the minimal demand in period k is at least $\overline{S}_k - \underset{\sim}{s}_{k+1}$, then $(\underline{s}_k, \underline{S}_k)$ is optimal for period k by Corollary 1. A special case of this result is established in [**12**, p. 545] for the stationary case.

The next example illustrates the application of Theorem 3(b).

EXAMPLE 2. Suppose unsatisfied demands are backlogged so that $v_j(y, t) = y - t$. Also let $a_{k+1} = \underset{\sim}{s}_{k+1}$ and let $\{a_j\}$ be defined recursively by $a_j - \eta_j = a_{j+1}$ for $j \leqq k$. Thus $a_i = \underset{\sim}{s}_{k+1} + \sum_{j=i}^{k} \eta_j$ for $i \leqq k$. Now let i be an integer for which

$$\overline{S}_j \leqq \underset{\sim}{s}_{k+1} + \sum_{t=j}^{k} \eta_t, \qquad j = i, i+1, \cdots, k. \tag{25}$$

Then the hypotheses of Theorem 3(b) are evidently satisfied. In particular, if $\eta_t = 0$ for $i \leqq t \leqq k$, then the optimal policy in periods i, $i+1, \cdots, k$ may be determined without evaluating $f_{k+1}(\cdot)$ if $\overline{S}_j \leqq \underset{\sim}{s}_{k+1}$ for $i \leqq j \leqq k$.

We remark that if (i)—(vii) hold and if $K_i = 0$ for all i, then we choose $\underline{s}_i$, $\overline{s}_i$, and $\overline{S}_i$ equal to $\underline{S}_i$ so $s_i = S_i = \underline{S}_i$ for all i by Theorem 1. In this event the hypothesis (vii) is equivalent to the hypothesis that (24) holds for all k. This observation together with

Corollary 1 establishes the next result which is known from [**10**], [**11**].

COROLLARY 2. *If* (i)—(vii) *hold and if* $K_i = 0$ *for all* i, *then* $\{(\underline{S}_i, \underline{S}_i)\}$ *is an optimal policy.*

4. **Applications and extensions.** In this section we discuss some applications and extensions of our results.

Applications of the basic lemmas. Lemmas 1 and 2 are useful in establishing bounds on the ordering regions and order quantities even where $-G_i(y)$ is not unimodal. We shall illustrate this point under the assumption that $K_i = 0$ for all i, leaving the other case to the reader. Suppose $G_i(y)$ appears as in Figure 2. The domain of $G_i(y)$ is divided into six regions labeled $1, 2, \cdots, 6$. If period i were considered by itself, and if the initial inventory in period i fell in an odd-numbered region, it would be optimal to order to the upper bound of that region, viz., to U_1, U_3, or U_5 as appropriate. If the initial inventory in period i fell in an even-numbered region, no order should be placed. If $i < n$, then the above policy need not be optimal for the n-period model. However, suppose y lies in an even-numbered region. Then by Lemma 1,

$$J_i(y') - J_i(y) \geqq G_i(y') - G_i(y) \geqq 0 \tag{26}$$

for all $y' \geqq y$ provided (v) holds so it is optimal not to order in period i for the n-period model. Notice also that the above inequality tells us that if the initial inventory level in period i is a and if it is optimal to order, then the inventory level after ordering must lie in the interval $[b, c]$.

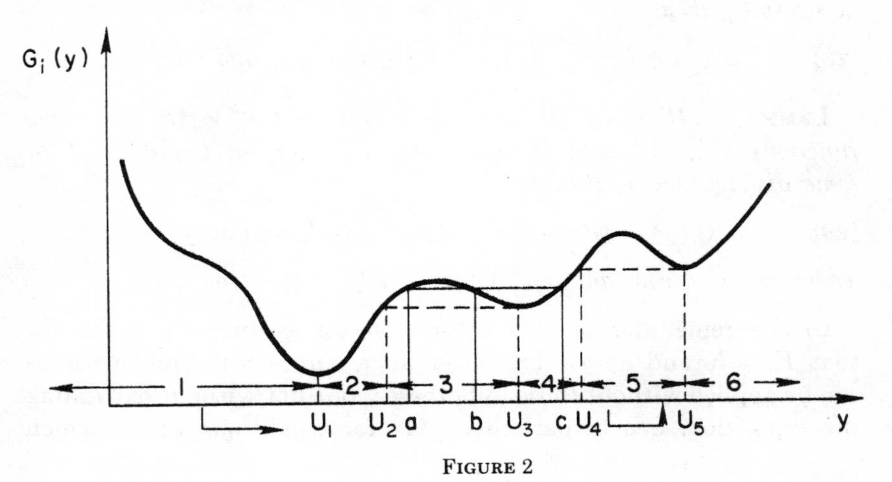

FIGURE 2

If (v) holds, if $v_j(U_1, t) \leqq U_1$ for $t \geqq \eta_j$, and if $G_j(y)$ is nonincreasing in y on $(-\infty, U_1]$ for $j \geqq i$, then by Lemma 2, $J_i(y)$ is minimized on $(-\infty, U_1]$ at $y = U_1$. Similarly from (26), $J_i(y)$ is minimized on $[U_1, \infty)$ at $y = U_1$. Combining these remarks we see that $J_i(y)$ is minimized on $(-\infty, \infty)$ at $y = U_1$. Hence, in region 1 it is optimal to order up to U_1.

Variation of the bounds over time. It is of interest to determine how s_i and S_i vary over time in relation to the variation of the cost functions and demand distributions. Although this appears to be quite difficult, we can instead examine how the bounds on s_i and S_i vary over time. Such studies are of interest in their own right and because they provide us with a tool for determining conditions under which the hypotheses (vii), (17)—(20), (24), (25) of our several results hold. Throughout this subsection we shall assume for simplicity that sufficient regularity conditions are imposed to permit differentiation and interchange of differentiation with integration where required.

As a preliminary we record several lemmas from [**11**]. Let I be a subset of the integers $1, \cdots, n$.

LEMMA 3. *If $\partial W_i(y,t)/\partial y$ is nonincreasing in $t \geqq \eta_i$ and $i \in I$ for each y, and if $\Phi_i(t) \geqq \Phi_j(t)$ for all t and $i, j \in I$, $i < j$, then*

$$G_i'(y) \geqq G_j'(y), \qquad i, j \in I, \ i < j, \text{ and all } y. \tag{27}$$

LEMMA 4. *If $W_i(y,t) = W_i^1(y-t) + W_i^2(t)$ for some functions W_i^1, W_i^2, if $dW_i^1(z)/dz$ is nonincreasing in $i \in I$ and nondecreasing in z, and if $\Phi_i(t) \geqq \Phi_j(t - b_{ij})$ for all t and $i, j \in I$, $i < j$, and some numbers b_{ij}, then*

$$G_i'(y) \geqq G_j'(y - b_{ij}), \qquad i, j \in I, \ i < j, \text{ and all } y. \tag{28}$$

LEMMA 5. *If $\lambda = 0$, if $W_i(y,t) = W^1(y-t) + W_i^2(t)$ for some functions W^1, W_i^2, and if $\Phi_i(t) = \Phi(t - \eta_i)$ for all t and $i \in I$ for some distribution Φ, then*

$$G_i(y) = G(y - \eta_i) + Q_i, \qquad i \in I, \text{ and all } y, \tag{29}$$

where Q_i is a constant and $G(y) = \int_{-\infty}^{\infty} W(y-t)\, d\Phi(t)$.

In the remainder of this subsection we assume for simplicity that $K_i = K$ and $\alpha_i = \alpha$ $(\leqq 1)$ for all i, and $\lambda = 0$. Our methods can be applied without these hypotheses, but not without expanding the exposition. See in particular [**11**] for conditions under which the hypotheses of Lemmas 3 and 4 are satisfied when $\lambda > 0$. There is also an analog of Lemma 5 when $\lambda > 0$.

Let $\mathscr{S}_i = (\underline{s}_i, \bar{s}_i, \underline{S}_i, \bar{S}_i)$, $B_{ij} = (b_{ij}, b_{ij}, b_{ij}, b_{ij})$, and $H_i = (\eta_i, \eta_i, \eta_i, \eta_i)$. Let $\mathscr{S} = (\underline{s}, \bar{s}, \underline{S}, \bar{S})$, where $\underline{s}, \bar{s}, \underline{S}, \bar{S}$ are defined for the function $G(\cdot)$ (see Lemma 5) in the usual way. For definiteness where $\mathscr{S}_i$ is not uniquely defined we choose it as follows. First pick the smallest possible $\underline{S}_i$. Then pick the smallest $\underline{s}_i, \bar{s}_i$, and $\bar{S}_i$. Do the same for $\mathscr{S}$. The following theorem, which is an easy consequence of Lemmas 3—5, describes how $\mathscr{S}_i$ varies over time in relation to the variation of W_i and Φ_i (as reflected in G_i) over time.

THEOREM 4. *Suppose* (i)—(iv) hold.
(a) *If* (27) *holds, then* $\mathscr{S}_i \leqq \mathscr{S}_j$ *for* $i, j \in I$, $i < j$.
(b) *If* (28) *holds, then* $\mathscr{S}_i - B_{ij} \leqq \mathscr{S}_j$ *for* $i, j \in I$, $i < j$.
(c) *If* (29) *holds, then* $\mathscr{S}_i - H_i = \mathscr{S}$ *for* $i \in I$.

Satisfying the hypotheses of the main results. In this subsection we use Theorem 4 to give conditions under which the important hypotheses (vii) and (24) of Theorem 1 and Corollary 1 respectively are satisfied. We begin by giving conditions under which (vii) holds.

It will be convenient in what follows to assume that there is an extended real number θ such that

(30) $\quad v_i(y, t) \leqq \max(\theta, y - \eta_i), \quad$ for $t \geqq \eta_i$, all i, and all y.

As an example, if $\eta_i \geqq 0$ and unsatisfied demands are backlogged so $v_i(y, t) = y - t$, then (30) holds with $\theta \geqq -\infty$. Alternatively if $\eta_i \geqq 0$ and if unsatisfied demands are lost, so $v_i(y, t) = \max(y - t, 0)$, then (30) holds with $\theta \geqq 0$. In applications θ should be chosen as small as possible. Thus $\theta = -\infty$ in the backlog case and $\theta = 0$ in the lost sales case.

The following result is a simple consequence of Theorem 4.

COROLLARY 3. *Suppose* (i)—(iv) *and* (30) *hold,* $I = \{1, 2, \cdots, n\}$, *and* $\underline{S}_i \geqq \theta$ *for all* $i > 1$.
(a) *If* (27) *holds and* $\eta_i \geqq 0$ *for all* $i < n$, *then* (vii) *holds.*
(b) *If* (28) *holds and* $\eta_i - b_{i,i+1} \geqq 0$ *for all* $i < n$, *then* (vii) *holds.*
(c) *If* (29) *holds and* $\eta_i \geqq 0$ *for all* $i \leqq n$, *then* (vii) *holds.*

The next corollary gives a condition ensuring that the hypothesis (24) of Corollary 1 holds.

COROLLARY 4. *If* (i)—(iv), (vii), (30) *hold, if* (28) *holds with* $I = \{k, k+1\}$, *if* $\underline{s}_{k+1} \geqq \theta$, *and if*

(31) $$\overline{S}_k - \underline{s}_k \leqq \eta_k - b_{k,k+1},$$

then (24) *holds*.

PROOF. $v_k(\overline{S}_k, t) \leqq \max(\theta, \overline{S}_k - \eta_k) \leqq \max(\theta, \underline{s}_k - b_{k,k+1}) \leqq \max(\theta, \underline{s}_{k+1}) = \underline{s}_{k+1} = \underline{s}_{k+1}$.

Stationary infinite horizon models. This paper is primarily concerned with a finite horizon model. If the model is stationary, i.e., $G_i, K_i, \alpha_i, \Phi_i$ are independent of i, then it is convenient to consider an infinite period version of the model. In this case fairly obvious modifications of Iglehart's results and proofs [1], [2] (see also [12, pp. 530-531]) for $0 \leqq \alpha \leqq 1$ show that if (i)—(vii) hold, there is an optimal (s, S) policy with the optimal choice of parameters being independent of time and satisfying (6), (7). Methods for computing these parameters are discussed in [8] and [12].

Restrictions on inventory levels. In some applications it may be desirable to limit the choice of the inventory y_i on hand and on order after ordering in period $i (= 1, 2, \cdots, n)$ to an interval $[\underline{y}_i, \overline{y}_i]$ say. The upper bound $\overline{y}_i$ might reflect limitations on storage space while the lower bound $\underline{y}_i$ could reflect a desire to limit the size of the backlogged demand. As a specific illustration, suppose demands occur over only the first $n - 1$ periods, so $D_i = 0,\ i \geqq n$. Then we may wish to require that no unsatisfied demand exist at the end of period $n + \lambda$.[7] This may be accomplished by setting $\underline{y}_n = 0$ so $y_n \geqq 0$. This implies $y_{n+\lambda} \geqq 0$ if $v_i(y, 0) \geqq 0$ for $y \geqq 0$ and $n \leqq i$.

In other applications it is natural to suppose that the demands are integers. Of course this restriction is already allowed in our formulation. However, in such cases it is usually necessary to impose the additional restriction that the order quantities and stock levels be integers. We shall now generalize our original model and results to provide for such integer restrictions and for bounds on the stock levels.

Let Y_i denote the nonempty set of admissible stock levels y_i on hand and on order after ordering in period i. Let $\underline{y}_i = \inf Y_i$ and $\overline{y}_i = \sup Y_i$. Let $\mathscr{D}_i$ denote the (Borel) set of possible values of the demand D_i in period i. Let X_{i+1} denote the nonempty set of possible values of the stock on hand and on order before ordering in period $i + 1$. We naturally impose the consistency condition that $v_i(y, t) \in X_{i+1}$ for all $y \in Y_i$ and $t \in \mathscr{D}_i$. In addition we suppose that if

[7] I am indebted to G. Lieberman for a discussion of this application.

the stock on hand and on order before ordering in period i is at least y_i in period i, then it is possible *not* to order in period i. Formally, we assume that $x \in X_i$ and $y_i \leqq x$ imply $x \in Y_i$, $i = 1, 2, \cdots, n$, where $X_1 \equiv \{x_1\}$. We also suppose that the domains of $f_i(\cdot)$, $G_i(\cdot)$, $J_i(\cdot)$, and $v_i(\cdot, \cdot)$ are respectively X_i, Y_i, Y_i, and $Y_i \times \mathscr{D}_i$. Moreover, we shall replace (i)—(iii) respectively by:

(i′) (i) holds and Y_i is closed;
(ii′) either (ii) holds or $\bar{y}_i < \infty$;
(iii′) either (iii) holds or $-\infty < \underline{y}_i$.

If (i′) holds, we may define

$$G_i^+(y) = G_i(\inf\{z \mid z > y, z \in Y_i\})$$

for $y < \bar{y}_i$ and $G_i^+(\bar{y}_i) = \infty$ if $\bar{y}_i < \infty$. Also

$$G_i^-(y) = G_i(\sup\{z \mid z < y, z \in Y_i\})$$

for $\underline{y}_i < y$ and $G_i^-(\underline{y}_i) = \infty$ if $-\infty < \underline{y}_i$. For example, if $Y_i = (-\infty, \infty)$, then $G_i^-(y) = G_i(y) = G_i^+(y)$, while if Y_i is the set of integers, then $G_i^-(y) = G_i(y-1)$ and $G_i^+(y) = G_i(y+1)$ for $y \in Y_i$.

If (i′)—(iii′) hold, there are a number $\underline{S}_i \in Y_i$ that minimizes $G_i(\cdot)$ on Y_i and numbers $\underline{s}_i$ $(\leqq \underline{S}_i)$ and $\overline{S}_i$ $(\geqq \underline{S}_i)$ such that

$$G_i(\overline{S}_i) \leqq G_i(\underline{S}_i) + \alpha_i K_{i+1} \leqq G_i^+(\overline{S}_i)$$

and

$$G_i(\underline{s}_i) \leqq G_i(\underline{S}_i) + K_i \leqq G_i^-(\underline{s}_i).$$

If in addition (vi) holds, there is a number $\bar{s}_i$, $\underline{s}_i \leqq \bar{s}_i \leqq \underline{S}_i$, such that

$$G_i(\bar{s}_i) \leqq G_i(\underline{S}_i) + K_i - \alpha_i K_{i+1} \leqq G_i^-(\bar{s}_i).$$

It is easy to check that the statements and proofs of Lemmas 1 and 2 remain valid in our new setup. Also Theorem 1 holds provided we replace (i)—(iii) by (i′)—(iii′) and replace (7) by

$$G_i^-(s_i) \geqq G_i(S_i) + (K_i - \alpha_i K_{i+1}). \tag{7′}$$

Only obvious modifications of the proof of Theorem 1 are required.

References

1. D. Iglehart, "Dynamic programming and stationary analysis of inventory problems" in *Multistage inventory models and techniques*, chapter 1 (see [7]).

2. ———, *Optimality of (s, S) policies in the infinite horizon dynamic inventory problem*, Management Sci. **9** (1963), 259-267.

3. S. Karlin, *Pólya type distributions*. II, Ann. Math. Statist. 28 (1957), 292.

4. ———, "One stage inventory models with uncertainty" in *Studies in the mathematical theory of inventory and production* edited by K. Arrow et al, Stanford Univ. Press, Stanford, Calif., 1958, chapter 8.

5. ———, *Dynamic inventory policy with varying stochastic demands,* Management Sci. **6** (1960), 231-258.

6. H. Scarf, "The optimality of (S, s) policies in the dynamic inventory problem" in *Mathematical methods in the social sciences* edited by K. Arrow et al, Stanford Univ. Press, Stanford, Calif., 1960, chapter 13.

7. H. Scarf, D. Gilford and M. Shelly (Editors), *Multistage inventory models and techniques,* Stanford Univ. Press, Stanford, Calif., 1963.

8. H. Scarf, "A survey of analytic techniques in inventory theory" in *Multistage inventory models and techniques,* chapter 7 (see [7]).

9. A. F. Veinott, Jr., "Optimal stockage policies with nonstationary stochastic demands" in *Multistage inventory models and techniques,* Chapter 4 (see [7]).

10. ———, *Optimal policy for a multiproduct, dynamic nonstationary inventory problem,* Management Sci. **12** (1965), 206-222.

11. ———, *Optimal policy in a dynamic, single product, nonstationary inventory model with several demand classes,* Operations Res. **13** (1965), 761-778.

12. A. F. Veinott, Jr. and H. Wagner, *Computing optimal (s, S) inventory policies,* Management Sci. **11** (1965), 525-552.

13. E. Zabel, *A note on the optimality of (S, s) policies in inventory theory,* Management Sci. **9** (1962), 123-125.

Stanford University
Stanford, California

IX
APPLIED PROBABILITY
AND
STATISTICS

Samuel Karlin[1]

Branching Processes

1. Discrete time branching processes. Historically the study of branching processes began with the Galton-Watson process formulated by Galton (1874) to describe a possible mechanism of extinction of distinguished English families. The model of Galton and Watson was used by R. A. Fisher (1922, 1930) to study the survival of the progeny of a mutant gene and to study random variations in gene frequencies. J. B. S. Haldane (1927) also applied the model to some questions in population genetics.

We define a 1-type discrete time Galton-Watson process (or simply a 1-type discrete time Markov branching process) to be a sequence $X_0, X_1, \cdots$ of integer valued random variables where

$$X_{n+1} = \sum_{r=1}^{X_n} \xi_r$$

and $\{\xi_r: r \geqq 1\}$ are independent identically distributed random variables with distribution

$$\Pr\{\xi_r = k\} = p_k, \quad k = 0, 1, 2, \cdots, \quad \sum_{k=0}^{\infty} p_k = 1, \tag{1}$$

X_0 represents the initial population size.

[1] Research supported in part at Stanford University under Contract N 0014-67-A-0112-0015.

We may interpret this structure as a model of population growth where in the initial population of X_0 elements, each individual gives birth, *independently of the others*, with probability p_k to k new individuals. The totality of direct descendants of the initial population constitutes the first generation whose size we denote by X_1. Each individual of the first generation independently bears a progeny whose size is governed by the probability distribution (1). The descendants produced constitute the second generation of size X_2. In general the nth generation is composed of descendants of the $(n-1)$st generation each of whose members independently produces k progeny with probabiltiy p_k, $k = 0, 1, 2, \cdots$. The population size of the nth generation is denoted by X_n. The X_n form a sequence of integer-valued random variables which generate a Markov chain.

Among the many scientific disciplines where 1-type Markov branching processes arise naturally, we list a few of the more prominent examples and some question of interest in these areas. Other examples will be presented subsequently.

(a) *Electron multipliers*. An electron multiplier is a device that amplifies a weak current of electrons. A series of plates are set up in the path of electrons emitted by a source. Each electron, as it strikes the first plate, generates a random number of new electrons, which in turn strike the next plate and produce more electrons, etc. Let X_0 be the number of electrons initially emitted, X_1 the number of electrons produced on the first plate by the impact due to the X_0 initial electrons; in general let X_n be the number of electrons emitted from the nth plate due to electrons emanating from the $(n-1)$st plate. The sequence of random variables $X_0, X_1, X_2, \cdots, X_n, \cdots$ constitutes a branching process.

(b) *Neutron chain reaction*. A nucleus is split by a chance collision with a neutron. The resulting fission yields a random number of new neutrons. Each of these secondary neutrons may hit some other nucleus producing a random number of additional neutrons, etc. In this case the initial number of neutrons is $X_0 = 1$. The first generation of neutrons comprises all those produced from the fission caused by the initial neutron. The size of the first generation is a random variable X_1. In general the population X_n at the nth generation is produced by the chance hits of the X_{n-1} individual neutrons of the $(n-1)$st generation.

(c) *Survival of family names*. The family name is inherited by sons only. Suppose that each individual has probability p_k of having k male offspring. Then from one individual there result the 1st, 2nd, $\cdots$, nth, $\cdots$ generations of descendants. We may investigate the distribution of such random variables as the number of descendants in the nth generation, or the probability that the family name will eventually become extinct. Such questions will be dealt with in the general analysis of branching processes of this section.

An important analytical tool in the theory of branching processes is the probability generating function

$$\phi(s) = \sum_{k=0}^{\infty} p_k s^k$$

where $\{p_k\}_{k=0}^{\infty}$ are the probabilities defined in (1). Assuming that the initial population consists of one individual, i.e. $X_0 = 1$, we develop an important functional relation for

$$\phi_n(s) = \sum_{k=0}^{\infty} \Pr\{X_n = k\} s^k.$$

In particular we show that

$$\phi_{n+1}(s) = \phi_n(\phi(s)) \tag{2}$$

where $\phi_0(s) = s$ and $\phi_1(s) = \phi(s)$.

For this consider

$$\begin{aligned}
\phi_{n+1}(s) &= \sum_{k=0}^{\infty} \Pr\{X_{n+1} = k\} s^k \\
&= \sum_{k=0}^{\infty} \sum_{j=0}^{\infty} \Pr\{X_{n+1} = k \mid X_n = j\} \Pr\{X_n = j\} s^k \\
&= \sum_{k=0}^{\infty} s^k \sum_{j=0}^{\infty} \Pr\{X_n = j\} \cdot \Pr\{\xi_1 + \cdots + \xi_j = k\} \\
&= \sum_{j=0}^{\infty} \Pr\{X_n = j\} \sum_{k=0}^{\infty} \Pr\{\xi_1 + \cdots + \xi_j = k\} s^k.
\end{aligned}$$

Since ξ_r $(r = 1, 2, \cdots, j)$ are independent, identically distributed random variables with common probability generating function $\phi(s)$, the sum $\xi_1 + \cdots + \xi_j$ has the probability generating function $[\phi(s)]^j$. Thus,

$$\phi_{n+1}(s) = \sum_{j=0}^{\infty} \Pr\{X_n = j\} [\phi(s)]^j.$$

But the right-hand side is just the generating function $\phi_n(\cdot)$ evaluated at $\phi(s)$. Thus

(3) $$\phi_{n+1}(s) = \phi_n(\phi(s)).$$

It follows, by induction, that for any $k = 0, 1, \cdots, n$

(4) $$\phi_{n+1}(s) = \phi_{n-k}(\phi_{k+1}(s)).$$

If instead of $X_0 = 1$ we assume $X_0 = i_0$ (constant), then

$$\phi_0(s) \equiv s^{i_0} \quad \text{and} \quad \phi_1(s) = [\phi(s)]^{i_0}.$$

We still have

$$\phi_{n+1}(s) = \phi_n(\phi(s))$$

but (4) no longer holds for $k = 0$.

With the help of (2), we can compute the expectation and variance of X_n. It is assumed henceforth, unless explicitly stated to the contrary, that $X_0 = 1$. We postulate that

$$m = EX_1 \quad \text{and} \quad \sigma^2 = \operatorname{Var} X_1 = E(X_1^2) - [E(X_1)]^2$$

exist and are finite.

Obviously, $EX_n = \phi_n'(1)$. Then differentiating (2) and setting $s = 1$ yields [since $\phi(1) = 1$] $\phi_{n+1}'(1) = \phi_n'(1)\,\phi'(1)$. Since $\phi'(1) = EX_1 = m$ we infer the formula

(5) $$EX_{n+1} = m^{n+1}.$$

To compute $\operatorname{Var} X_{n+1}$ we use $\phi_{n+1}''(1)$ and a simple induction argument to obtain

$$\begin{aligned} \operatorname{Var} X_{n+1} &= \sigma^2 m^n \cdot (m^{n+1} - 1)/(m - 1) \qquad &&\text{if } m \neq 1, \\ &= (n + 1)\,\sigma^2 &&\text{if } m = 1. \end{aligned}$$

Thus the variance increases (decreases) geometrically if $m > 1$ $(m < 1)$, and linearly if $m = 1$. This behavior is characteristic of many results for branching processes.

We now illustrate the role of the basic relation (2) in the calculation of extinction probabilities; i.e. $\Pr\{X_0 = 0 \text{ for some } n\}$. Obviously whenever $X_n = 0$, $X_k = 0$ for all $k > n$.

Note first that extinction never occurs if the probability that an individual gives birth to no offspring is zero, i.e. when $p_0 = 0$. Thus in investigating the probability of extinction we will assume $0 < p_0 < 1$. Let

$$q_n = \Pr\{X_n = 0\} = \phi_n(0).$$

Then by formula (2)

$$q_{n+1} = \phi_{n+1}(0) = \phi(\phi_n(0)) = \phi(q_n). \tag{6}$$

Since $\phi(s)$ is a strictly increasing function on $0 \leqq s \leqq 1$ (it is a power series with nonnegative coefficients and $p_0 < 1$) and $q_1 = \phi_1(0) = p_0 > 0$, $q_2 = \phi(q_1) > \phi(0) = q_1$. Assume that $q_n > q_{n-1}$; then $q_{n+1} = \phi(q_n) > \phi(q_{n-1}) = q_n$. This shows inductively that $q_1, q_2, \cdots q_n, \cdots$ is a monotone increasing sequence bounded by 1. Hence

$$\pi = \lim_{n\to\infty} q_n$$

exists and $0 < \pi \leqq 1$. Since $\phi(s)$ is continuous, for $0 \leqq s \leqq 1$, letting $n \to \infty$ in (6) yields

$$\pi = \phi(\pi). \tag{7}$$

Since q_n is defined as the probability of extinction at or prior to the nth generation, we infer that π is the probability of eventual extinction and (7) shows that π is a root of the equation

$$\phi(s) = s. \tag{8}$$

We now extablish the result that π is the smallest positive root of (8). Let s_0 be a positive root of (8). Then $q_1 = \phi(0) < \phi(s_0) = s_0$. Assume $q_n < s_0$. Then by (6), $q_{n+1} = \phi(q_n) < \phi(s_0) = s_0$. Thus we infer by induction that $q_n < s_0$ holds for all n. It follows that $\pi = \lim q_n \leqq s_0$, validating the assertion that π is the smallest positive root of (8).

A modification of the above analysis shows that

$$\lim_{n\to\infty} \phi_n(s) = \pi \qquad \text{for } 0 \leqq s < 1. \tag{9}$$

The fact that the $\phi_n(s)$ converge to the constant function π on $0 \leqq s < 1$ implies that in the series

$$\phi_n(s) = \sum_{k=0}^{\infty} \Pr\{X_n = k\} s^k$$

the first coefficient

$$\Pr\{X_n = 0\} \qquad \text{converges to } \pi \text{ as } n \to \infty,$$

and all the other coefficients

$$\Pr\{X_n = k\} \qquad \text{converge to 0 as } n \to \infty \text{ for } k = 1, 2, \cdots.$$

Hence, regardless of the actual value of $m = EX_1 > 1$, the probability that the nth generation will consist of any positive finite number of individuals tends to zero as $n \to \infty$, while the probability of extinction tends to π. In this circumstance we say the $X_n \to \infty$ as $n \to \infty$ with probability $1 - \pi$.

We close this section by noting the interesting property that the conditional expectation of X_{n+r} (r a positive integer), given X_n, is $m^r \cdot X_n$, i.e. $E(X_{n+r} | X_n) = m^r X_n$. To prove this we first consider the case $r = 1$:

$$E\{X_{n+1} | X_n\} = E\left\{ \sum_{j=1}^{X_n} \xi_j \,\middle|\, X_n \right\} = X_n E\xi_j = mX_n.$$

We now assume the stated relation for r and prove the formula for $r + 1$. Thus

$$\begin{aligned} E\{X_{n+r+1} | X_n\} & \\ &= E\{E[X_{n+r+1} | X_{n+r}, X_{n+r-1}, \cdots, X_n] | X_n\} \\ &= E\{E[X_{n+r+1} | X_{n+r}] | X_n\}, \end{aligned}$$

where we use the Markov nature of $\{X_n\}$. But $E[X_{n+r+1} | X_{n+r}] = X_{n+r} \cdot m$ and by the induction hypothesis, $E(mX_{n+r} | X_n) = m^{r+1} X_n$. Thus we have

$$(10) \quad E\{X_{n+r} | X_n\} = X_n m^r \quad \text{for} \quad r = 0, 1, 2, \cdots, \quad n = 0, 1, 2, \cdots.$$

Now consider the random variables

$$W_n = X_n / m^n, \qquad n = 0, 1, 2, \cdots.$$

Then on the basis of (10), we have

$$(11) \quad E\{W_{n+r} | W_n\} = \frac{1}{m^{n+r}} E\{X_{n+r} | X_n\} = \frac{1}{m^{n+r}} \cdot X_n \cdot m^r = W_n$$

which shows that $\{W_n\}_{n=0}^{\infty}$ is a martingale. We refer the reader to Karlin [**10**] and Harris [8] for more extensive discussion of these questions.

Examples.

(i) Let $\phi(s) = p_0 + p_1 s$, $0 < p_0 < 1$. The associated branching process is a pure death process. In each period each individual independently dies with probability p_0 and survives with probability $1 - p_0 = p_1$.

(ii) Consider the example where each individual produces N or 0 direct descendants with probabilities p or q, respectively.

Thus $p_0 = q$, $p_N = p$, and $p_k = 0$ for $k \neq 0, N$. Then

$$\phi(s) = q + ps^N. \tag{12}$$

(iii) Each individual may have k offspring where k has a binomial probability distribution with parameters N and p. Then

$$\phi(s) = (q + ps)^N. \tag{13}$$

(iv) In population biology we often assume that an individual gene can give birth to normal progeny as well as a mutant type which begins its own line of descent. It is frequently assumed that the probability of a mutant gene having k direct descendants $(k = 0, 1, 2, \cdots)$ is governed by a Poisson distribution with mean $\lambda = 1$. Then $\phi(s) = e^{s-1}$ and $\pi = 1$.

The rationale behind this choice of distribution is as follows. In many populations a large number of zygotes (fertilized eggs) are produced, only a small number of which grow to maturity. The events of fertilization and maturation of different zygotes obey the law of independent binomial trials. The number of trials (i.e. number of zygotes) is large so that the actual number of mature progeny follows the Poisson distribution. This is a corollary of the principle of rare events commonly invoked to justify the Poisson approximation. It seems quite appropriate in the model of population growth of a rare mutant gene. If the mutant gene carries a biological advantage (or disadvantage), then the probability distribution is taken to be the Poission distribution with mean $\lambda > 1$ or < 1. Specifically,

$$\phi(s) = e^{\lambda(s-1)} \tag{14}$$

and $0 < \pi < 1$ if and only if $\lambda > 1$.

In a heterogeneous population of mutant genes we may assume that the probability distribution of the number of offspring is a Poisson distribution, but with the mean also a random variable.

For example, we may have a large geographical area in which for each subarea a branching process characterized by the probability generating function of a Poisson distribution with parameter λ is taking place. We assume, furthermore, that the value of λ varies depending on the subarea and its distribution over the whole area is that of a gamma. Formally we postulate that the probability of a mutant gene having exactly k direct descendants is given by

$$p_k = e^{-\lambda}\lambda^k/k!, \qquad k = 0, 1, 2, \cdots,$$

where λ itself is a random variable distributed according to a gamma distribution with the density function

$$f(\lambda) = \frac{(q/p)^\alpha \lambda^{\alpha-1}}{\Gamma(\alpha)} \exp\left(-\frac{q}{p}\lambda\right) \qquad \text{for } \lambda \geqq 0,$$

$$= 0 \qquad \text{otherwise,}$$

where q, p, α are positive constants and $q + p = 1$. The probability of an individual having k offspring, if we average with respect to the values of the parameter λ, is

$$\Pr\{\xi = k\} = \int_0^\infty \Pr\{\xi = k \mid \lambda\} f(\lambda)\, d\lambda.$$

Thus the generating function is

$$\phi(s) = \sum_{k=0}^\infty \Pr\{\xi = k\} s^k = \left(\frac{q/p}{(q/p) + 1 - s}\right)^\alpha = \left(\frac{q}{1 - ps}\right)^\alpha.$$

This we recognize as the probability generating function of the negative binomial distribution.

(v) In Examples (ii)—(iv) no closed-form expressions are known for the nth-generation probability generating function $\phi_n(s)$. The final example studied below is amenable to a rather complete analysis. Specifically, we will compute the nth-generation probability generating function. Let

$$p_k = bc^{k-1}, \qquad k = 1, 2, \cdots,$$

and

$$p_0 = 1 - \sum_{k=1}^\infty p_k,$$

where $b, c > 0$ and $b + c < 1$. Then

$$p_0 = 1 - b\sum_{k=1}^\infty c^{k-1} = 1 - \frac{b}{1-c} = \frac{1 - b - c}{1 - c},$$

and the corresponding probability generating function is

$$\phi(s) = 1 - \frac{b}{1-c} + bs\sum_{k=1}^\infty (cs)^{k-1} = \frac{1 - (b + c)}{1 - c} + \frac{bs}{1 - cs}. \tag{15}$$

Notice that $\phi(s)$ has the form of a linear fractional transformation

$$f(s) = \frac{\alpha + \beta s}{\gamma + \delta s}, \qquad \alpha\delta - \beta\gamma \neq 0, \tag{16}$$

which may be iterated to yield

$$\begin{gathered} \phi_n(s) = 1 - m^n \left(\frac{1 - s_0}{m^n - s_0} \right) + \frac{m^n [(1 - s_0)/(m^n - s_0)]^2 s}{1 - [(m^n - 1)/(m^n - s_0)] s}, \\ s_0 = \frac{1 - b - c}{c(1 - c)}, \quad m \neq 1, \quad m = \frac{b}{(1 - c)^2}. \end{gathered} \tag{17}$$

Then the probabilities of extinction at the nth generation are

$$\Pr\{X_n = 0\} = \phi_n(0) = 1 - m^n \left(\frac{1 - s_0}{m^n - s_0} \right).$$

Note that this expression converges to s_0 as $n \to \infty$ if $m > 1$ and to 1 if $m \leqq 1$. See S. Karlin [**10**] for details.

Multi-type Galton-Watson process. A first step in generalizing the simple Galton-Watson process is the consideration of processes involving several types of objects. For example, in cosmic ray cascades, electrons produce photons and photons may produce electrons. In this situation we have a 2-type branching process in which the number of each type of particle is counted. It is also common practice to divide the range of energy levels of cosmic ray particles into a finite number of portions and let each portion be identified with a type.

In biological populations there may exist a finite number of possible mutant forms each evolving according to the generating function prescription presented in the following discussion. Each mutant form is then identified as a type and we consider the vector process of all types evolving simultaneously.

Associated with type i is the probability generating function

$$f^{(i)}(s_1, \cdots, s_p) = \sum_{r_1, \cdots, r_p = 0}^{\infty} p^{(i)}(r_1, \cdots, r_p)\, s_1^{r_1}, \cdots, s_p^{r_p},$$
$$|s_1| \leqq 1, \cdots, |s_p| \leqq 1, \quad i = 1, \cdots, p,$$

where $p^{(i)}(r_1, \cdots, r_p)$ is the probability that a single object of type i has r_1 children of type 1, r_2 children of type 2, $\cdots, r_p$ of type p.

We introduce the vector notation $s = (s_1, \cdots, s_p)$.

Let $f_n^{(i)}(s)$ denote the nth-generation probability generating function arising from one individual of type i. Analogous to (2) we have

$$f_{n+1}^{(i)}(s) = f^{(i)}(f_n^{(1)}(s), f_n^{(2)}(s), \cdots, f_n^{(p)}(s)), \quad f_0^{(i)}(s) = s_i,$$
$$n = 0, 1, \cdots, \quad i = 1, \cdots, p.$$

Let $Z_n = (Z_n^{(1)}, \cdots, Z_n^{(p)})$ denote the vector representing the population size of p types in the nth generation. The analog of (10) is

$$E(Z_{n+m} \mid Z_n) = Z_n M^m,$$

where $M = \| m_{ij} \|_{i,j=1}^p$ is the matrix of first moments:

$$m_{ij} = E(Z_1^{(j)} \mid Z_0 = e_i) = \frac{\partial f^{(i)}}{\partial s_j}(1, 1, \cdots, 1), \qquad i, j = 1, \cdots, p,$$

and e_i denotes the vector with 1 for the ith component and zero otherwise.

We now state the analogue of (8) for p types. We will assume $m_{ij} > 0$ for all i, j. (It suffices to have $m_{ij}^{(n)} > 0$ for some n and all i, j.) Let $\pi^{(i)}$ be the extinction probability if initially there is one object of type i $(i = 1, \cdots, p)$; that is,

$$\pi^{(i)} = \Pr\{Z_n = 0 \quad \text{for some} \quad n \mid Z_0 = e_i\}.$$

The vector $(\pi^1, \cdots, \pi^p)$ is denoted by π. Let $\mathbf{1}$ denote the vector $(1, 1, \cdots, 1)$.

THEOREM 1. *Let $m_{ij} > 0$ for all i, $j = 1, \cdots, p$ and let ρ denote the eigenvalue of largest absolute value of the matrix M. If $\rho \leqq 1$, then $\pi = \mathbf{1}$. If $\rho > 1$, then $0 \leqq \pi \ll \mathbf{1}$ (i.e. $\pi_i < 1$ for all i) and π satisfies the equation*

$$\pi^{(i)} = f^{(i)}(\pi), \qquad i = 1, \cdots, p.$$

We refer the reader to Harris [8] or Karlin [**10**] for proofs.

2. Continuous time branching processes. The branching processes dealt with in §1 are limited in that generation times are fixed. Although some phenomena, particularly experimental trials, fit this situation, most natural reproductive processes occur continuously in time. It is therefore of interest to formulate a continuous time version of simple branching processes.

We determine a continuous time Markov branching process with state variable $X(t) = \{$number of particles at time t, given

$X(0) = 1\}$ by specifying the infinitesimal probabilities of the process. Let

$$\delta_{1k} + a_k h + o(h), \qquad k = 0, 1, 2, \cdots, \tag{18}$$

represent the probability that a single particle will split producing k particles (or objects) during a small time interval $(t, t+h)$ of length h. In (18) δ_{1k} denotes, as customary, the Kronecker delta symbol, and we assume that $a_1 \leqq 0$, $a_k \geqq 0$ for $k = 0, 2, 3, \cdots$, and

$$\sum_{k=0}^{\infty} a_k = 0. \tag{19}$$

We further postulate that individual particles act independently of each other, always governed by the infinitesimal probabilities (18).

Another way to express the infinitesimal transitions is to differentiate between the time until a split occurs and the nature of the split. Thus each object lives a random length of time following an exponential distribution with mean $\lambda^{-1} = a_0 + a_2 + a_3 + \cdots$. On completion of its lifetime, it produces a random number D of descendants of like objects, where the probability distribution of D is

$$\Pr\{D = k\} = \frac{a_k}{a_0 + a_2 + a_3 + \cdots}, \qquad k = 0, 2, 3, \cdots. \tag{20}$$

The lifetime and progeny distribution of separate individuals are independent and identically distributed.

For a simple example of a continuous time branching process we put $a_2 = \lambda$, $a_0 = \mu$, $a_1 = -(\lambda + \mu)$, and $a_k = 0$ otherwise. Then $(\lambda + \mu)^{-1}$ can be interpreted as the mean time for a birth or death event; $\lambda/(\lambda + \mu)$ $(\mu/(\lambda + \mu))$ is the probability of a birth (death) under the condition that an event has happened. The stochastic process so obtained, whose state variable is population size, can also be recognized as a linear growth birth and death process.

Now let $P_{ij}(t)$ denote the probability that the population of size i at time zero will be of size j at time t, or in symbols $P_{ij}(t) = \Pr\{X(t+s) = j | X(s) = i\}$. As the notation indicates, this probability depends only on the elapsed time, i.e. the process has stationary transition probabilities. We introduce the generating function

$$\phi(t; s) = \sum_{j=0}^{\infty} P_{1j}(t)\, s^j. \tag{21}$$

Since individuals act independently, we have the fundamental relation

$$\sum_{j=0}^{\infty} P_{ij}(t)\, s^j = [\phi(t; s)]^i. \tag{22}$$

The formula (22) characterizes and distinguishes branching processes from other continuous time Markov chains. It expresses the property that different individuals (i.e., particles) give rise to independent realizations of the process uninfluenced by the pedigrees evolving owing to the other individuals present. In other words, the population $X(t; i)$ evolving in time t from i initial parents is the same, probabilistically, as the combined sum of i populations each with one initial parent.

Introducing the generating function

$$u(s) = \sum_{k=0}^{\infty} a_k s^k, \tag{23}$$

we assert the continuous time analog of equation (8) for the calculation of the probability of extinction.

THEOREM 2. *The probability of extinction q is the smallest nonnegative root of the equation*

$$u(s) = 0. \tag{24}$$

Hence, $q = 1$ if and only if $u'(1) \leqq 0$.

A more precise result of the probability of no extinction by time t is given by

THEOREM 3. *Consider a continuous time branching process $X(t)$ determined by the infinitesimal generating function*

$$u(s) = \sum_{k=0}^{\infty} a_k s^k \quad \textit{where} \quad \sum_{k=0}^{\infty} a_k = 0.$$

Suppose that $u''(1) < \infty$ and that $u'(1) < 0$. Then the probability of no extinction by time t tends to zero at an exponential rate according to

$$\lim_{t \to \infty} \frac{1 - P_{10}(t)}{\exp[u'(1)\, K(0)] \exp[u'(1)\, t]} = 1 \tag{25}$$

where

$$K(s) = \int_1^s \left[\frac{1}{u(x)} - \frac{1}{u'(1)(x-1)} \right] dx + \frac{\log(1-s)}{u'(1)}.$$

Moreover, the random variable $X(t)$ *conditioned by* $X(t) > 0$ *has a limit distribution whose probability generating function is given by*

$$g(z,t) = \frac{\sum_{k=1}^{\infty} \Pr\{X(t) = k | X(0) = 1\} z^k}{1 - \Pr\{X(t) = 0\}} \tag{26}$$
$$\rightarrow 1 - \exp\left[u'(1) \int_0^z \frac{dx}{u(x)} \right] \quad \text{as } t \rightarrow \infty.$$

We also state without proof the following limit theorem which corresponds to the cases $u'(1) = 0$ and $u'(1) > 0$.

THEOREM 4. (i) *Suppose* $u'(1) = 0, u'''(1) < \infty$. *Then*

$$\Pr\{X(t) > 0 | X(0) = 1\} \sim \frac{2}{u''(1)} \frac{1}{t}, \qquad t \rightarrow \infty,$$

and

$$\lim_{t \rightarrow \infty} \Pr\left\{ \frac{2X(t)}{u''(1)\, t} > \lambda \,\middle|\, X(t) > 0 \right\} = e^{-\lambda}, \qquad \lambda > 0.$$

(ii) *If* $u'(1) > 0$ *and* $u''(1) < \infty$, *then*

$$Z(t) = \frac{X(t)}{\exp[u'(1)\, t]}$$

has a limit distribution as $t \rightarrow \infty$.

See Karlin [**10**, Chapter 11] and Harris [**8**] for details.

Two-type continuous time branching process. Consider two different types of particles which we will call type 1 and type 2 particles, respectively. A continuous time branching Markov process for two types of particles will be determined by appropriately specifying the infinitesimal probabilities. Explicitly, we postulate that each particle of type i $(i = 1, 2)$ may at any time, independent of its past and independent of the history or present state of any of the other particles of either type, convert during a small time interval $(t, t + h)$ into k_1 and k_2 particles of types 1 and 2, respectively, with probabilities

$$\delta_{1k_1} \delta_{0k_2} + a^{(1)}_{k_1,k_2} h + o(h)$$

(δ_{ij} denotes the familiar Kronecker delta symbol) for a single parent of type 1 and

$$\delta_{0k_1}\delta_{1k_2} + a^{(2)}_{k_1,k_2}h + o(h)$$

for a single parent of type 2 $(k_1, k_2 = 0, 1, 2, \cdots)$. Note that we are again postulating time homogeneity for the transition probabilities in that the constants $a^{(i)}_{k_1,k_2}$ are time independent. The parameters obey the restrictions

$$\begin{aligned} &a^{(1)}_{1,0} \leqq 0, && a^{(2)}_{0,1} \leqq 0, \\ &a^{(1)}_{k_1,k_2} \geqq 0 && \text{for all } k_1, k_2 = 0, 1, 2, \cdots \\ & && \text{except } k_1 = 1, \ k_2 = 0, \\ &a^{(2)}_{k_1,k_2} \geqq 0 && \text{for all } k_1, k_2 = 0, 1, 2, \cdots \\ & && \text{except } k_1 = 0, \ k_2 = 1, \end{aligned}$$

and

$$\sum_{k_1,k_2=0}^{\infty} a^{(i)}_{k_1,k_2} = 0, \qquad i = 1, 2.$$

We introduce the pair of infinitesimal generating functions

$$u^{(i)}(s_1, s_2) = \sum_{k_1,k_2=0}^{\infty} a^{(i)}_{k_1,k_2} s_1^{k_1} s_2^{k_2}, \qquad i = 1, 2 \quad (|s_1| \leqq 1, \quad |s_2| \leqq 1).$$

Let $P_{k_1,k_2;j_1,j_2}(t)$ be the probability that a population of k_1 objects of type 1 and k_2 objects of type 2 present at time 0 will be transformed into a population consisting of j_1 objects of type 1 and j_2 objects of type 2 over a time period of length t. Since the infinitesimal probabilities are necessarily time homogeneous, we define the probability generating function

$$\phi^{(1)}(t; s_1, s_2) = \sum_{j_1,j_2=0}^{\infty} P_{1,0;j_1,j_2}(t)\, s_1^{j_1} s_2^{j_2}.$$

$$\phi^{(2)}(t; s_1, s_2) = \sum_{j_1,j_2=0}^{\infty} P_{0,1;j_1,j_2}(t)\, s_1^{j_1} s_2^{j_2}.$$

Then as in the model of the one-type branching process, it follows that

$$\sum_{j_1,j_2=0}^{\infty} P_{k_1,k_2;j_1,j_2}(t)\, s_1^{j_1} s_2^{j_2} = [\phi^{(1)}(t; s_1, s_2)]^{k_1}[\phi^{(1)}(t; s_1, s_2)]^{k_2} \tag{27}$$

$$(k_1, k_2 = 0, 1, 2, \cdots).$$

In fact, (27) can be regarded as the defining relation of a continuous time two-type branching process. In other words any transition probability matrix function satisfying (27) is said to generate a two-type continuous time Markov branching process. The Markov character of the process is summarized in the Chapman-Kolmogorov equation:

$$P_{k_1,k_2;j_1,j_2}(t+\tau) = \sum_{l_1,l_2=0}^{\infty} P_{k_1,k_2;l_1,l_2}(t)\, P_{l_1,l_2;j_1,j_2}(\tau). \tag{28}$$

Then from (27) and (28),

$$\phi^{(i)}(t+\tau; s_1, s_2) = \phi^{i}(t; \phi^{(1)}(\tau; s_1, s_2), \phi^{(2)}(\tau; s_1, s_2)) \qquad i = 1, 2. \tag{29}$$

We next offer some applications and examples of two-type continuous time branching processes.

EXAMPLE 1.Our first example involves a branching process with immigration. We consider the one-type continuous time branching process and enlarge its scope by allowing, in addition to branching, some migration of particles into the system. Recall that

$$\delta_{1k} + a_k h + o(h), \qquad k = 0, 1, 2, \cdots,$$

represents the probability that a particle will convert into k particles during a small time interval $(t, t+h)$ independent of its past and of all other particles. Let us superimpose immigration into the population as follows. Specifically, let

$$\delta_{0k} + b_k h + o(h), \qquad k = 0, 1, 2, \cdots,$$

denote the probability, independent of the present or past history of the population, that k particles of the same kind immigrate and merge with the population during the time interval $(t, t+h)$. Note that the parameters a_k as well as the parameters b_k are assumed to be independent of the precise time t that the conversion or the immigration takes place. In other words the associated infinitesimal transition probabilities per individual are time homogeneous. For the a_k and b_k we impose the conditions

$$a_1 \leqq 0, \qquad b_0 \leqq 0, \qquad a_k \geqq 0 \quad \text{for } k = 0, 2, 3, \cdots,$$
$$b_k \geqq 0 \quad \text{for } k = 1, 2, 3, \cdots,$$

$$\sum_{k=0}^{\infty} a_k = \sum_{k=0}^{\infty} b_k = 0.$$

Let

$$
\begin{aligned}
P_k(t) &= \Pr\{\text{population at time } t \text{ is size } k \text{ if there were} \\
&\qquad \text{no particles at time } t = 0\} \\
&= \Pr\{X(t) = k \mid X(0) = 0\}, \qquad k = 0, 1, 2, \cdots,
\end{aligned} \tag{30}
$$

and denote its generating function by

$$
\phi(t; s) = \sum_{k=0}^{\infty} P_k(t)\, s^k. \tag{31}
$$

Our objective is to evaluate $P_k(t)$ or, if this is not feasible, to ascertain some of its properties.

We introduce the infinitesimal generating functions

$$
u(s) = \sum_{k=0}^{\infty} a_k s^k \quad \text{and} \quad v(s) = \sum_{k=0}^{\infty} b_k s^k.
$$

It is possible to cast the one-type continuous time branching process with immigration in the form of a two-type continuous time branching process. This is done as follows. Assume that we have two different types of particles, types 1 and 2, with infinitesimal probabilities of conversion which we will now specialize, as described at the start of this section.

The idea underlying the identification runs as follows. We have available two types of particles: the first is real, while the second is of a fictitious nature. Real particles upon termination of their lifetime (which is of random duration distributed according to an exponential law with parameter $\lambda^{-1} = a_0 + a_2 + a_3 + \cdots$) create k new real particles with probability $\lambda \cdot a_k$ $(k = 0, 2, 3, \cdots)$. A fictitious particle also lives a random length of time (exponentially distributed with parameter $\tilde{\lambda}^{-1} = b_1 + b_2 + \cdots$) and at the end of its life produces l real particles and one further fictitious particle with probability $\tilde{\lambda} \cdot b_l$ $(l = 1, 2, 3, \cdots)$. Notice that $\sum_{l=1}^{\infty} \tilde{\lambda}\, b_l = 1$. The progeny of fictitious particles account for the immigration factor. Thus, we set

$$
\begin{aligned}
a^{(1)}_{k_1, k_2} &= a_{k_1} && \text{if } k_2 = 0, \\
&= 0 && \text{if } k_2 \neq 0, \\
a^{(2)}_{k_1, k_2} &= b_{k_1} && \text{if } k_2 = 1, \\
&= 0 && \text{if } k_2 \neq 1.
\end{aligned}
$$

Then in accordance with the notation of the beginning of this section we have

$$u^{(1)}(s_1, s_2) = \sum_{k1,k2=0}^{\infty} a_{k1,k2}^{(1)} s_1^{k1} s_2^{k2} = \sum_{k1=0}^{\infty} a_{k1} s_1^{k1},$$

$$u^{(2)}(s_1, s_2) = \sum_{k1,k2=0}^{\infty} a_{k1,k2}^{(2)} s_1^{k1} s_2^{k2} = s_2 \sum_{k1=0}^{\infty} b_{k1} s_1^{k1}.$$

Thus

$$u^{(1)}(s_1, s_2) = u(s_1), \qquad u^{(2)}(s_1, s_2) = s_2 v(s_1).$$

In the special case under consideration the generating functions satisfy the differential equations

$$\frac{\partial \phi^{(i)}(t; s_1, s_2)}{\partial t} = \frac{\partial \phi^{(i)}(t; s_1, s_2)}{\partial s_1} u(s_1) + \frac{\partial \phi^{(i)}(t; s_1, s_2)}{\partial s_2} s_2 v(s_1), \quad i = 1, 2 \tag{32}$$

and

$$\frac{\partial \phi^{(1)}(t; s_1, s_2)}{\partial t} = u(\phi^{(1)}(t; s_1, s_2)), \tag{33}$$

$$\frac{\partial \phi^{(2)}(t; s_1, s_2)}{\partial t} = [\phi^{(2)}(t; s_1, s_2)] v(\phi^{(1)}(t; s_1, s_2)). \tag{34}$$

Now we will relate the probabilities $P_{0,1;\, j1, j2}(t)$ of the two-type process to the probabilities defined in (30). In accordance with the meaning ascribed to the two types of particles, the initial state $(0, 1)$ signifies that we start at time 0 with no real particles but with the presence of a potential immigrant signified by a fictitious particle. By the very meaning of the symbols we obviously have

$$\begin{aligned} P_{0,1;\, j1, j2}(t) &= P_{j1}(t) \qquad \text{if } j_2 = 1, \\ &= 0 \qquad \text{if } j_2 \neq 1, \end{aligned}$$

and hence

$$\phi^{(2)}(t; s_1, s_2) = s_2 \phi(t; s_1). \tag{35}$$

Then from (32) we obtain

$$\frac{\partial \phi(t; s)}{\partial t} = \frac{\partial \phi(t; s)}{\partial s} u(s) + \phi(t; s) v(s), \tag{36}$$

where we have written s in place of s_1. The initial condition here is

$$\phi(0; s) = 1. \tag{37}$$

Denoting the solution of (33) by $f(t; s)$, where s has taken the place of s_1 and s_2 is suppressed, we can rewrite (34) as

$$\frac{\partial \phi(t; s)}{\partial t} = \phi(t; s)\, v(f(t; s)). \tag{38}$$

With the initial condition (37), the solution of (38) is

$$\phi(t; s) = \exp\left[\int_0^t v(f(\tau; s))\, d\tau\right].$$

EXAMPLE 2. *Embedding of urn schemes in continuous time multidimensional Markov branching processes.* We start with W_0 white and B_0 black balls in an urn. A draw consists of the following operations: (i) pick a ball at random from the urn, (ii) notice its color C and return the ball to the urn, (iii) add α balls of the color C and β balls of the opposite color. Let (W_n, B_n) denote the composition of the urn after n successive draws. The stochastic process $\{(W_n, B_n),\ n = 0, 1, 2, \cdots\}$ is usually referred to as Bernard Friedman's urn scheme, and in the special case of $\beta = 0$ we have a classical Pólya urn scheme.

These urn schemes can be embedded in Markov branching processes in the following manner. Consider a two-type continuous time Markov branching process defined by

$$a_1 = a_2 = 1, \qquad h_1(s) = s_1^{\alpha+1} s_2^{\beta},$$
$$h_2(s) = s_1^{\beta} s_2^{\alpha+1}, \qquad X_1(0) = W_0, \quad X_2(0) = B_0,$$

where the infinitesimal generating functions are now given by $u_i(s) = a_i[h_i(s) - s_i]$, $i = 1, 2$, $0 < a_i < \infty$, $s = (s_1, s_2)$, and $h_i(s)$ is a probability generating function.

Let $\tau_n(\omega)$ denote the nth time a particle split occurs. At time $\tau_n + 0$ there are $X_1(\tau_n)$ particles of type 1 and $X_2(\tau_n)$ of type 2, respectively. Since $a_1 = a_2 = 1$ all particles irrespective of type have exponential life times with mean one. Furthermore, they act independently. Hence the next split will be of type 1 with probability

$$X_1(\tau_n)/(X_1(\tau_n) + X_2(\tau_n))$$

and of type 2 with probability

$$X_2(\tau_n)/(X_1(\tau_n)+X_2(\tau_n)).$$

After a split we lose the particle that splits but since this creates $(\alpha+1)$ new particles of its own type and β of the opposite type, we get a net addition of α particles of the type that split and β of the opposite type. This implies that the stochastic mechanism yielding the movement from $X(\tau_n)=(X_1(\tau_n), X_2(\tau_n))$ to $X(\tau_{n+1})$ is the same as that of (W_n, B_n) to (W_{n+1}, B_{n+1}). Furthermore, both processes are Markov and are thus equivalent in the sense that their finite dimensional distributions are identical. It is in this sense that we say the Friedman and Pólya urn schemes are embeddable in the continuous time Markov branching process $\{X(t), t \geqq 0\}$.

The embedding approach is a powerful tool for establishing limit theorems such as when $\alpha=\beta$ then

$$\frac{W_n}{W_n+B_n} \xrightarrow[n\to\infty]{} Y \text{ in law 1}$$

where Y is Beta-distributed with parameters $(W_0/\alpha, B_0/\alpha)$, and $\{(W_n, B_n), n=0,1,2,\cdots\}$ is a Pólya urn scheme with parameters (W_0, B_0, α). When $\alpha\neq\beta$ and $(\alpha-\beta)/(\alpha+\beta)>\frac{1}{2}$, then using a classical Martingale convergence theorem, it follows that $W_n/(W_n+B_n)\to\frac{1}{2}$ a.s. See K. Athreya and S. Karlin [2] and K. Athreya [1] for an extensive discussion of this and more general multi-dimensional limit theorems.

We conclude this section by mentioning that there is a considerable literature on branching processes which do not require that individual particles have a fixed or exponentially distributed lifetime. These age dependent processes, as well as branching processes where the progeny distribution is also time and population size dependent, are described in Harris [8] (see also references therein).

3. **Branching and some compounded stochastic processes.** We begin with a discussion of immigration and its role in describing the evolution of populations by considering a simple model in which a single type is evolving from an initial population whose growth behavior follows the laws of a continuous time Markov process. Moreover, in addition to the inherent growth of the population, there is an influx of immigrants of the same type which contribute further descendants whose growth behavior is

governed by the laws of the same Markov process. The arrival pattern of immigrants into the colony is generally also a stochastic process. We formulate the process, for concreteness, as a model of growth of bacterial populations.

Consider a bacterial colony of n_0 individuals. Assume that each bacterium, independent of the others, produces descendents which in turn produce further offspring, etc. The population growth evolving from a single bacterium describes a continuous time Markov process. Let $F(s,t)$ be the probability generating function of the population size at time t derived from a single bacterium. Clearly, the population size at time t due to the existing colony of size n_0 at time 0 is a random variable which has the generating function $[F(s,t)]^{n_0}$. Assume, furthermore, that immigration of new bacteria occurs at times t_j, $j = 1, 2, \cdots, N$. Each immigrant will evolve a population following the same laws of reproduction as the original n_0 bacteria, independent of them and each other. The population size at time t derived from an immigrant arriving at time t_j has the probability generating function $F(s, t - t_j)$. The total population size at time t has the probability generating function

$$[F(s,t)]^{n_0} \prod_{j=1}^{N} F(s,\, t - t_j),$$

since each bacterium creates independent families. Assume, however, that immigration occurs not at fixed times t_j, but that the t_j constitute events of a Poisson process with parameter r. We want to calculate the probability generating function of the total population size in terms of $F(s,t)$ and r.

The immigration times t_j are random variables and their number $N(t)$ during the time interval $[0,t]$ is also a random variable whose distribution function is Poisson with parameter rt. Let $Y_j(t,t_j)$ denote the population size at time t derived from a single immigrant into the colony at time t_j for $j = 1, 2, \cdots, N(t)$. Then

$$Y(t) = \sum_{j=1}^{N(t)} Y_j(t, t_j)$$

is the population size at time t evolved from immigrants arriving during the time epoch $[0,t]$.

The probability generating function of $Y(t)$ can be computed by conditioning on $N(t)$ in the usual way. This leads to the expression

$$E[s^{Y(t)}] = \sum_{k=0}^{\infty} E[s^{Y(t)} | N(t) = k] \Pr\{N(t) = k\}. \tag{39}$$

We know that the joint distribution in $[0, t]$ of the arrival times t_j, $j = 1, 2, \cdots, N(t)$, given the number of arrivals $N(t) = k$, is the same as the distribution of the order statistics of k independent, uniformly distributed, random variables over $[0, t]$. Thus

$$\begin{aligned} E[s^{Y(t)} | N(t) = k] &= \frac{k!}{t^k} \int_0^t dt_1 \int_{t_1}^t dt_2 \cdots \int_{t_{k-1}}^t dt_k E[s^{\sum_{j=1}^k Y_j(t,t_j)}] \\ &= \frac{1}{t^k} \int_0^t dt_1 \int_0^t dt_2 \cdots \int_0^t dt_k E[s^{\sum_{j=1}^k Y_j(t,t_j)}], \end{aligned}$$

because the integrand is a symmetric function of $t_1, t_2, \cdots, t_k$. Further, since different immigrants create independent histories, we get

$$E[s^{\sum_{j=1}^k Y_j(t,t_j)}] = \prod_{j=1}^k E[s^{Y_j(t,t_j)}] = \prod_{j=1}^k F(s, t - t_j).$$

Therefore

$$\begin{aligned} E[s^{Y(t)} | N(t) = k] &= \frac{1}{t^k} \int_0^t dt_1 \int_0^t dt_2 \cdots \int_0^t dt_k \prod_{j=1}^k F(s, t - t_j) \\ &= \frac{1}{t^k} \prod_{j=1}^k \int_0^t F(s, t - t_j)\, dt_j \\ &= \left\{ \frac{1}{t} \int_0^t F(s, t - \tau)\, d\tau \right\}^k. \end{aligned}$$

Inserting this formula into (39) and taking account of the fact that $N(t)$ is a Poisson process, we obtain

$$\begin{aligned} E[s^{Y(t)}] &= \sum_{k=0}^{\infty} \left\{ \frac{1}{t} \int_0^t F(s, t - \tau)\, d\tau \right\}^k e^{-rt} \frac{(rt)^k}{k!} \\ &= \exp\left\{ r \int_0^t [F(s, t - \tau) - 1] d\tau \right\}. \end{aligned} \tag{40}$$

Hence, the probability generating function of the total population size at time t is

$$G(s,t) = [F(s,t)]^{n_0} \exp\left\{ r \int_0^t [F(s,t-\tau) - 1] d\tau \right\}. \tag{41}$$

EXAMPLE. As an example we assume that each individual bacterium follows a growth law described by the Yule process $X(t)$ of parameter $\beta > 0$. Then the size of the population at time t derived from a single bacterium at time $t = 0$ is governed by the probability law

$$P_k(t) = \Pr\{X(t) = k \mid X(0) = 1\},$$

where

$$P_k(t) = e^{-\beta t}(1 - e^{-\beta t})^{k-1}, \qquad k = 1, 2, \cdots.$$

Equivalently the generating function of the Yule process is given by

$$F(s,t) = e^{-\beta t} \sum_{k=1}^{\infty} (1 - e^{-\beta t})^{k-1} s^k = \frac{se^{-\beta t}}{1 - (1 - e^{-\beta t})s}.$$

Since we assumed that each immigrant follows the same law of growth as the original population, the generating function of the population size emanating from immigrant bacteria, in accordance with (40), has the form

$$\begin{aligned} E[s^{Y(1)}] &= \exp\left\{ r \int_0^t \frac{se^{-\beta(t-r)} d\tau}{1 - s + se^{-\beta(t-\tau)}} - rt \right\} \\ &= \exp\left\{ \frac{-r}{\beta} \log[1 - s + se^{-\beta t}] - rt \right\} \\ &= e^{-rt}(1 - s + se^{-\beta t})^{-r/\beta} = e^{-rt}[1 - (1 - e^{-\beta t})s]^{-r/\beta}. \end{aligned}$$

If we take account of the original population of bacteria, then by (41) the generating function of population size at time t evolving from the initial population in addition to the immigrant population is

$$G(s,t) = \exp[-(r + \beta n_0)t][1 - (1 - e^{-\beta t})s]^{-(n_0 + r/\beta)}.$$

The expected population size at time t is $(\partial/\partial s) G(s,t)|_{s=1}$, which reduces to

$$(n_0 + r/\beta)(e^{\beta t} - 1).$$

Higher-order moments can also be evaluated by further differentiation of the generating function.

A more sophisticated immigration structure arises very naturally in ecological and genetic questions concerned with the continued formation and growth of mutant populations. Karlin and McGregor [2] proposed the following rather flexible model to describe the fluctuations in the number of mutant lines over time.

We assume that new mutant lines arise over time according to a general stochastic process called the input process. (The origin of new lines may be ascribed to either migration or mutation forces.) The specific mutant populations are then assumed to fluctuate in size according to another stochastic process which may also include a description of interaction between different populations.

The simplest assumption one can make about the growth of individual populations is that they evolve independently of one another according to the laws of a continuous time Markov chain $\mathscr{P}$ with stationary transition probability function

$$P_{ij}(t), \qquad i,j = 0,1,2,\cdots.$$

The following examples of input processes $\{I(t), t > 0\}$ are of particular interest in this context.

(i) $I(t)$ is a Poisson process with parameter ν or, more generally, a variable time Poisson process of intensity parameter $\nu(t)$. In this case, the probability of a new mutant line coming into existence during the time interval $(t, t+h)$ is $\nu(t)h + o(h)$, $h \to 0$, while the probability of no line being created is $1 - \nu(t)h + o(h)$, $h \to 0$. Moreover, the number of mutant lines formed during disjoint time intervals are independent random variables. The dependence of $\nu(t)$ on time reflects the possibility of changing environmental conditions.

(ii) $I(t)$ is a renewal process, i.e. the times between the successive starts of new mutant lines are independent positive random variables with common distribution function $G(\xi)$ $(0 < \xi < \infty)$. The times of creation of new mutant lines are taken to occur at times

$$\xi_1, \xi_1 + \xi_2, \cdots, \xi_1 + \cdots + \xi_n, \cdots$$

where $\xi_1, \xi_2, \cdots$ are independent observations from $G(\xi)$.

(iii) $I(t)$ is a general continuous time increasing point process.

Some examples of growing processes $\mathscr{P}$ which are particularly pertinent are as follows:

(i) Birth and death processes with individual infinitesimal birth and death rates $\{\lambda_n\}$ and $\{\mu_n\}$ respectively, and state 0 an absorbing state. See Karlin and McGregor [13]—[15].

(ii) Continuous time Markov branching processes with non-positive infinitesimal mean change of population size corresponding to an initial single parent.

Having described the general mutation growth models, we formulate some particular questions to be investigated.

1) Specifying the input process $\{I(t),\ t>0\}$ and the individual growing process $\mathscr{P}$ for each mutant type, we want to determine the distribution function of the number N_t^* of different mutant lines existing at time t. More generally, what is the nature of the whole process $\{N_t^*,\ t>0\}$?

2) Determine the distribution functions of the random variables $N_k^*(t)$, the number of mutant populations with exactly k members at time t. Describe the vector process $N(t)=\{N_0(t), N_1(t), N_2(t), \cdots\}$, $t>0$, and the nature of specific functionals of this process.

3) Determine the random time required for all current mutant populations to disappear.

In the case where the input process is Poisson, the solution of problem 2) is very simple and essentially classical. We have the following theorem:

THEOREM 4. *Let the input process be generalized Poisson with intensity parameter $\nu(t)$. Let $\mathscr{P}$ be a continuous time Markov chain with transition probability matrix $P_{ij}(t)$. Let $N_t^*(k)$ denote the number of mutant populations of size k existing at time t. Then the random process $N_t^*(k)$, $k=1,2,\cdots$, are independent Poisson processes with joint generating function*

$$\phi(t, z_1, z_2, \cdots) = \sum_{\bar{r}} z_1^{r_1} z_2^{r_2} \cdots \Pr\{N_t^*(i) = r_i,\ i = 1, 2, \cdots\}$$

$$= \exp\left[\sum_{k=1}^{\infty} (z_k - 1) \int_0^t P_{1k}(t-\tau)\,\nu(\tau)\,d\tau\right].$$

Subject to the hypotheses of the above theorem, it follows that

$$E(N_t^*(k)) = \int_0^t P_{1k}(t-\tau)\,\nu(\tau)\,d\tau. \tag{42}$$

If $\nu(t)=\nu$ is constant, the equilibrium distribution of $N_t^*(k)$ $(N^*(k) = \lim_{t\to\infty} N_t^*(k))$ is Poisson with parameter

$$E(N^*(k)) = \nu \int_0^\infty P_{1k}(\tau)\,d\tau$$

provided the integral exists, which is certainly the case if 0 is an absorbing state. The expected number of alleles N^* in the stationary case is obviously

$$E(N^*) = \nu \int_0^\infty \sum_{k=1}^\infty P_{1k}(\tau)\, d\tau = \nu \int_0^\infty [1 - \Omega(\tau)]\, d\tau, \tag{43}$$

where $\Omega(\tau) = 1 - \sum_{k=1}^\infty P_{1k}(\tau)$ is the distribution function of the extinction time for a particular mutant line generated by a single initial parent. Notice that $E(N^*)$ is finite if and only if the distribution function $\Omega(t)$ has finite mean. It is possible for $E(N^*(k))$ to be finite while $E(N^*)$ is infinite. This situation is of interest, and we will discuss some aspects of this phenomenon later.

The formula (43) is also valid when the input process is a renewal process, provided we interpret ν as the reciprocal of the mean interarrival time. Higher moments of $N_t^*(k)$ can be obtained easily by iterating certain recursion relations. These calculations are also accessible when the input process is a renewal process or generalized Poisson process.

We next discuss briefly the situation where the input process is Poisson with parameter ν and the growing process $\mathscr{P}$ is such that each mutant type ultimately becomes lost with probability 1, but the expected time until extinction is ∞. Specifically we will assume that the probability distribution $F(t)$ of the time until extinction (starting with a single initial parent) has the asymptotic growth behavior

$$1 - F(t) = \sum_{k=1}^\infty P_{1k}(t) \sim \frac{C}{t^\alpha} \quad \text{as } t \to \infty \tag{44}$$

where $0 < \alpha \leqq 1$, and C is a constant.

For example, if $\mathscr{P}$ is a classical birth and death process with infinitesimal parameters

$$\lambda_n = n\lambda, \qquad \mu_n = n\lambda, \qquad n > 0,$$

then it is known (see Karlin and McGregor [**15**]) that

$$1 - F(t) \sim 1/\lambda t \quad \text{as } t \to \infty.$$

In this case we can easily determine the limiting growth behavior of $N_t^*(k)$ and $N_t^* = \sum_{k=1}^\infty N_t^*(k)$ as $t \to \infty$. Explicitly, we obtain that $N_t^*(k)$ is Poisson distributed with mean $(\lambda/k)[\lambda t/(1 + \lambda t)]^k$. Hence

$$\lim_{t\to\infty} E(N_t^*(k)) = \lambda/k.$$

Thus the number of populations with k members is of mean size λ/k.

What is more striking is the fact that the total number of existing mutant types N_t has an asymptotic normal distribution of mean $\log(1+\lambda t)$ and variance $\log(1+\lambda t)$. The precise result is as follows:

THEOREM 5. *If the input process is Poisson with parameter* ν *and* $\mathscr{P}$ *is a continuous time Markov process such that the distribution function* $F(t)$ *of the time until extinction of a newly created mutant satisfies*

$$1 - F(t) \sim C/t^{\alpha} \quad \textit{for } t \to \infty \quad (C \textit{ is a constant})$$

where $0 < \alpha \leqq 1$, *then*

$$N_t^* = \sum_{k=1}^{\infty} N_t^*(k)$$

has an asymptotic normal distribution (as $t \to \infty$*) with mean and variance*

$$E(N_t^*) \sim \frac{c\nu t^{1-\alpha}}{1-\alpha}, \qquad \mathrm{Var}(N_t^*) \sim \frac{c\nu t^{1-\alpha}}{1-\alpha}, \quad \textit{when } 0 < \alpha < 1$$

and

$$E(N_t) \sim c\nu \log t, \qquad \mathrm{Var}\, N_t \sim c\nu \log t, \quad \textit{when } \alpha = 1.$$

As an indication of the problems associated with functionals of the vector process $N(t)$, we consider the special example

$$S(t) = \{\text{number of different sizes of mutant populations at time } t\}$$

which can be interpreted as a measure of population fluctuation. The determination of exact asymptotic formulas for $ES(t)$ and $\mathrm{Var}\, S(t)$ as $t \to \infty$ leads to some rather surprising results. For example, if the input process is Poisson with parameter $\lambda = 1$ and the growing process $\mathscr{P}$ is a null recurrent or transient birth and death process whose infinitesimal parameters satisfy

$$\pi_n = \frac{\lambda_0 \lambda_1 \cdots \lambda_{n-1}}{\mu_1 \mu_2 \cdots \mu_n} \sim D n^{\gamma - 1}, \qquad \frac{1}{\lambda_n \pi_n} \sim C n^{\beta - 1}, \quad n \to \infty,$$

C, D, $\gamma > 0$, and $\beta + \gamma = 1$, then

$$ES(t) \sim t \int_0^{\infty} \left(1 - \exp\left[-\int_0^1 g(0, w; u)\, du\right]\right) dw, \tag{45}$$

$$\operatorname{Var} S(t) \sim t \int_0^\infty \left(\exp\left[- \int_0^1 g(0, w; u)\, du \right] \right. \tag{46}$$
$$\left. - \exp\left[-2 \int_0^1 g(0, w; u)\, du \right] \right) dw$$

as $t \to \infty$, where

$$g(0, w; u) = c_1 \frac{w^{\gamma - 1}}{u^{1-\beta}} \exp\left(- c_2 \frac{w}{u} \right)$$

and $g(x, w; u)$ is the transition density of a Bessel diffusion process which is the limit in the sense of weak convergence of the original birth and death process with the normalization $X(tu)/t$. The verification of these formulas requires a local limit theorem for the convergence of the transition density of $X(tu)/t$ to that of the limiting diffusion. See B. Singer [**19**] for detailed proofs and further discussion of results of this nature.

The vector process $N(t)$ and some related functionals also appear naturally in an infinite urn scheme which has some curious properties. Suppose n balls are thrown independently at a fixed *infinite* array of cells with probability p_k of hitting the kth cell.

$N_k(n) = \{$number of balls in the kth cell after n tosses$\}$ is then identified with the previously mentioned mutant population model where the input process is deterministic and the growth process is generated by throwing balls in urns.

Some functionals of particular interest are

$S(n)$ = number of occupied cells after n throws,
$U(n)$ = number of cells containing an odd number of balls,
$C(n) = \sum_{r=1}^\infty C_{n,r}$,

where

$$C_{n,r} = np_r \quad \text{if } N_r(n) = 0,$$
$$= 0 \quad \text{if } N_r(n) > 0,$$

which can be interpreted modulo a factor n as the percentage of cells not occupied after n tosses.

A novel interpretation of $U(n)$ occurs in Spitzer [**20**] in the following context. Consider an infinite sequence of light bulbs and an infinite sequence of numbers $\{p_k\}$, $p_k > 0$, $\sum_{k=1}^{\infty} p_k = 1$. At time 0 all the light bulbs are "off". At each unit time $t = 1, 2, 3, \cdots$, therefore one of the bulbs is selected (the kth with probability p_k) and its switch is turned. Thus it goes on if it was off and off if it was on. The number of light bulbs on at time n is precisely $U(n)$.

The functionals $S(n)$, $U(n)$ and $C(n)$ give rise to central limit theorems as indicated in the following theorem.

THEOREM 6. *Let $\{p_k\}$ be such that $p_k \geqq p_{k+1}$, $k = 1, 2, \cdots$, and assume that*

$$\alpha(x) = \max\{j : p_j \geqq 1/x\} = x^{\gamma} L(x) \quad \textit{for } x \to \infty,$$

$0 < \gamma \leqq 1$, *where $L(x)$ is slowly varying. Then*

$$\text{(i)} \qquad \Pr\left\{\frac{S(n) - ES(n)}{\sqrt{B(n)}} \leqq x\right\} \to \Phi(x), \quad n \to \infty,$$

$$\text{(ii)} \qquad \Pr\left\{\frac{U(n) - EU(n)}{\frac{1}{2}\sqrt{M(4n)}} \leqq x\right\} \to \Phi(x), \quad n \to \infty,$$

$$\text{(iii)} \qquad \Pr\left\{\frac{C(n) - EC(n)}{\sqrt{\operatorname{Var} C(n)}} \leqq x\right\} \xrightarrow[n \to \infty]{} \Phi(x),$$

where

$$B(n) = \Gamma(1-\gamma)(2^{\gamma} - 1) n^{\gamma} L(n), \ 0 < \gamma < 1, \qquad L^*(n) = \int_0^{\infty} \frac{L(nx)\,dx}{1+x},$$
$$= nL^*(n), \quad \gamma = 1,$$

$$M(n) = \Gamma(1-\gamma) n^{\gamma} L(n), \quad 0 < \gamma < 1,$$
$$= nL^*(n), \quad \gamma = 1,$$

$$\Phi(x) = \frac{1}{\sqrt{2\pi}} \int_{-\infty}^{x} e^{-y/2}\,dy.$$

It is interesting to note, however, that there are simple examples of $\{p_k\}$ which do not lead to a central limit theorem for $S(n)$; for example, if $p_k = 1/2^k$, $k = 1, 2, \cdots$. Then

$$\alpha(x) \sim c \log x \text{ as } x \to \infty, \qquad \operatorname{Var} S(n) < \infty \text{ for } n = 1, 2, \cdots,$$

and in fact,

$$S(n) - ES(n) \xrightarrow{\text{law}} W,$$

a nondegenerate random variable which is difficult to identify but whose moments can be calculated explicitly. We refer the reader to Karlin [11] for a detailed discussion of the infinite urn.

A considerable generalization of the previously mentioned branching models is presently undergoing intensive investigation in the framework of what we might call compounded branching stochastic processes. In this setting one considers particles moving in a given topological space according to the laws of a stochastic process and also giving rise to progeny or dying according to a specified branching law. In other words, we refer to processes which involve the superposition of branching and general stochastic motion.

An example of this more general structure and of the variety of limit theorems which appear can be seen by looking at branching Brownian motion. For this we consider a particle executing Brownian motion in N-dimensional Euclidean space for an exponential length of time and then splitting into k Brownian motions with probability p_k. Each new particle begins its evolution at the position of the original particle when the split occurred and evolves independent of all other particles according to the motion of its parent (i.e. Brownian motion for an exponential length of time and splitting into a random number of particles).

Then let

$Z_t^D =$ number of Brownian motions in a compact region D at time t,

$Z_t =$ number of Brownian motions which have come into existence up to time t.

Assume $F(s) = \sum_{k=0}^{\infty} p_k s^k$ satisfies $F'(1) > 1$ and $F''(1) < \infty$. Then we have

$$\Pr\left[\lim_{t\to\infty} (2\pi t)^{N/2}\frac{Z_t^D}{Z_t} = |D| \,\middle|\, \lim_{t\to\infty} Z_t = \infty\right] = 1$$

(see S. Watanabe [21] for details).

More generally we can assume a particle undergoing a diffusion begins its evolution at $x_0 \in D$, a bounded region in Euclidean N-space, and splits after a general continuously distributed time into a random number of particles which again evolve independently of each other and behave like the parent.

Some questions of particular interest can be formulated in the following manner:

(1) If the boundary absorbs particles, describe the process $\{N(t),\ t > 0\}$ where

$$N(t) = \text{number of particles alive at time } t$$

and determine the distribution of the first hitting time of the boundary by any particle. This hitting time represents the instant at which absorption first begins.

(2) If particles can enter a region at random times starting from the interior (continuous formation of new particles in addition to branching and diffusion) and become absorbed upon hitting the boundary, we have a problem involving annihilation and creation of mass which leads to nonlinear semigroups and hence to problems in nonlinear differential equations. In this context we refer the reader to Fujita [5] for a treatment of exploding solutions of the Cauchy problem for $u_t = \Delta u + u^{1+\alpha}$ and its probabilistic interpretation. For a general discussion of this problem see Nagasawa and Sirao [18].

Questions of the above type also have an interpretation in neutron transport theory. Particles are assumed to move in a medium and collide with an atomic nucleus causing a fission process. New particles are generated which again move in the medium and split according to the branching law of the parent. Branching plus stochastic motion of the particles is thus involved. Usually the motion is in a confined region and the boundaries can be absorbing.

We can pose the following special problems:

(1) Find nontrivial conditions on the branching law, stochastic motion, and the boundary of a region such that $N(t)$ = number of particles in existence at time t can blow up in finite time. Determine the distribution of the explosion time.

(2) Assume that at the times of split both the number of particles and their energy may be random variables. This leads to a multi-type process such as

$$N(E_i, D_j; t) = \text{number of particles with energy in } E_i \text{ located in region } D_j \text{ at time } t,$$

where $i = 1, 2, \cdots, n$; $j = 1, 2, \cdots, n$. A question of particular interest is the limit behavior of $\{N(E_i, D_j; t)\}$ as $t \to \infty$.

4. **Direct product branching processes.** In this section we introduce a class of finite state Markov chains of special structure which include many cases of interest in applications to population genetics, ecology and possibly other disciplines. Our discussion includes many of the idealized models proposed by S. Wright [22] and R. Fisher [4] to investigate the fluctuation of gene frequency under the influence of mutation, migration, selection and genetic drift.

We begin by developing the relationship between the theory of branching processes and various frequency models. To this end consider two populations of type a and A, respectively, each independently multiplying according to 1-type Galton-Watson processes. The two dimensional process unfolds as a sequence of pairs of random variables $Z_n = \{X_n, Y_n\}$ where $X_n(Y_n)$ denotes the number of a-types (A-types) in the nth generation. In the above formulation, the components X_n and Y_n generate independent branching processes so that Z_n is the direct product process. Let the generating functions of the number of offspring of a-types and A-types per individual be $f(s)$ and $g(s)$, respectively. Suppose initially we have a population of i a-types and j A-types. After one generation the joint probability distribution of the populations of a- and A-types has a generating function $H(s,t) = f^i(s)g^j(t)$. The joint probability that the first generation contains k a-types and a total population of M individuals of either type is the coefficient of $s^k t^M$ in the generating function $f^i(st)g^j(t)$. Symbolically,

$$\Pr\{X_1 = k, X_1 + Y_1 = M \mid X_0 = i, Y_0 = j\} \\ = \text{coefficient of } s^k t^M \text{ in } f^i(st)g^j(t). \tag{47}$$

Moreover, the probability that the offspring generation includes k a-types conditioned that the total progeny is M is computed by the formula

$$\Pr\{X_1 = k \mid X_0 = i, Y_0 = j, X_1 + Y_1 = M\} \\ = \frac{\text{coefficient of } s^k t^M \text{ in } f^i(st)g^j(t)}{\text{coefficient of } t^M \text{ in } f^i(t)g^j(t)}. \tag{48}$$

If we specialize so that

$$f(s) = e^{\lambda(s-1)}, \qquad g(s) = e^{\mu(s-1)} \quad (\lambda > 0, \mu > 0)$$

then (48) becomes

$$\binom{M}{k}\left(\frac{i\lambda}{i\lambda + j\mu}\right)^k \left(\frac{j\mu}{i\lambda + j\mu}\right)^{M-k} \tag{49}$$

When $\lambda \neq \mu$ and $M = N = i + j$ then (49) is precisely the transition probability matrix of Wright's selection model where $\lambda/\mu - 1 = \sigma$ is the relative selection coefficient.

Mutation, selection and migration factors are subsumed in the following general construction. We postulate that each individual of type a can produce offspring of both types. We denote its generating function as $f(s,t)$. Similarly, we assume that an individual of type A may produce individuals of both types and let $g(s,t)$ designate the generating function of the progeny. Let $h(s,t)$ represent the generating function of the number of a- and A-types immigrating into the system during each period. Let $(X_n, Y_n) = Z_n$, $n = 0, 1, \cdots$, denote the resulting two-dimensional branching process. The generating function of Z_1 for the initial condition $X_0 = i$, $Y_0 = j$ is

$$[f(s,t)]^i[g(s,t)]^j h(s,t). \tag{50}$$

The transition probability matrix obtained by conditioning that the population size has a fixed size is calculated in the usual way. We get

$$\begin{aligned} \Pr\{X_1 = k \mid X_0 = i, Y_0 = j; X_1 + Y_1 = M\} &= P(k \mid i,j; M) \\ &= \frac{\text{coefficient of } s^k t^M \text{ in } f^i(st,t) g^j(st,t) h(st,t)}{\text{coefficient of } t^M \text{ in } f^i(t,t) g^j(t,t) h(t,t)}. \end{aligned} \tag{51}$$

For the special choice $F(s,t) = \exp\{\lambda[(1-\alpha_i)s + \alpha_1 t] - 1\}$, $g(s,t) = \exp\{\mu[(1-\alpha_2)t + \alpha_2 s - 1]\}$ and $h(s,t) \equiv 1$, the expression (51) provided $i + j = N = M$ reduces to Wright's transition probability matrix for a model involving mutation and selection. If we take $h(s,t) = \exp[a(s-1) + b(t-1)]$, keeping $f(s,t)$ and $g(s,t)$ unaltered and $M = N = i + j$ then (51) becomes

$$\begin{aligned} P_{ik}(N) = \binom{N}{k} & [(1-\alpha_1)\lambda i + \mu(N-i)\alpha_2 + a]^k \\ & \times \frac{[\alpha_i \lambda i + (1-\alpha_2)\mu(N-i) + b]^{N-k}}{[\lambda i + \mu(N-i) + a + b]^N}. \end{aligned} \tag{52}$$

(Here N is a fixed parameter representing the constant population size.)

The formula (51) when $i + j = N = M$ by various specifications of $f(s,t)$, $g(s,t)$ and $h(s,t)$ provides a variety of interesting transition probability matrices P_{ik} associated with Markov chains on the state space $(0, 1, 2, \cdots, N)$.

We record some examples of (51) which arise in different biological situations.

A. If $f(s) = g(s) = (1 - p + ps)^\kappa$, $\quad i + j = M = N$ then

$P(r \mid i,j; M) = \Pr\{r\ a\text{-types and } M - r\ A\text{-types} \mid \text{total of } M \text{ offsprings}\}$

$$(53) \qquad = \binom{i\kappa}{r}\binom{j\kappa}{M-r}\Big/\binom{(i+j)\kappa}{M} \qquad r = 0, 1, \cdots, M$$

where the initial population composition consisted of i a-types and j A-types. Notice that this formula is independent of p.

B. Suppose the a and A types reproduce according to negative binomial distributions with parameter p; specifically,

$$f(s) = (1-p)^{\alpha}/(1-ps)^{\alpha}, \qquad g(t) = (1-p)^{\beta}/(1-pt)^{\beta}.$$

Then (48) becomes

$$(54) \qquad P(r\,|\,i,j;M) = \frac{\binom{i\alpha+r-1}{r}\binom{j\beta+M-r-1}{M-r}}{\binom{i\alpha+j\beta+M-1}{M}}.$$

Example A with κ equal to 2 occurs as a model proposed by Kimura describing polysomic inheritance. The state variable of this process is the number of mutant subunits. Specifically each chromosome consists of N subunits and suppose a mutation has occurred in one of them. The subunits duplicate to produce $2N$ which divide at random into two daughter chromosomes of N subunits. A single line of descent is observed. The state in each generation designates the number of mutant subunits contained in the cell. The transition probabilities are given by

$$P_{nm} = \binom{2n}{m}\binom{2N-2n}{N-m}\Big/\binom{2N}{N}$$

and this is the same as (53) for $n = i$, $m = r$, $i + j = N = M$ and $\kappa = 2$.

Multi-type models. The formulation of the preceding theory for the case of any number of types is direct. We consider a multi-type branching process with p types of individuals labeled respectively, $A_1, A_2, \cdots, A_p$. Suppose each individual of type A_k in one generation yields progeny of all types whose generating function is given by

$$(55) \qquad f_k(s_1, s_2, \cdots, s_p) \qquad k = 1, 2, \cdots, p.$$

Individuals are assumed to act independently. Let $[X_1(n), X_2(n), \cdots, X_p(n)]$ denote the associated branching process where $X_k(n)$ represents the number of A_k-type at the start of the nth generation.

The probability generating function of the progeny after one generation is

$$f_1^{i_1}(s_1, \cdots, s_p)\, f_2^{i_2}(s_1, \cdots, s_p) \cdots f_p^{i_p}(s_1, \cdots, s_p)\, h(s_1, \cdots, s_p) \tag{56}$$

where $X_k(0) = i_k$ $(k = 1, 2, \cdots, p)$ and $h(s_1, \cdots, s_p)$ denotes the generating function of the various types immigrating into the system.

As in the case of two types, the Markov chain arising by fixing the population size has an interpretation as a frequency model, described as follows. The state space consists of all p-tuples of nonnegative integers $\bar{k} = (k_1, k_2, \cdots, k_p)$ obeying the constraint $\sum_{\nu=1}^{p} k_\nu = N$. The transition probability matrix is constructed from the branching process as follows: let $\bar{k} = (k_1, k_2, \cdots, k_p)$, $\bar{l} = (l_1, l_2, \cdots, l_p)$. Then

$$\begin{aligned} P_{\bar{k},\bar{l}} = \Pr \Big\{ & X_1(1) = l_1,\ X_2(1) = l_2, \cdots, X_p(1) = l_p \,|\, X_\nu(0) \\ & = k_\nu \ (\nu = 1, \cdots, p),\ \sum_{\nu=1}^{p} X_\nu(0) = N,\ \sum_{\nu=1}^{p} X_\nu(1) = N \Big\} \\ = & \frac{\text{coefficient } s_1^{l_1} s_2^{l_2} \cdots s_p^{l_p} \text{ in } f_1^{k_1}(\bar{s}) f_2^{k_2}(\bar{s}) \cdots f_p^{k_p}(\bar{s})\, h(\bar{s})}{\text{coefficient } t^N \text{ in } f_1^{k_1}(\bar{t}) f_2^{k_2}(\bar{t}) \cdots f_p^{k_p}(\bar{t})\, h(\bar{t})} \end{aligned} \tag{57}$$

where $\bar{s} = (s_1, s_2, \cdots, s_p)$ and $\bar{t} = (t, t, \cdots, t)$.

In the special case

$$\begin{aligned} f_i(\bar{s}) &= \exp\left\{ \lambda_i \left(\sum_{\nu=1}^{p} \alpha_{i\nu} s_\nu - 1 \right) \right\}, \qquad i = 1, 2, \cdots, p, \\ h(\bar{s}) &= \exp\left\{ \sum_{\nu=1}^{p} c_\nu (s_\nu - 1) \right\}, \end{aligned} \tag{58}$$

$$\begin{aligned} \alpha_{i\nu} \geqq 0, \quad \sum_{\nu=1}^{p} \alpha_{i\nu} = 1, \quad \lambda_i > 0 \ (\nu, i = 1, \cdots, p), \\ c_\nu > 0 \ (\nu = 1, \cdots, p), \end{aligned} \tag{59}$$

the transition probability matrix (57) reduces to

$$P_{\bar{i},\bar{k}} = \frac{\binom{N}{l_1, l_2, \cdots, l_p} \prod_{\nu=1}^{p} \left[\sum_{i=1}^{p} k_i \lambda_i \alpha_{i\nu} + c_\nu \right]^{l_\nu}}{\left[\sum_{i,\nu} k_i \lambda_i \alpha_{i\nu} + c_\nu \right]^{N}} \tag{60}$$

where

$$\binom{N}{l_1, l_2, \cdots, l_p} = \frac{N!}{l_1!\, l_2! \cdots l_p!}.$$

The parameters occurring in (60) are to be interpreted as follows:

$\alpha_{i\nu}$ represents the chance that an A_i-type individual after birth will mutate into an A_ν-type individual, λ_i represents the relative selection (= fitness) coefficient of an A_i-type, c_i represents the average rate at which A_i-type individuals are immigrating into the population.

The probability matrix (60) is plainly the multi-type version of Wright's gene frequency Markov chain stochastic model allowing for mutation, migration and selection.

Eigenvalues of Markov chains of frequency models. The eigenvalues and eigenvectors of the transition matrices of the frequency models mentioned above play a significant role in describing the limiting behavior of the associated Markov chains.

To underscore the essential ideas we list the simplest class of examples. Consider the model of two types (labeled A and a). Let the probability generating function of the offspring population be $f(s)$, the same for A and a individuals.

The associated Markov chain has a transition probability matrix $P = \| P_{ik} \|$ where

$$P_{ik} = \frac{\text{coefficient } t^k s^N \text{ of } f^i(ts) f^{N-i}(s)}{\text{coefficient } s^N \text{ of } f^N(s)}, \qquad i, k = 0, 1, \cdots, N. \tag{61}$$

THEOREM 7. *Let $f(s)$ be a probability generating function of a nonnegative integer valued random variable, i.e.*

$$f(s) = \sum_{r=0}^{\infty} c_r s^r, \qquad c_r \geqq 0, \quad \sum_{r=0}^{\infty} c_r = 1.$$

The eigenvalues of the matrix (61) *are*

$$\lambda_0 = 1, \ \lambda_1 = 1, \ \lambda_r = \frac{\text{coefficient } s^{N-r} \text{ of } f^{N-r}(s)\, [f'(s)]^r}{\text{coefficient } s^N \text{ of } f^N(s)}, \qquad r = 2, 3, \cdots, N. \tag{62}$$

If $c_0 \cdot c_1 \cdot c_2 > 0$ *then*

$$1 > \lambda_2 > \lambda_3 > \cdots > \lambda_N > 0. \tag{63}$$

[Actually much weaker assumptions suffice to guarantee the validity of (63).]

Moreover, the right eigenvector α_r (apart from a constant factor) corresponding to λ_r $(r = 2, 3, \cdots, N)$ is associated with a polynomial $Q_r(z)$ of degree r such that

$$\alpha_r = [Q_r(0), Q_r(1), \cdots, Q_r(N)] \qquad r = 2, \cdots, N. \tag{64}$$

Two linearly independent right eigenvectors associated with $\lambda_0 = \lambda_1 = 1$ are

$$\alpha_0 = \left(1, 1 - \frac{1}{N}, 1 - \frac{2}{N}, \cdots, 0\right), \quad \alpha_1 = \left(0, \frac{1}{N}, \frac{2}{N}, \cdots, \frac{N}{N}\right). \tag{65}$$

The left eigenvectors β_r (apart from a constant factor) are of the form

$$\beta_r = \left[\binom{N}{0}(-1)^0 R_{N-r}(0), \binom{N}{1}(-1)^1 R_{N-r}(1), \cdots, \right.$$

$$\left. \binom{N}{i}(-1)^i R_{N-r}(i), \cdots, \binom{N}{N}(-1)^N R_{N-r}(N)\right], \tag{66}$$

$$r = 2, \cdots, N$$

where $R_l(z)$ is a polynomial of degree l. Also

$$\beta_0 = (0, 0, \cdots, 0, 1), \qquad \beta_1 = (1, 0, 0, \cdots, 0). \tag{67}$$

The eigenvectors $\{\alpha_n\}_{n=0}^N$ and $\{\beta_m\}_{m=0}^N$ constitute a biorthogonal system. In terms of the biorthogonal set of eigenvectors displayed we have the representation

$$P_{ij}^t = \sum_{r=0}^N \lambda_r^t \alpha_r(i) \beta_r(j) \tag{68}$$

where $\alpha_r(i)$ is the ith component of the rth right eigenvector and $\beta_r(j)$ is the jth component of the rth left eigenvector as listed above after Theorem 7. (Here t is a nonnegative integer and α_r and β_r are biorthogonal.) It is useful to separate the two terms of (68) for $\lambda_0 = \lambda_1 = 1$ so it has the form

$$P_{ij}^t = \alpha_0(i)\beta_0(j) + \alpha_i(i)\beta_1(j) + \sum_{r=2}^N \lambda_r^t \alpha_r(i)\beta_r(j). \tag{69}$$

The sum in (69) goes to zero at the rate λ_2^t $(\lambda_2 < 1)$. Moreover, inspection of the explicit expression of $\beta_0(j)$ and $\beta_1(j)$ [cf. (67)] reveals that P_{ij}^t is precisely the sum term when $0 < j < N$. It follows that

$$\lim_{t\to\infty} P_{ij}^t = \alpha_0(i)\beta_0(j) + \alpha_1(i)\beta_1(j)$$

$$(70) \qquad \begin{aligned} &= 0 \quad \text{if } j \neq 0, N, \\ &= (N-i)/N \quad \text{if } j = 0, \\ &= i/N \quad \text{if } j = N, \end{aligned}$$

and the rate of convergence is geometric of order λ_2, i.e. λ_2 is the "rate of approach to homozygosity". Equivalently, the probability that the system is not in a homozygous state (0 or N) behaves like λ_2^t as $t \to \infty$. Furthermore, since $\lambda_3 < \lambda_2$ we see that

$$(71) \qquad \lim_{t\to\infty} P_{ij}^t / \lambda_2^t = \alpha_2(i)\beta_2(j), \quad i, j \neq 0, N.$$

The expression (71) can be interpreted to the effect that the limiting probability of being in state j, given $j \neq 0, N$ is

$$(72) \qquad \beta_2(j) \Big/ \sum_{j=1}^{N-1} \beta_2(j).$$

More generally, for p-types we consider the Markov chain with transition probability matrix

$$(73) \qquad P_{\bar{i},\bar{k}} = \frac{\text{coefficient of } t_1^{k_1} t_2^{k_2} \cdots t_p^{k_p} \text{ in } \prod_{\nu=1}^{p} f^{i_\nu}(t_\nu)}{\text{coefficient of } t^N \text{ in } f^N(t)}.$$

The state space consists of the integral points of the simplex

$$\Delta_p = \left\{ \bar{i} = (i_1, \cdots, i_p) \mid i_\nu \text{ integers} \geqq 0, \ \sum_{\nu=1}^{p} i_\nu = N \right\}.$$

The eigenvalues $\{\lambda_r\}$ where $\lambda_r > \lambda_{r+1}$ for $r \geqq 1$ have a probabilistic interpretation according to the following theorem.

THEOREM 8. *The rate of absorption (fixation in a single pure type) is λ_2, i.e. if $\bar{i}$ and $\bar{j}$ are not vertices, then*

$$P_{\bar{i}\bar{j}}^t \sim C_{\bar{i}\bar{j}} \lambda_2^t \quad \textit{as } t \to \infty$$

where $C_{\bar{i}\bar{j}}$ is a constant depending on $\bar{i}$ and $\bar{j}$ but not on t.

(ii) *The rate at which the population loses all but k (unspecified) types ($k \leqq p$) is λ_k. Equivalently the probability that the population at the ith generation includes at least k types is $\sim C_{\bar{i}} \lambda_k^t$ (C_i is a constant* *depending on the initial state but not on t). In particular the probability that the population contains all types at the ith generation decreases to zero at the rate λ_p^t.*

For an extensive discussion of the probabilistic interpretation of eigenvalues and eigenvectors in the more general models we refer the reader to Karlin and McGregor [**16**] and [**17**].

We close this section by mentioning a class of limiting processes associated with frequency models of the above type. In particular we consider the transition probability matrix defined by

(74)

$$P_{ij} = \frac{\text{coefficient of } s^j\omega^{N-j} \text{ in } f^i((1-\alpha_2)s + \alpha_2\omega)\, f^{N-i}(\alpha_1 s + (1-\alpha_1)\omega)}{\text{coefficient of } t^N \text{ in } f^N(t)},$$

where $i, j = 0, 1, 2, \cdots, N$. Here α_1 and α_2 represent mutation probabilities in the previously described sense.

THEOREM 9. *Let $X(t; N)$ denote a sequence of Markov processes with associated transition probability matrix (74). Assume that $f'(1) = 1$ and $f'''(1) < \infty$. (The condition $f'(1) = 1$ involves no essential loss of generality and amounts to merely choosing appropriate units for measuring population size.)*

(i) *If $\alpha_1 = \gamma_1/N$, $\alpha_2 = \gamma_2/N$ $(\gamma_1, \gamma_2 > 0)$, then $Y_N(t) = X([Nt]; N)/N$ ($[h]$ denotes the greatest integer in h) converges (in the sense of weak convergence of stochastic processes) to the diffusion process $Z(t)$ on the state space $0 \leqq \xi \leqq 1$ whose associated backward differential equation has the form*

$$\frac{\partial u}{\partial t} = \sigma\frac{\xi(1-\xi)}{2}\frac{\partial^2 u}{\partial \xi^2} + [\gamma_1 - (\gamma_1+\gamma_2)\xi]\frac{\partial u}{\partial \xi}, \qquad 0 \leqq \xi \leqq 1,$$

where $\sigma = f''(1)$. The boundaries 0 and 1 act as reflecting barriers.

(ii) *If $\alpha_1 = \gamma_1/N$, $\alpha_2 = \gamma_2/N$ $(\gamma_1, \gamma_2 > 0)$, then $Y_N^{(d)}(t) = X([N^d t]; N)/N^d$ (d is a parameter, $0 < d < 1$) converges to the diffusion process $Z^{(d)}(t)$ on the state space $(0, \infty)$ whose associated backward differential equation has the form*

$$\frac{\partial u}{\partial t} = \sigma\frac{\xi}{2}\frac{\partial^2 u}{\partial \xi^2} + \gamma_1\frac{\partial u}{\partial \xi}, \qquad 0 < \xi < \infty.$$

The boundaries 0 and ∞ are natural boundaries if $\gamma_1 \geqq 1$, and 0 is a reflecting barrier if $0 < \gamma_1 \leqq 1$.

(iii) *If $\alpha_1 = \gamma_1/N$, $\alpha_2 = \gamma_2/N^d$, then $Y_N^{(d)}(t) = X([N^d t]; N)/N^d$*

$(0 < d < 1)$ converges to the diffusion process $\tilde{Z}^{(d)}(t)$ whose backward differential equation has the form

$$\frac{\partial u}{\partial t} = \sigma \frac{\xi}{2} \frac{\partial^2 u}{\partial \xi^2} + (\gamma_1 - \gamma_2 \xi) \frac{\partial u}{\partial \xi} .$$

The boundary 0 acts as a reflecting barrier.

The first limit process is concerned with fluctuations of bona fide frequencies of the a-type among the population of both types. A unit of time for the limit process corresponds to N renewals of the population in the approximating process. The second and third limit process refer to the case where the times of observations are more frequent (every N^d generations). Moreover, now the relative frequency of a is approximately N^d/N, which is essentially zero. In this process fluctuations of the number of a individuals of order N^d are observed.

The difference between (ii) and (iii) is the magnitude of the mutation rate α_2.

The limit processes exhibited in Theorem 9 are all of diffusion type. The next theorem describes a different sort of limit process valid as $N \to \infty$.

THEOREM 10. *Let the conditions of Theorem 9 be satisfied. Let $P_{ij}(N)$ denote the transition probability matrix of $X(t; N)$.*

If $\alpha_1 = \gamma_1/N$, $\alpha_2(N) \to \alpha_2^*$, $0 \leqq \alpha_2^* < 1$, then

$$\lim_{N \to \infty} P_{ij}(N) = P_{ij}^* \quad \text{for all } i, j = 0, 1, \cdots$$

and

$$\sum_{j=0}^{\infty} P_{ij}^* s^j = f^i((1 - \alpha_2^*)s + \alpha_2^*) e^{\gamma_1 (s-1)};$$

that is, the limit process $X^*(t)$ of $X(t; N)$ is a branching process with Poisson immigration. We interpret $X^*(t)$ as the number of individuals of the rare mutant a-type in an effectively infinite population of A-individuals.

The results of the above theorem are typical of a wide variety of limit processes connected with the Markov chains of the transition probability matrices (57).

The theorems announced above testify to the rich structure the class of induced Markov chains (57) possesses.

References

1. K. Athreya, *Limit theorems for multitype continuous time Markov branching processes and some classical urn schemes,* Ph. D. Thesis, Stanford University, 1967.

2. K. Athreya and S. Karlin, *Embedding of classical urn schemes in continuous time Markov branching processes and some applications,* to appear in Ann. Math. Statist.

3. R. A. Fisher, *On the dominance ratio,* Proc. Roy. Soc. Edinburgh **42** (1922), 321-341.

4. ———, *The genetical theory of natural selection,* Dover, New York, 1958.

5. H. Fujita, *On the blowing up of solutions of the Cauchy problem for* $u_t = \Delta u + u^{1+\alpha}$, J. Faculty Sci., Univ. of Tokyo, Ser. 1, **13,** Part 2 (1966), 109-124.

6. F. A. Galton and H. W. Watson, *On the probability of extinction of families,* J. Anthropol. Inst. Great Britain and Ireland **4** (1874), 138-144.

7. J. B. S. Haldane, *A mathematical theory of natural and artificial selection. Part* V: *Selection and mutation,* Proc. Cambridge Philos. Soc. **23** (1927), 838-844.

8. T. E. Harris, *The theory of branching processes,* Springer-Verlag, Berlin, 1963.

9. N. Ikeda, M. Nagasawa and S. Watanabe, *Branching Markov processes,* (to appear).

10. S. Karlin, *A first course in stochastic processes,* Academic Press, New York, 1966.

11. ———, *Central limit theorems for certain infinite urn schemes,* J. Math. Mech. **17** (1967), 373-402.

12. S. Karlin and J. McGregor, *The number of mutant forms maintained in a population,* Proc. 5th Berkeley Sympos. Prob. and Stat. (1967), 415-438.

13. ———, *The differential equations of birth and death processes and the Stieltjes moment problem,* Trans. Amer. Math. Soc. **85** (1957), 489-546.

14. ——— *The classification of birth and death processes,* Trans. Amer. Math. Soc. **86** (1957), 366-400.

15. ———, *Linear growth birth and death processes,* J. Math. Mech. **4** (1958), 643-662.

16. ———, *Direct product branching processes and related Markoff chains,* Proc. Nat. Acad. Sci. **51** (1964), 598-602.

17. ———, "Direct product branching processes and related induced Markoff chains, I. Calculation of rates of approach to homozygosity" in *Bernoulli, Bayes, Laplace Anniversary Volume,* Springer-Verlag, Berlin, 1965, pp. 111-145.

18. M. Nagasawa and T. Sirao, *Probabilistic treatment of blowing up of solutions for a non-linear integral equation* (to appear).

19. B. Singer, *Some limit theorems for Markov chains and related occupancy problems,* Ph.D. Thesis, Stanford University, 1967.

20. F. Spitzer, *Random walks,* Van Nostrand, New York, 1965.

21. S. Watanabe, *On a limit theorem for branching Brownian motion,* (to appear).

22. S. Wright, *Evolution in Mendelian populations,* Genetics **16** (1931), 97.

DEPARTMENT OF MATHEMATICS
STANFORD UNIVERSITY
STANFORD, CALIFORNIA

Donald L. Iglehart

Diffusion Approximations in Applied Probability[1]

1. **Introduction and summary.** For many problems in applied probability it is difficult to obtain explicit expressions for the distributions of random quantities of interest. In some problems however, approximations to these distributions can be obtained from limit theorems as a particular physical parameter approaches a limit. These approximations are similar in spirit to the normal approximation, for sums of independent random variables, resulting from the central limit theorem. Our purpose in this expository paper is to sketch the general approach to these limit theorems and approximations and to mention a number of techniques which have proved to be useful in various probabilistic models.

The starting point for us is a given sequence of stochastic processes $\{X_n(k) : k = 0, 1, \cdots\}$ for $n = 1, 2, \cdots$ with each process defined on its own probability space $(\Omega_n, \mathscr{F}_n, P_n)$. We have indicated a discrete time parameter for these processes, however, this is not crucial and, in fact, we shall later consider processes with a continuous time parameter. The state space of our processes can be either discrete or continuous and either real-valued or vector-valued. The index n for the sequence is meant to correspond to the physical parameter mentioned above. In some models, however, the sequence index will not be associated with the positive

[1] This work was supported by the Office of Naval Research, Contract Nonr-401 (55) at Cornell University.

integers. With this set-up given, our aim is to scale the time parameter and to translate and scale the space variable in such a manner that the resulting processes converge to a limit process as n goes to infinity. In general we shall seek sequences $\{a_n\}$, $\{b_n\}$, and $\{c_n\}$ to form the processes[2]

$$Y_n(t) = \frac{X_n([a_n t]) - b_n}{c_n}, \qquad t \geqq 0, \quad n = 1, 2, \cdots.$$

The a_n's and c_n's will be positive real numbers, generally tending to infinity as $n \to \infty$. However, the b_n's will be vectors if the X_n's are. Notice that the $Y_n(\cdot)$ processes will be constant for stretches of length a_n^{-1} because of the discrete nature of the X_n's and hence have discontinuous paths. There is an alternative way to define the $Y_n(\cdot)$ processes which leads to continuous path functions. This latter approach has certain advantages and will be introduced later.

There are a number of modes of convergence for the $Y_n(\cdot)$ processes which are of interest. The simplest of these is to require convergence in distribution of the value of the $Y_n(\cdot)$ processes at a fixed value of t; i.e.

$$\lim_{n \to \infty} P_n\{ Y_n(t) \leqq x\} = P\{ Y(t) \leqq x\} \quad \text{for all } t \geqq 0$$

and all x for which the right-hand side is continuous, where $Y(\cdot)$ is a limit process defined on a probability space $(\Omega, \mathscr{F}, P)$. Often the limit process, $Y(\cdot)$, is a diffusion; i.e. a strong Markov process with continuous path functions. The classical central limit theorem is of this type, where $Y(\cdot)$ is Brownian motion. Next we might be interested in showing the convergence of the finite-dimensional distributions (f.d.d.) of the $Y_n(\cdot)$ processes to the corresponding distributions of the $Y(\cdot)$ process. In other words, for every $k > 1$ and $0 \leqq t_1 < t_2 < \cdots < t_k$ we would like to show that

$$\lim_{n \to \infty} P_n\{ Y_n(t_1) \leqq x_1, \cdots, Y_n(t_k) \leqq x_k\}$$
$$= P\{ Y(t_1) \leqq x_1, \cdots, Y(t_k) \leqq x_k\},$$

for all x_j for which the right-hand side is continuous.

A third mode of convergence which is important for applied problems is weak convergence of a sequence of probability measures defined as follows. Let S be a metric space and $\mathscr{S}$ be the smallest

[2] The symbol $[x]$ denotes the integer part of x.

Borel field containing the open sets of S. If ν_n and ν are probability measures on $\mathscr{S}$ and if the

$$\lim_{n\to\infty} \int_S f\,d\nu_n = \int_S f\,d\nu$$

for every bounded, continuous, real-valued function f on S, then we say ν_n converges weakly to ν and write $\nu_n \Rightarrow \nu$.

For our problems the measures ν_n will be generated by the $Y_n(\cdot)$ processes and the measure ν by $Y(\cdot)$. In most cases the metric space S will be $C[0,1]$, the space of continuous functions on $[0,1]$ with the metric of uniform convergence, or a multi-dimensional analog. Weak convergence is important because it implies convergence in distribution of certain functionals of the original processes. Sometimes in applications the distribution of the functional is more important than that of the original process.

Once these limit theorems have been obtained we can consider using the limit distributions as an approximation to the distribution of $X_n(\cdot)$ when the parameter n is sufficiently "large." Of course, the question of how large n must be before a "good" approximation is obtained introduces the question of rates of convergence for the limit theorems. Little work of an analytical nature has been done on the rate of convergence issue; however, for some models numerical results are available.

The protype of these limit theorems and their resulting approximations is the convergence of sums of independent, identically distributed (i.i.d.) random variables to Brownian motion. These results will be sketched in §2. Models from queueing theory will be taken up in §§3, 4, and 5. §6 will be concerned with the multi-urn Ehrenfest model. A quality control model will be discussed in §7. Finally, in §8 brief mention will be made of other work on the convergence of processes in applied probability.

2. **Sums of random variables and Brownian motion.** Let $X_1, X_2, \cdots$ be a sequence of i.i.d. random variables defined on the product probability space $(\Omega, \mathscr{F}, P)$ and having mean 0 and variance 1. We shall denote the partial sums by $S_i = X_1 + \cdots + X_i$ and set $S_0 = 0$. The appropriate sequence of processes to consider is defined by

$$Y_n(t) = S_{[nt]}/\sqrt{n}, \qquad 0 \leqq t \leqq 1, \quad n = 1, 2, \cdots.$$

The fact that we have restricted the time parameter t to the unit interval is not important, but does make the exposition easier.

Observe that for a fixed value of t the number of jumps of the $Y_n(\cdot)$ process is of order n, the mean value of each jump is 0, and the variance of each jump is n^{-1}. Thus the jumps are occurring very often and their heights are becoming small as n grows large. Hence it is natural to hope that as $n \to \infty$ the paths of $Y_n(\cdot)$ will become continuous.

Alternatively, we could consider a sequence of processes defined by

$$Z_n(t) = S_k/\sqrt{n} + (nt - k)\, X_{k+1}/\sqrt{n}, \qquad kn^{-1} \leqq t \leqq (k+1)n^{-1}$$

for $k = 0, 1, \cdots, n-1$. Notice that $Z_n(t) = Y_n(t)$ for t of the form kn^{-1}, $k = 0, \cdots, n$, and is defined by linear interpolation for other values of t. Hence the paths of $Z_n(\cdot)$ are automatically continuous and this is convenient when considering weak convergence of probability measures.

We shall be interested in showing that the sequences $\{Y_n(\cdot)\}$ and $\{Z_n(\cdot)\}$ converge, as discussed in §1, to Brownian motion, $Y(\cdot)$, on the unit interval. As a check to see that we are on the right track, we calculate the infinitesimal mean and variance of $Y_n(t)$ per unit time. Denoting these means and variances by $m_n(y)$ and $\sigma_n^2(y)$, we have

$$\begin{aligned} m_n(y) &= nE\{Y_n(t + 1/n) - Y_n(t) \mid Y_n(t) = y\} \\ &= nE\, X_{[nt]+1}/\sqrt{n} = 0 \end{aligned}$$

and

$$\begin{aligned} \sigma_n^2(y) &= nE\{[Y_n(t + 1/n) - Y_n(t)]^2 \mid Y_n(t) = y\} \\ &= E\{X_{[nt]+1}^2\} = 1. \end{aligned}$$

This infinitesimal mean and variance agrees with those of Brownian motion which provides a useful check before proceeding to a rigorous analysis.

The proof of the convergence of the one-dimension distributions is simply the central limit theorem; i.e.

$$\lim_{n \to \infty} P\{Y_n(t) \leqq x\} = (2\pi t)^{-1/2} \int_{-\infty}^{x} \exp\left\{-\frac{s^2}{2t}\right\} ds.$$

The normal distribution on the right-hand side is also the distribution of the position of Brownian motion at time t. To show convergence of the f.d.d. the Lévy continuity theorem for characteristic functions can be used. The case $k = 2$ embodies the general argument which goes as follows. Let $\phi_n(s_1, s_2; t_1, t_2)$ be the joint characteristic function of $Y_n(t_1)$ and $Y_n(t_2)$. Then

$$\begin{aligned}\phi_n(s_1, s_2; t_1, t_2) &= E\left[\exp\left\{\frac{i(s_1+s_2)}{\sqrt{n}} S_{[nt_1]} + \frac{is_2}{\sqrt{n}}(S_{[nt_2]} - S_{[nt_1]})\right\}\right] \\ &= E\left[\exp\left\{\frac{i(s_1+s_2)}{\sqrt{n}} S_{[nt_1]}\right\}\right] \\ &\quad \cdot E\left[\exp\left\{\frac{is_2}{\sqrt{n}} S_{[nt_2]-[nt_1]}\right\}\right].\end{aligned}$$

Letting $n \to \infty$ we see that

$$\phi_n(s_1, s_2; t_1, t_2) \to \exp\{-(s_1+s_2)^2 t/2 - s_2^2(t_2 - t_1)/2\},$$

which is the joint characteristic function of Brownian motion at times t_1 and t_2. Similar arguments can be used to show the convergence of the f.d.d. of the $Z_n(\cdot)$ processes.

To establish weak convergence of the measures associated with $\{Y_n(\cdot)\}$ or $\{Z_n(\cdot)\}$ two steps are required. The first step is the convergence of the f.d.d. which we have demonstrated above. Secondly, the probability that the approximating processes can have large fluctuation between points at which they are determined by their f.d.d. must be shown to be small. The notation of weak convergence is intimately related to the so-called invariance principles. An invariance principle for sums of i.i.d random variables states roughly that the limit of the distribution of various functionals of the S_i's is independent of the common distribution of the X_i's, provided the mean is 0 and variance 1. Such an invariance principle was first given by Erdos and Kac [**7**] and later generalized by Donsker [**5**] and Billingsley [**2**]. To carry out the second step mentioned above for $\{Y_n(\cdot)\}$ (respectively, $\{Z_n(\cdot)\}$) the reader should consult Donsker [**5**] (Billingsley [**2**]). Other important references for weak convergence of sums of i.i.d. random variables are Prokhorov [**28**], Skorokhod [**33**], and Itô-McKean [**13**].

It is important to note that while the approximating processes could all be defined on a single probability space the limiting Brownian motion was defined on another probability space. Thus with this set-up it makes no sense to speak of convergence with probability one or in quadratic mean, for example. However, when the X_i's are Bernoulli random variables, Knight [**21**] has succeeded in defining the approximating processes and the limit process on a single probability space and then showing probability one convergence.

3. **Waiting time for the queue** $GI/G/1$. In this section we shall consider the distribution of the waiting time in the single-server queue with general independent input and general service time. Our objective is to demonstrate the usefulness of the notion of weak convergence in studying the asymptotic behavior of the waiting time as the traffic intensity ρ goes to 1. This discussion is based on work of Prokhorov [**29**], [**30**].

Let W_n be the waiting time (time before service begins) for the nth customer. We shall denote the service time of the nth customer by v_n and the interval between the arrival times of the nth and $(n+1)$th customers by u_n. Then if $W_1 = 0$,

$$(1) \quad F_{n+1}(x) = \Pr\{W_{n+1} \leqq x\} = \Pr\{S_1 \leqq x, S_2 \leqq x, \cdots, S_n \leqq x\},$$

where $X_n = v_n - u_n$ and $S_i = \sum_{j=1}^{i} X_j$. This result was first derived by Lindley (1952); see Prabhu [**28**] for a comprehensive discussion of the $GI/G/1$ queue. The usual independence assumptions regarding $\{v_n\}$ and $\{u_n\}$ imply that the X_i's are independent. Since the distribution of W_{n+1} is seen to be equal to the distribution of the maximum functional on the process of partial sums, S_i, it is natural to hope for limit theorems which lead to the maximum functional on Brownian motion.

Consider now a sequence of such queueing processes indexed by δ_i in which $E\{u_n^{(\delta i)}\} = 1/(1-\delta_i)$, $E\{v_n^{(\delta i)}\} = 1$, and hence $\rho_i = 1 - \delta_i$. Thus $E\{X_n^{(\delta i)}\} = -\delta_i/(1-\delta_i)$. We shall treat the case $\delta_i > 0$ (or $\rho_i < 1$) here, although for many of the arguments the sign of δ_i is not important. Our goal now is to obtain limit theorems for $W_{ni}^{(\delta i)}$ as $\delta_i \searrow 0$ and $n_i \to \infty$. There are a number of possible limit theorems depending on whether $n_i\delta_i^2$ converges to zero, a positive constant, or plus infinity.

Theorem 3.1 of Prokhorov [**29**] is the principal tool used in obtaining these limits theorems. The set-up for the theorem is as follows. A double sequence

$$X_{n,1}, X_{n,2}, \cdots, X_{n,k_n}$$

of random variables is given which are independent for each n and subject to the asymptotic negligibility condition

$$\lim_{n\to\infty} \max_{1 \leqq k \leqq k_n} \Pr\{|X_{n,k}| > \epsilon\} = 0 \quad \text{for all } \epsilon > 0.$$

The first two moments satisfy $E\{X_{n,k}\} = 0$, $\sigma_{n,k}^2 = \sigma^2[X_{n,k}] > 0$, and $\sum_{k=1}^{k_n} \sigma_{n,k}^2 = 1$. Let the partial sums be denoted by $S_{n,0} = 0$ and $S_{n,k} = \sum_{j=1}^{k} X_{n,j}$ for $1 \leqq k \leqq k_n$. We let $t_{n,k} = \sigma^2[S_{n,k}]$ and construct the continuous path function $Y_n(t)$ which is piece-wise linear with vertices at the points $(t_{n,k}, S_{n,k})$. The measure induced by $Y_n(t)$ on $C[0,1]$ we shall denote by P_n and that induced by Brownian motion by P. Then the theorem is as follows.

THEOREM. *A necessary and sufficient condition for P_n to converge weakly to P is that*

$$\lim_{n\to\infty} \sum_{k=1}^{k_n} \int_{|x|>\lambda} x^2 dF_{n,k}(x) = 0 \tag{2}$$

for all $\lambda > 0$, where $F_{n,k}(x) = \Pr\{X_{n,k} \leqq x\}$.

First consider the case where $n_i\delta_i^2 \to 0$ as $i \to \infty$. We shall assume that $\sigma^2[X_j^{(\delta i)}] = \sigma_i > 0$, $\sigma_i \to \sigma$, and that

$$\int_{|x|\geqq z} x^2 dK^{(\delta_i)}(x) \to 0 \tag{3}$$

uniformly in δ_i as $z \to \infty$, where $K^{(\delta i)}$ is the distribution function of $X_n^{(\delta i)}$. The latter condition assures us that condition (2) is satisfied for our queueing problem. From (1) we have for $0 \leqq x < \infty$

$$\begin{aligned} F_{n_i+1}^{(\delta_i)}(x\sigma_i\sqrt{n_i}) &= \Pr\{W_{n_i+1}^{(\delta_i)} \leqq x\sigma_i\sqrt{n_i}\} \\ &= \Pr\{S_k^{(\delta_i)}/\sigma_i\sqrt{n_i} \leqq x,\ 1 \leqq k \leqq n_i\}. \end{aligned} \tag{4}$$

One can easily check to see that the conditions of the above theorem hold and hence

$$\Pr\{S_k^{(\delta i)}/\sigma_i\sqrt{n_i} \leqq x + kE\{X_1^{(\delta i)}\}/\sigma_i\sqrt{n_i},\ 1 \leqq k \leqq n_i\} \tag{5}$$

converges as $i \to \infty$ to the maximum functional of Brownian motion, namely

$$P\left\{\max_{0\leqq t\leqq 1} Y(t) \leqq x\right\},$$

where $Y(t)$ is the path function of Brownian motion. This probability is known to be $2\Phi(x) - 1$, the truncated normal distribution. Since we have assumed that $n_i\delta_i^2 \to 0$, the terms

$$\frac{kE\{X_1^{(\delta_i)}\}}{\sigma_i\sqrt{n_i}} = -\frac{k\delta_i}{\sigma_i\sqrt{n_i}(1-\delta_i)} \to 0 \quad \text{as } i \to \infty, \tag{6}$$

for $k = 1, 2, \cdots, n_i$. Hence putting together (5) and (6) yields

$$F_{n_i+1}^{(\delta_i)}(x\sigma_i\sqrt{n_i}) \to 2\Phi(x) - 1.$$

This result is of course also true in the case where $\delta_i = 0$.

Next we assume $n_i\delta_i^2 \to \tau$, $0 < \tau < \infty$. Using (1) again we have

$$F_{n_i+1}\left(\frac{x}{\delta_i}\right) = \Pr\left\{\frac{S_k^{(\delta_i)} - kE[X_1^{(\delta_i)}]}{\sigma_i\sqrt{n_i}} \leqq \frac{x - (k/n_i)\cdot n_iE[X_1^{(\delta_i)}]\delta_i}{\sigma_i\sqrt{n_i}\delta_i},\ i \leqq k \leqq n_i\right\}.$$

By assumption $-n_iE[X_1^{(\delta_i)}]\delta_i \to \tau$ and $\sigma_i\sqrt{n_i}\delta_i \to \sigma\sqrt{\tau}$. Using the weak convergence again we have

$$F_{n_i+1}^{(\delta_i)}\left(\frac{x}{\delta_i}\right) \to P\left\{\max_{0\leqq t\leqq 1}\left[Y(t) - \frac{t\tau}{\sigma\sqrt{\tau}}\right] \leqq \frac{x}{\sigma\sqrt{\tau}}\right\}. \tag{7}$$

By a change in time scale the last probability can also be written as

$$P\left\{\max_{0\leqq t\leqq \tau}\left[Y(t) - \frac{t}{\sigma}\right] \leqq \frac{x}{\sigma}\right\}. \tag{8}$$

Finally, consider the case where $n_i\delta_i^2 \to \infty$. We would expect the limit of $F_{n_i+1}^{(\delta_i)}(x/\delta_i)$ to be the expression in (8) with τ replaced by ∞. This is in fact true, although an additional argument employing Kolmogorov's inequality is required. For a fixed $\delta > 0$, the stationary distribution of the Markov chain $\{W_n^{(\delta)}\}$ exists and is given by

$$F^{(\delta)}(x) = \Pr\left\{\sup_{k\geqq 1} S_k^{(\delta)} \leqq x\right\}.$$

In the course of deriving (8) with τ replaced by ∞, Prokhorov also shows that

$$F^{(\delta_i)}\left(\frac{x}{\delta_i}\right) \to P\left\{\max_{0\leqq t<\infty}\ [Y(t) - t\sigma^{-1}] \leqq \mathbf{x}\sigma^{-1}\right\} = 1 - \exp\left(-\frac{2x}{\sigma^2}\right). \tag{9}$$

The last equality was derived by Darling and Siegert [**4**]. The result given in (9) was first derived by Kingman [**19**]. For an expository account of Kingman's work in this area of so-called "heavy traffic" the reader should consult Kingman [**20**].

From this work of Prokhorov's it should be clear that the notion of weak convergence is an important one for applied probability. For other work in this spirit consult Viskov [**37**], Viskov and Prokhorov [**38**], and Borovkov [**3**].

4. **The many server queue and telephone trunking problem.** In this section we shall discuss the many server queue and telephone trunking problem (infinite server queue) with Poisson arrivals and exponential service time. As usual the service times are independent of the arrival process and no server is idle if a customer is waiting. If there are n servers and we let $X_n(t)$ denote the number of customers waiting or being served at time t, then it is well known that $X_n(t)$ is a birth and death process. As such, the transition probabilities, $p_{ij}(t) = P\{X_n(t+\tau) = j \mid X_n(\tau) = i\}$, can be expressed by the integral representation of Karlin and McGregor [**15**]. If we knew $p_{ij}(t)$ explicitly, we would in principle be able to calculate all the distributions of interest for this model. Unfortunately, it is exceedingly difficult to compute $p_{ij}(t)$ when n is larger than 3 or 4.

Motivated by this difficulty, we shall seek a diffusion approximation for $X_n(t)$ when n is large. We shall follow the discussion of the author (Iglehart [**10**]). Naturally, if we keep the expected interarrival time and the expected service time fixed, the behavior of the many server queue approaches that of the telephone trunking problem. While the transition probabilities for the telephone trunking problem are well known, we would have lost the characteristics of the original problem. In fact, the traffic intensity ρ, defined to be the ratio of the arrival rate to n times the service rate, would tend to zero whereas the processes being approximated have a positive traffic intensity. Therefore, we shall let the arrival rate become large with the number of servers, and keep the service time constant in such a manner that ρ is maintained at a fixed value less than one. To be specific, we let $X_n(t)$ be the birth and death process with parameters

$$\begin{aligned}\lambda_i^{(n)} &= n\rho,\\ \mu_i^{(n)} &= i, \quad i \leqq n,\\ &= n, \quad n > i,\end{aligned}$$

for $i = 0, 1, \cdots$ and $n = 1, 2, \cdots$, where $0 < \rho < 1$.

As a heuristic aid to setting up the appropriate approximating processes, consider the imbedded jump process of $X_n(t)$. The expected displacement in one jump starting at state $i \leqq n$ is $(n\rho - i)/(n\rho + i)$ which is nonnegative if $i \leqq n\rho$ and negative if $i > n\rho$. For $i > n$ the expected displacement is always negative. In other words, the state $[n\rho]$ is an "equilibrium point" of the process. Thus for our approximating processes it is natural to consider the fluctuations of $X_n(t)$ about $[n\rho]$, measured in an appropriate scale. We shall set

$$Y_n(t) = (X_n(t) - n\rho)/(n\rho)^{1/2}, \qquad n = 1, 2, \cdots.$$

Again as a guide to the limiting diffusion process, we shall calculate the infinitesimal mean and variance of $Y_n(t)$. The infinitesimal mean is defined in terms of the jump process $\tilde{Y}_n(k)$, say, as

$$m_n(y) = E\{\tilde{Y}_n(k+1) - \tilde{Y}_n(k) \mid \tilde{Y}_n(k) = y\}/E\{\text{holding time in } y\}.$$

In our case, if we let $\alpha_n(y) = [n\rho + (n\rho)^{1/2}y]$, then

$$m_n(y) = \frac{1}{(n\rho)^{1/2}} \frac{\lambda^{(n)}_{\alpha_n(y)} - \mu^{(n)}_{\alpha_n(y)}}{\lambda^{(n)}_{\alpha_n(y)} + \mu^{(n)}_{\alpha_n(y)}} \cdot (\lambda^{(n)}_{\alpha_n(y)} + \mu^{(n)}_{\alpha_n(y)}).$$

We now take n so large that $\alpha_n(y) < n$ (which is possible since $\rho < 1$). Thus

$$m_n(y) = (1/(n\rho)^{1/2})\{n\rho - [n\rho + (n\rho)^{1/2}y]\} \to -y \quad \text{as } n \to \infty. \tag{10}$$

The infinitesimal variance is similarly defined as

$$\sigma_n^2(y) = \frac{E\{(\tilde{Y}_n(k+1) - \tilde{Y}_n(k))^2 \mid \tilde{Y}_n(k) = y\}}{E\{\text{holding time in } y\}}. \tag{11}$$

Hence,

$$\sigma_n^2(y) = \frac{1}{(n\rho)} \frac{\lambda^{(n)}_{\alpha_n(y)} + \mu^{(n)}_{\alpha_n(y)}}{\lambda^{(n)}_{\alpha_n(y)} + \mu^{(n)}_{\alpha_n(y)}} \cdot (\lambda^{(n)}_{\alpha_n(y)} + \mu^{(n)}_{\alpha_n(y)})$$

$$= \frac{1}{n\rho}(n\rho + [n\rho + (n\rho)^{1/2}y]) \to 2 \quad \text{as } n \to \infty.$$

The limit process should then be governed by the backward equation

$$\frac{\partial u}{\partial t} = \frac{\partial^2 u}{\partial x^2} - x\frac{\partial u}{\partial x}, \qquad -\infty < x < \infty,$$

which is recognized as the equation of the Ornstein-Uhlenbeck diffusion process.

At this point we are in need of a technique for showing that the distribution of $Y_n(t)$ converges to the distribution of the Ornstein-Uhlenbeck process $Y(t)$, say. If we could explicitly obtain the representation for $p_{ij}(t)$, we might use analytic techniques involving orthogonal polynomials to achieve our result. While the parameters $\{\pi_j\}$ and the orthogonal polynomials $\{Q_i(x)\}$ can be easily obtained, the measure $\psi(x)$ is difficult to characterize. In fact, the difficulty in obtaining this representation provided our initial motivation for looking for a diffusion approximation. Fortunately, there is a general theory due to Stone [**34**], [**35**] which gives necessary and sufficient conditions for the weak convergence of a sequence of birth and death processes (or random walks, or diffusions) to a limiting diffusion process. Although the general set-up and notation required to discuss Stone's results in detail are too involved for this paper, perhaps a few heuristic remarks would be helpful.

The processes considered by Stone (random walks, birth and death processes, and diffusions) enjoy the property that points in the state space are not jumped over by the process, i.e. possible transitions can only occur to neighboring states in the discrete case and the path functions are continuous in the continuous state space case. This property results in the infinitesimal operator of the semigroup of transition functions being a local operator. The infinitesimal operator is essentially determined (aside from any boundary conditions which may have to be imposed) by the infinitesimal mean and variance of the process. Since convergence of the infinitesimal operators of a sequence of processes implies convergence of the processes, it seems natural to assume that convergence of the infinitesimal means and variances would also imply convergence of the processes. This is in fact true, except the infinitesimal operator of a semigroup is not determined until the boundary conditions (lateral conditional in Feller's [8] terminology) have been specified. Hence to obtain convergence of the processes we need to assume that the behavior of the sequence of processes in the neighborhood of a boundary point converges to that of the limit process. Finally, we need to check that the state space of the approximating processes becomes dense in the state space of the limit process.

In our example of the many server queue we have shown in (10) and (11) that the infinitesimal mean and variance converge uniformly in every compact interval of the line. Furthermore, the state space of $Y_n(t)$ becomes dense in $(-\infty, +\infty)$, the state space of the limit process, as $n \to \infty$. These are the essential facts required to apply Stone's results (cf. Iglehart [**10**]). The conclusion is that the processes $Y_n(t)$ converge weakly to $Y(t)$ provided $X_n(0) = [n\rho + (n\rho)^{1/2}y]$ for any real y.

To illustrate how the representation for $p_{ij}(t)$ can be used to obtain a limiting diffusion, we shall consider the telephone trunking problem. In this model $X_n(t)$, the number of busy channels, is a birth and death process with parameters

$$\lambda_j^{(n)} = nc, \qquad \mu_j^{(n)} = j$$

for $j = 0, 1, \cdots$ and $n = 1, 2, \cdots$, where $c > 0$. The representation (Karlin and McGregor [15]) in this case is

$$p_{ij}(t) = \pi_j \sum_{k=0}^{\infty} e^{-kt} c_i(k; nx) c_j(k; nc) \frac{e^{-nc}(nc)^k}{k!},$$

where $\{c_k(x; a)\}$ are the Poisson-Charlier polynomials, orthogonal with respect to the measure $e^{-a}a^x/x!$ on $x = 0, 1, \cdots$ and $\pi_j = (nc)^j/j!$. By showing that the Poisson-Charlier polynomials, properly normalized, converge to the Hermite polynomials (this is not surprising considering the fact that the Hermite polynomials are orthogonal with respect to the normal density and the implications of the central limit theorem), it is not hard to show that for $\alpha_n(x) = [(nc)^{1/2}x + nc]$, $p_{\alpha n(x), \alpha n(y)}(t)$ is asymptotic (when normalized) to the transition density for the Ornstein-Uhlenbeck process. This convergence provides a local limit theorem for the convergence of the processes $(X_n(t) - nc)/(nc)^{1/2}$ to the Ornstein-Uhlenbeck process.

For another example of such a local limit theorem obtained from the integral representation for $p_{ij}(t)$ in the case of the Ehrenfest urn model, the reader should consult Karlin and McGregor [17].

5. **Luchak's queueing model in heavy traffic.** Luchak [27] studied a single-server queueing model in which customers arrive according to a Poisson process with rate $\lambda > 0$ and require a random number, N, of phases of service, each phase being exponential with parameter μ. Luchak considered the transient behavior of the phase length process, $Q(t)$, which is the number of phases present in the system at time t including the phase in service. His result is given in terms of a rather unwieldy transform.

Recently Lalchandani [22] considered in his thesis the behavior of $Q(t)$ in heavy traffic (i.e., as the traffic intensity approaches one). We shall give a brief outline of his method which seems to have some general interest.

The distribution of N is discrete $(P[N = n] = c_n,\ n = 1, 2, \cdots)$ with finite mean a and second moment b. With these parameters the traffic intensity $\rho = \lambda a/\mu$. Consider now a sequence of queueing systems indexed by n $(n \geqq 1)$ for which the arrival and service parameters are λ_n and μ_n. For the corresponding traffic intensity

ρ_n to tend to 1, we choose, as an example, $\lambda_n = n$ and $\mu_n = a(n + \sqrt{n})$. If $Q_n(t)$ denotes the phase length process for the nth system, Lalchandani shows that the distribution of $X_n(t)$, defined as

$$X_n(t) = (Q_n(t) - n)/[(a+b)n]^{1/2},$$

converges as $n \to \infty$ to the distribution of Brownian motion at time t with initial state y, provided $Q_n(0) = [((a+b)n)^{1/2} \cdot y + n]$.

While the $Q_n(t)$ process is a continuous parameter Markov chain, it is not a birth and death process and hence we do not have available the special techniques for such processes. Consider, however, the discrete parameter jump chain $\tilde{Q}_n(k)$, corresponding to $Q_n(t)$. In one step the chain $\tilde{Q}_n(k)$ goes up j with probability $c_j \lambda_n/(\lambda_n + \mu_n)$ and down 1 with probability $\mu_n/(\lambda_n + \mu_n)$, provided $\tilde{Q}_n(k) > 0$. If $\tilde{Q}_n(k) = 0$, then the only transitions are up j with probability c_j. Except for the troublesome state 0, the $\tilde{Q}_n(k)$ process is a process of sums of independent random variables. However, when $\rho_n \nearrow 1$, we would not expect the queue to be idle often, and thus it is not too suprising that the limit process for $\tilde{X}_n([(\lambda_n + \mu_n)t])$ is Brownian motion, where $\tilde{X}_n(k) = (\tilde{Q}_n(k) - n)/[(a+b)n]^{1/2}$. One of the crucial steps in showing this convergence in distribution is the study of the behavior of $P\{\tilde{Q}_n(j) = 0\}$.

Once the convergence of the suitably translated and scaled jump chain is obtained it is not difficult to show that the $X_n(t)$ process converges. If we consider the time points at which jumps in the $Q_n(t)$ process occur, it is clear that these time points are essentially a renewal process with exponential lifetimes having parameter $(\lambda_n + \mu_n)$. Very occasionally, however, the queue is empty and the exponential parameter is λ_n. Define $N_n(t)$ to be the number of jumps of the $Q_n(\cdot)$ process in the interval $[0, t]$. Then the mean and variance of $N_n(t)$ are shown to behave like $(\lambda_n + \mu_n)t$ as $n \to \infty$. Now a conditional probability argument can be used in which the $P\{X_n(t) \leqq x \mid N_n(t) = j\}$ is identified with the $P\{\tilde{X}_n(j) \leqq x\}$. Finally, in this manner Lalchandani shows that the distribution of $X_n(t)$ converges to the appropriate normal distribution.

While the argument outlined above used the specific structure of the Markov chain $Q_n(t)$, it seems likely that the idea of looking first for limit processes of the jump chain should have greater applicability.

6. **The multi-urn Ehrenfest model.** In the multi-urn Ehrenfest model, N balls are distributed among $d+1$ $(d \geqq 2)$ urns. If we label the urns $0, 1, \cdots, d$, then the system is said to be in state $\mathbf{i} = (i_1, i_2, \cdots, i_d)$ when there are i_j balls in urn j $(j = 1, 2, \cdots, d)$

and[3] $N - \mathbf{1}\cdot\mathbf{i}$ balls in urn 0. At discrete epochs a ball is chosen at random from one of the $d+1$ urns; each of the N balls has probability $1/N$ of being selected. The ball chosen is removed from its urn and placed in urn i $(i = 0, 1, \cdots, d)$ with probability p^i, where the p^i's are elements of a given vector $(p^0, \mathbf{p})$, satisfying $p^i > 0$ and $\sum_{i=0}^{d} p^i = 1$. We shall let $\mathbf{X}_N(k)$ denote the state of the system after the kth such rearrangement of balls. In this section we shall discuss some limit theorems which were obtained by the author (Iglehart [**11**]) for the sequence of processes $\{\mathbf{X}_N(k) : k = 0, \cdots, N\}$ as N tends to infinity. For the classical Ehrenfest model ($d = 1$, $p^0 = p^1 = 1/2$) Kac [**14**] showed that the distribution of $(X_N([Nt]) - N/2)/(N/2)^{1/2}$ converges as $N \to \infty$ to the distribution of the Ornstein-Uhlenbeck process at time t having started at y_0 at $t = 0$, provided $X_N(0) = [(N/2)^{1/2} y_0 + N/2]$. Recently, Karlin and McGregor [**17**] obtained a similar result for the continuous time version of the model with $d = 2$; in this version the random selection of balls is done at the occurrence of events of an independent Poisson process.

A preliminary calculation indicates that the process $\{\mathbf{X}_N(k) : k = 0, \cdots, N\}$ is attracted to the pseudo-equilibrium state $N\mathbf{p}$ and that states far from $N\mathbf{p}$ will only occur rarely. Thus it is natural to consider the fluctuations of $\mathbf{X}_N(k)$ about $N\mathbf{p}$ measured in an appropriate scale. For our purposes the appropriate processes to consider are $\{\mathbf{Y}_N(k) : k = 0, \cdots, N\}$, where

$$\mathbf{Y}_N(k) = (\mathbf{X}_N(k) - N\mathbf{p})/N^{1/2}.$$

Next we define a sequence of stochastic processes $\{\mathbf{y}_N(t) : 0 \leqq t \leqq 1\}$ which are continuous, linear on the intervals $((k-1)N^{-1}, kN^{-1})$, and satisfy $\mathbf{y}_N(kN^{-1}) = \mathbf{y}_N(k)$ for $k = 0, 1, \cdots, N$. In other words, we let

$$\mathbf{y}_N(t) = \mathbf{Y}_N(k) + (Nt - k)(\mathbf{Y}_N(k+1) - \mathbf{Y}_N(k))$$

if $kN^{-1} \leqq t \leqq (k+1)N^{-1}$. Throughout this discussion we shall let[4] $\mathbf{X}_N^i(0) = [N^{1/2} y_0^i + Np^i]$, where $\mathbf{y}_0 = (y_0^1, \cdots y_0^d)$ is an arbitrary, but fixed, element of[5] R^d. With this initial condition and the Markov

[3] The vector $\mathbf{1}$ has all its components equal to 1 and $\mathbf{x}\cdot\mathbf{y}$ is the usual scalar product.

[4] It will always be understood that N is sufficiently large so that $0 \leqq X_N^i(0) \leqq N$ for all $i = 1, 2, \cdots, d$, where $X_N^i(\cdot)$ is the ith component of the vector $\mathbf{X}_N(\cdot)$.

[5] R^d is d-dimensional Euclidean space.

structure of the model, the processes $\{\mathbf{X}_N(k) : k = 0, \cdots, N\}$ for $N = 1, 2, \cdots$ can be defined on a probability triple $(\Omega_N, \mathscr{F}_N, P_N)$. We shall let $C_d[0,1]$ denote the product space of d copies of $C[0,1]$, the space of continuous functions on $[0,1]$ with the topology of uniform convergence, and endow $C_d[0,1]$ with the product topology. The topological Borel field of $C_d[0,1]$ will be denoted by $\mathscr{C}_d$. Clearly, the transformation taking the sequence $\{\mathbf{X}_N(k) : k = 0, \cdots, N\}$ into $\{\mathbf{y}_N(t) : 0 \leqq t \leqq 1\}$ is measurable and induces a probability measure on $\mathscr{C}_d$. We shall denote this induced measure by $\mu_N(\cdot\,; \mathbf{y}_0)$.

The principal result of Iglehart [11] is that $\mu_N(\cdot\,; \mathbf{y}_0) \Rightarrow \mu(\cdot\,; \mathbf{y}_0)$ as $N \to \infty$, where $\mu(\cdot\,; y_0)$ is the probability measure on $\mathscr{C}_d$ of a d-dimensional diffusion process, $\mathbf{y}(\cdot)$, starting at the point $\mathbf{y}_0$. The limit process $\mathbf{y}(\cdot)$ is a d-dimensional analog of the Ornstein-Uhlenbeck process whose distribution at time t is a multivariate normal with mean vector $e^{-t}\mathbf{y}_0$ and covariance matrix $\mathbf{\Sigma}$, where the elements of $\mathbf{\Sigma}$ are

$$\begin{aligned} \sigma_{ij} &= (1 - e^{-2t}) p^i (1 - p^i), \qquad i = j, \\ &= -(1 - e^{-2t}) p^i p^j, \qquad i \neq j. \end{aligned}$$

To obtain the weak convergence of the measures μ_N to μ we must first show the convergence of the corresponding f.d.d. We shall only be able to sketch the proof here. For the convergence of the distribution of $\mathbf{y}_N(t)$ we can consider the distribution of $\mathbf{Y}_N([Nt])$, since $| Y_N^i([Nt]) - y^i(t) | \leqq N^{-1/2}$ for $i = 1, \cdots, d$ with probability one. The method of characteristic functions and the Lévy continuity theorem is used. If we let [6] $\psi_N(\mathbf{s}; k) = E_N\{\exp\{i\mathbf{s} \cdot \mathbf{Y}_N(k)\}\}$, then by using a standard conditional probability argument and obvious asymptotic expansions we show that

$$\psi_N(\mathbf{s}, k+1) = g_N(\mathbf{s}) \psi_N(\mathbf{h}_N(\mathbf{s}, 1), k)$$

for $k = 0, 1, \cdots, N-1$, where

$$\begin{aligned} g_N(\mathbf{s}) &= \exp\{-N^{-1}\mathbf{s}'\mathbf{A}\mathbf{s} + o(N^{-1})\}. \\ \mathbf{A} &= (1 - e^{-2t})^{-1}\mathbf{\Sigma}, \\ \mathbf{h}_N(\mathbf{s}, 1) &= (1 - N^{-1} + o(N^{-1}))\mathbf{s} \quad \text{as } N \to \infty, \end{aligned}$$

and the terms $o(N^{-1})$ are uniform for $\mathbf{s}$ in a compact set of R^d and independent of k. From this result, a simple iteration, one shows that

[6] The symbol $E_N\{\cdot\}$ denotes expectation with respect to P_N.

$$\psi_N(\mathbf{s}, k) = \prod_{j=0}^{k-1} g_N[\mathbf{h}_N(\mathbf{s}, j)]\psi_N[\mathbf{h}_N(\mathbf{s}, k), 0] \tag{12}$$

for $k = 1, 2, \cdots, N$, where $\mathbf{h}_N(\mathbf{s}, p) = \mathbf{s}$ and

$$\mathbf{h}_N(\mathbf{s}, j) = \mathbf{h}_N[\mathbf{h}_N(\mathbf{s}, j-1), 1] \quad \text{for } j \geqq 1.$$

Now letting $k = [Nt]$ in (12) and taking logarithms we obtain

$$\lim_{N\to\infty} \ln \psi_N(\mathbf{s}, [Nt]) = -(1/2)\,\mathbf{s}'\,\boldsymbol{\Sigma}\,\mathbf{s} + ie^{-t}\mathbf{y}_0 \cdot \mathbf{s},$$

which is the characteristic function of $\mathbf{y}(t)$. The convergence of the f.d.d. is shown by a similar argument. To complete the proof of weak convergence, a combination of the methods of Stone [**34**] and Billingsley [**2**] are used.

The method outlined above to show convergence of the f.d.d. can be used for a variety of related urn models, some of which are associated with queueing problems. These results will appear in future publications.

7. **Weak convergence of a sequence of quickest detection problems.** Consider a production process which is in one of two states, a good state and a bad state, which correspond to being in control and out of control. Production begins with the process in control, and after each item is produced there is a probability π of the process going out of control. A statistical control procedure is desired which will enable one to detect the fact that the process is out of control in some optimal manner. This model of a production process was first introduced by Girshick and Rubin [**9**] and later discussions of the problem are due to Shiryaev [**32**], Taylor [**36**], and Bather [**1**]. Most of the work carried out in these papers deals with a continuous time analog in which a Brownian motion process with mean 0 has a drift of 1 introduced after some independent exponential time. The corresponding optimal statistical control procedures are then derived and proposed as good rules for controlling the discrete processes. The passage from discrete to continuous and back to discrete has never been carried out in a rigorous way. It turns out that the notions of weak convergence are exactly what is required.

Our discussion, based on the paper of Iglehart and Taylor [**12**], begins with a sequence of truncated processes which can be easily described as follows. In the truncated problem of length n ($\geqq 2$), the process produces n independent items and goes out of control at a random time T_n ($\leqq n$). All items produced at or before time T_n are assumed to have a random quality with distribution function F_0 (having mean 0 and variance 1). All items produced after T_n possess quality given by $F_1(x) = F_0(x - 1)$, i.e. the process when out of control shifts the d.f. F_0 by 1 unit. The distribution of T_n is given by

$$\begin{aligned} \Pr\{T_n = j\} &= (\pi/n)(1 - \pi/n)^{j-1}, \qquad j = 1, 2, \cdots, n - 1, \\ &= (1 - \pi/n)^{n-1}, \qquad j = n, \end{aligned}$$

which is simply a geometric distribution with parameter π/n truncated at $n - 1$ with the remaining mass lumped at $j = n$. If $T_n = n$, then all n items produced have quality given by F_0. The process is just turned off after n items have been produced, and if $T_n = n$ it would be out of control, but this is then irrelevant.

We introduce two sequences of measures on $\mathscr{C}$ (Borel sets of $C[0,1]$) as follows. Let $X_1, X_2, \cdots$ be a sequence of independent random variables with d.f. F_0 and define

$$X_n(k) = \left(\sum_{i=1}^{k} X_i\right) \Big/ n^{1/2}$$

for $k = 1, 2, \cdots, n$ and $X_n(0) = 0$. Then the paths $x_n(t)$ in $C[0,1]$ are obtained by setting $x_n(t) = X_n(k)$ for $t = k/n$, $k = 0, 1, \cdots, n$ and by linear interpolation for other values of t. Let μ_n denote the measure induced on $\mathscr{C}$ by $x_n(\cdot)$. Next define the continuous paths $(\theta_n(\cdot))$ on $[0,1]$ by

$$\begin{aligned} \theta_n(t) &= 0 \qquad t \leqq T_n/n, \\ &= (t - T_n/n) \qquad t > T_n/n. \end{aligned}$$

The measures induced by $\theta_n(\cdot)$ we denote by λ_n. Clearly, the observed process of production corresponds to $y_n(t) = x_n(t) + \theta_n(t)$.

From the work of Prokhorov **[29]** we know that $\mu_n \Rightarrow \mu$, where μ is Wiener measure for paths starting at 0. We now introduce the measure λ on $\mathscr{C}$ induced by the process $\theta(\cdot)$ on $[0,1]$ defined by

$$\begin{aligned} \theta(t) &= 0 \qquad t \leqq T, \\ &= (t - T) \qquad t > T, \end{aligned}$$

where the random variable T has an exponential density $f_T(t)$ with parameter π for $0 \leqq t < 1$ and assumes the value 1 with probability $e^{-\pi}$. Using characteristic functions it is easy to show that the f.d.d. of λ_n converge to those of λ. Furthermore, since the support of all the measures $\{\lambda_n\}$ is contained in the compact set $M \subset C[0,1]$ given by

$$M = \{x : x(t) = 0,\ t \leqq t_0;\ x(t) = t - t_0,\ t > t_0;\ \text{for some } t_0 \in [0,1]\},$$

it follows from Prokhorov [**29**] that the family of measures is tight and thus $\lambda_n \Rightarrow \lambda$. Finally, since the measures induced by $y_n(t)$ are simply the convolutions $\lambda_n * \mu_n$, an additional argument shows that $\lambda_n * \mu_n \Rightarrow \lambda * \mu$.

Girshick and Rubin [**9**] show that the optimal control (under a cost structure which we won't mention) is to stop the process when the posteriori probability that the process will be out-of-control for the next item produced, given the history of observations, exceeds a certain level. It is more convenient to consider a monotone function of this posteriori probability which maps each path of the process $\{y_n(t) : 0 \leqq t \leqq 1\}$ into $C[0,1]$. This sequence of mappings (one for each n) is continuous and converges uniformly on compact sets of $C[0,1]$. Another result of Prokhorov [**29**] implies that the measures induced by these monotone functions converge weakly. Finally, this last result implies that the distribution of the optimal stopping times and the optimal costs converge. Thus we have established in a rigorous manner the relationship between the discrete models and the continuous analog.

8. **Other work on convergence of processes in applied probability.** In this final section we shall mention briefly some other work on convergence of processes. The area of applied probability in which diffusion approximations have been most widely used is population genetics. This work was initiated by Fisher and Wright in the 1930's. We have not discussed any of these applications since a comprehensive review is available by Kimura [**18**]. A subsequent paper which treats diffusion approximations in genetics from the point of view discussed here is Karlin and McGregor [**16**].

In branching processes several papers have recently appeared which deal with convergence of processes. These papers are Lamperti [**24**], Lamperti and Ney [**26**], and Lamperti [**25**].

Distribution-free statistics such as those of the Kolmogorov-Smirnov and Cramér-von-Mises types can be defined as functionals on the sequence of empirical stochastic processes. Convergence of these processes has been studied by Doob, Donsker, and Prokhorov. For an excellent summary of this work, complete references, and further extensions the reader should consult Pyke [**31**]. Two additional papers which deal with order statistics and related random variables are Dwass [**6**] and Lamperti [**23**].

References

1. J. Bather, *On a quickest detection problem*, Ann. Math. Statist. **38** (1967), 711-724.

2. P. Billingsley, *The invariance principle for dependent random variables*, Trans. Amer. Math. Soc. **83** (1956), 250-268.

3. A. Borovkov, *Some limit theorems in the theory of mass service*, Theor. Probability Appl. **9** (1964), 550-565 (English transl.).

4. D. Darling and A. Siegert, *The first passage problem for a continuous Markov process*, Ann. Math. Statist. **24** (1953), 624-639.

5. M. Donsker, *An invariance principle for certain probability limit theorems*, Mem. Amer. Math. Soc. no. 6, 1951.

6. M. Dwass, *Extremal processes*, Ann. Math. Statist. **35** (1964), 1718-1725.

7. P. Erdös and M. Kac, *On certain limit theorems in the theory of probability*, Bull. Amer. Math. Soc. **52** (1946), 292-302.

8. W. Feller, *Generalized second order differential operators and their lateral conditions*, Illinois J. Math. **1** (1957), 459-504.

9. M. Girshick and H. Rubin, *A Bayes approach to a quality control model*, Ann. Math. Statist. **23** (1952), 114-125.

10. D. Iglehart, *Limit diffusion approximations for the many server queue and the repairman problem*, J. Appl. Prob. **2** (1965), 429-441.

11. ———, *Limit theorems for the multi-urn Ehrenfest model*, Tech. Report no. 19, Department of Operations Research, Cornell University, 1967.

12. D. Iglehart and H. Taylor, *Weak convergence for a sequence of quickest detection problems*, Tech. Report no. 30, Department of Operations Research, Cornell University, 1967.

13. K. Itô and H. McKean, Jr., *Diffusion processes and their sample paths*, Springer-Verlag, Berlin, 1965.

14. M. Kac, *Random walk and the theory of Brownian motion*, Amer. Math. Monthly **54** (1947), 369-391.

15. S. Karlin and J. McGregor, *Many server queueing processes with Poisson input and exponential service times*, Pacific J. Math. **8** (1958), 87-118.

16. ———, "On some stochastic models in genetics" in *Stochastic models in medicine and biology* edited by J. Gurland, Univ. of Wisconsin Press, Madison, Wis., pp. 245-279.

17. ———, *Ehrenfest urn models*, J. Appl. Prob. **2** (1965), 352-376.

18. M. Kimura, *Diffusion models in population genetics*, J. Appl. Prob. **1** (1964), 177-232.

19. J. Kingman, *On queues in heavy traffic*, J. Roy. Statist. Soc. Ser. B **24** (1962), 383-392.

20. ———, *The heavy traffic approximation in the theory of queues*, Proc. Sympos. Congestion Theory, Univ. of North Carolina Press, Chapel Hill, N. C., 1965, pp. 137-159.

21. F. Knight, *On the random walk and Brownian motion*, Trans. Amer. Math. Soc. **103** (1962), 218-228.

22. A. Lalchandani, *Some limit theorems in queueing theory*, Tech. Report no. 29, Department of Operations Research, Cornell University, 1967.

23. J. Lamperti, *On extreme order statistics*, Ann. Math. Statist. **35** (1964), 1726-1737.

24. ———, *Limiting distributions for branching processes,* Proc. Fifth Berkeley Sympos. Math. Statist. and Prob. (to appear).

25. ———, *The limit of a sequence of branching processes,* Z. Wahrscheinlichkeitstheorie Verw. Gebiete **7** (1967), 271-288.

26. J. Lamperti and P. Ney, *Conditioned branching processes and their limiting diffusions* (to appear).

27. G. Luchak, *The continuous time solution of the equations of the single channel queue with a general class of service time distributions by the method of generating functions,* J. Roy. Statist. Soc. Ser. B **20** (1958), 176-181.

28. N. Prabhu, *Queues and inventories,* Wiley and Sons, New York, 1965.

29. Yu. Prokhorov, *Convergence of random processes and limit theorems in probability theory,* Theor. Probability Appl. **1** (1956), 157-214 (English transl.).

30. ———, *Transient phenomena in processes of mass service,* Litovsk. Mat. Sb. **3** (1963), 262-290, 1963, no. 1 (in Russian).

31. R. Pyke, *Spacings,* J. Roy. Statist. Soc. B **27** (1965), 395-449.

32. A. Shiryaev, *On optimum methods in quickest detection problems,* Theor. Probability Appl. **8** (1963), 22-46 (English transl.).

33. A. Skorokhod, *Limit theorems for stochastic processes,* Theor. Probability Appl. **1** (1956), 262-290 (English transl.).

34. C. Stone, *Limit theorems for birth and death processes and diffusion processes,* Ph. D. Thesis, Stanford University, 1961.

35. ———, *Limit theorems for random walks, birth and death processes, and diffusion processes,* Illinois J. Math. **7** (1963), 638-660.

36. H. Taylor, *Statistical control of a Gaussian process,* Technometrics **9** (1967), 29-41.

37. O. Viskov, *Two asymptotic formulas in the theory of queues,* Theor. Probability Appl. **9** (1964), 158-159 (English transl.).

38. O. Viskov and Yu. Prokhorov, *The probability of loss calls in heavy traffic,* Theor. Probability Appl. **9** (1964), 92-96 (English transl.).

Stanford University

Richard E. Barlow

Reliability Theory[1]

Closure under the formation of coherent structures. Perhaps the most common structures for reliability consideration are the *series structures*

$$c_1 \; c_2 \; \cdots \; c_n$$

with Boolean structure function $\phi(\underline{x}) = x_1 x_2 \cdots x_n$, where the *indicator variable*

$$x_i = 1 \quad \text{if the } i\text{th component, } c_i, \text{ works,}$$
$$= 0 \quad \text{otherwise,}$$

and the *parallel structures*

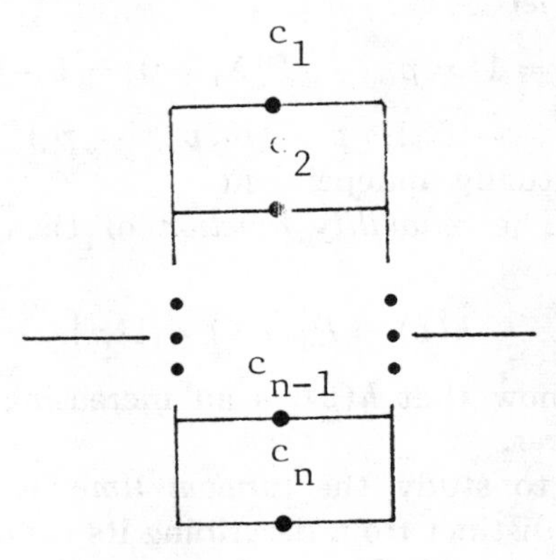

[1] This is an *expository* account of certain key results in reliability theory and is not meant as a general introduction to the subject.

with Boolean structure function

$$\phi(\underline{x}) = x_1 \vee x_2 \vee \cdots \vee x_n,$$

where $x \vee y = 1$ if and only if $x = 1$ or $y = 1$, or $x = 1$ and $y = 1$. Also important are the k out of n structures which work if any k or more of the n components work. The 2 out of 3 structure, for example, can also be represented in terms of series and parallel structures if we allow replication, i.e.

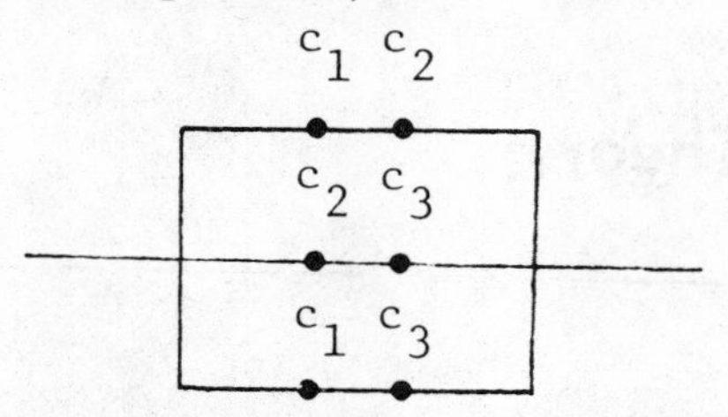

The Boolean structure function for this special structure is

$$\phi(\underline{x}) = x_1x_2 \vee x_2x_3 \vee x_1x_3.$$

More generally, we have the following:

DEFINITION. A *coherent structure* is a couple (C, ϕ) consisting of

(1) A set of components $C = \{c_1, c_2, \cdots, c_n\}$.

(2) A Boolean function ϕ defined on vectors $\underline{x} = (x_1, \cdots, x_n)$ of binary indicator variables and satisfying

(i) $\phi(\underline{0}) = 0$ and $\phi(\underline{1}) = 1$;

(ii) $\underline{x} \leqq \underline{y}$ (coordinatewise) implies $\phi(\underline{x}) \leqq \phi(\underline{y})$.

Let X_i be a binary random variable corresponding to the ith component and let

$$P[X_i = 1] = p_i, \qquad P[X_i = 0] = 1 - p_i = q_i.$$

Let $\underline{X} = (X_1, X_2, \cdots, X_n)$, $\underline{p} = (p_1, p_2, \cdots, p_n)$ and assume that the X_i's are mutually independent.

DEFINITION. The *reliability function* of the coherent structure (C, ϕ) is

$$h(\underline{p}) = P[\phi(\underline{X}) = 1 \mid \underline{p}].$$

It is easy to show that $h(\underline{p})$ is an increasing function of $\underline{p}$ for coherent structures.

We now wish to study the *random time* at which a coherent structure fails as distinct from describing its condition at a *specified time*. To do this, let $T_1, T_2, \cdots, T_n$ denote the failure times of components. The reliability of ith component at time t is

$$\bar{F}_i(t) = P\{T_i > t\}.$$

Let

$$X_i(t) = 1 \quad \text{if } T_i > t,$$
$$= 0 \quad \text{otherwise,}$$

and let T be the time to failure of the structure. Then $T > t$ if and only if $\phi[\underline{X}(t)] = 1$ where $\underline{X}(t) = (X_1(t), X_2(t), \cdots, X_n(t))$. The reliability at time t of the structure is

$$\bar{F}(t) = 1 - F(t) = P[T > t] = P\{\phi[\underline{X}(t)] = 1\} = h[\underline{\bar{F}}(t)]$$

where $\underline{\bar{F}}(t) = (\bar{F}_1(t), \bar{F}_2(t), \cdots, \bar{F}_n(t))$.

Suppose we build a coherent structure from stochastically independent components whose failure times follow an exponential law, i.e.

$$F_i(t) = 1 - \exp(-\lambda_i t) \quad \text{for } \lambda_i, t > 0.$$

If the structure is a series structure then the lifetime of the structure, T, again has an exponential distribution. However, in the parallel case it is easy to verify that T is *not* exponentially distributed. What can we say in general about the properties of the distribution of T? Birnbaum, Esary and Marshall [**5**] have actually characterized this class of distributions. It is perhaps easiest to describe this class in terms of the failure rate function. Let F be a distribution on $(0, \infty)$ with density f and

$$r(t) = f(t)/(1 - F(t)) \quad \text{for } t \geqq 0.$$

Then, intuitively, $r(t)\,dt$ is the conditional probability of failure in $(t, t + dt)$ given survival to time t. Note that

$$\bar{F}(t) = 1 - F(t) = \exp\left[-\int_0^t r(u)\,du\right].$$

DEFINITION. A distribution F such that $F(0) = 0$ is called *IFRA* (for increasing failure rate average) if and only if

$$-\log \bar{F}(t)/t$$

is nondecreasing in $t \geqq 0$.

If F has a density f, then it is easy to verify that F is IFRA iff $(1/t)\int_0^t r(u)\,du$ is nondecreasing in $t \geqq 0$. Note that exponential distributions are IFRA.

THEOREM 1 (BIRNBAUM, ESARY AND MARSHALL). *The IFRA class of distributions is closed under the formation of coherent structures.*

Furthermore, the closure under coherent structures of the exponential class of distributions is dense in the IFRA class with respect to limits in distribution, i.e.

$$\{\text{IFRA}\}^{\text{CS}} = \{\text{IFRA}\} = \{\exp\}^{\text{CS,LD}}.$$

We omit the proof to this theorem. The key to the proof of this theorem is an inequality which is of independent interest.

THEOREM 2 (BIRNBAUM, ESARY AND MARSHALL). *If h is the reliability function of a coherent structure and ψ is defined on $[0,1]$, by either*

(i) $\psi(u) = -u \log u$,

(ii) $\psi(u) = -(1-u)\log(1-u)$ *(the dual of* (i)),

or

(iii) $\psi(u) = u(1-u)$,

then the inequality

$$\sum_{i=1}^{n} \frac{\partial h(\underline{p})}{\partial p_i} \psi(p_i) \geqq \psi[h(\underline{p})] \tag{1}$$

holds for all $\underline{p}$ vectors.

If $p_1 = p_2 = \cdots = p_n = p$, then we use the notation $h(p) \equiv h(\underline{p})$. If, in addition, $\psi(u) = u(1-u)$, then

$$\frac{dh(p)}{dp} = \sum_{i=1}^{n} \frac{\partial h(\underline{p})}{\partial p_i}$$

and (1) becomes

$$pq\,(dh(p)/dp) > h(p)\,[1 - h(p)]$$

for $0 < p < 1$. (The strict inequality can be shown for p in the open interval using the fact that all components are assumed essential.) If $h(p_0) = p_0$, then

$$h'(p_0) > \frac{h(p_0)\,[1 - h(p_0)]}{p_0(1-p_0)} = 1,$$

i.e. at a crossing point of $h^*(p) \equiv p$ by $h(p)$ we see that $h(p)$ is increasing and has slope >1 so that we have the situation in Figure 1. Since $h(p)$ is increasing it can cross $h(p) = p$ at most once and from below. The usefulness of this result follows from the fact that a redundant structure with reliability function $h(p)$ will have higher reliability than a single component for component reliability $p > p_0$. This result was first discovered by Moore and Shannon [8] for two terminal networks.

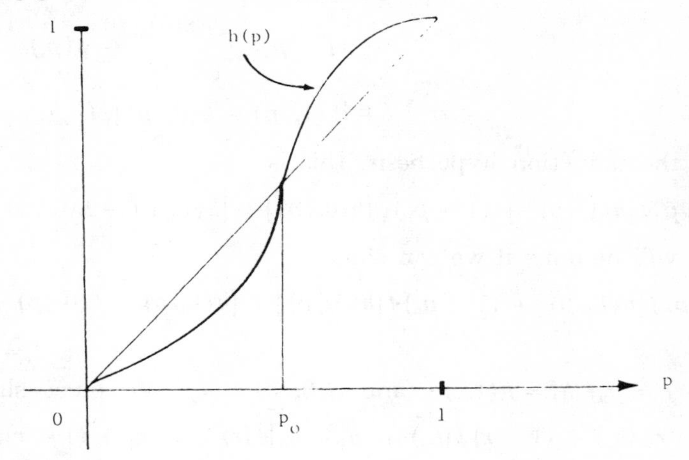

FIGURE 1.

PROOF OF THEOREM 2. The proof is by induction. For $n = 1$, $\phi(x) = x$ with $h(p) = p$ and we have equality in (1).

Now we assume the theorem is true for $n - 1$. We claim (1) is true for $h(1_n, \underline{p})$. Either $\phi(1_n, \underline{x})$ is coherent as a function of the $n - 1$ vector $\underline{x}$ or $\phi(1_n, \underline{x}) \equiv 1$. In either case $h(1_n, \underline{p})$ satisfies (1).

We claim (1) is true for $h(0_n, \underline{p})$. Either $\phi(0_n, \underline{x})$ is coherent or $\phi(0_n, \underline{x}) \equiv 0$. In either case, $h(0_n, \underline{p})$ satisfies (1).

Now $\phi(\underline{x}) = x_n \phi(1_n, \underline{x}) + (1 - x_n)\phi(0_n, \underline{x})$ and

$$h(\underline{p}) = E[\phi(\underline{X})] = p_n h(1_n, \underline{p}) + (1 - p_n) h(0_n, \underline{p}).$$

Also,

$$\partial h(\underline{p}) / \partial p_n = h(1_n, \underline{p}) - h(0_n, \underline{p}).$$

Hence,

$$\text{(2)} \qquad \sum_{i=1}^{n} \frac{\partial h(\underline{p})}{\partial p_i} \psi(p_i) = \sum_{i=1}^{n-1} \frac{\partial h(\underline{p})}{\partial p_i} \psi(p_i) + [h(1_n, \underline{p}) - h(0_n, \underline{p})] \psi(p_n).$$

Now we substitute into (2) using

$$h(\underline{p}) = p_n h(1_n, \underline{p}) + (1 - p_n) h(0_n, \underline{p})$$

so that

$$\sum_{i=1}^{n} \frac{\partial h(\underline{p})}{\partial p_i} \psi(p_i) = p_n \sum_{i=1}^{n-1} \frac{\partial h(1_n, \underline{p})}{\partial p_i} \psi(p_i) + (1 - p_n) \sum_{i=1}^{n-1} \frac{\partial h(0_n, \underline{p})}{\partial p_i} \psi(p_i) + [h(1_n, \underline{p}) - h(0_n, \underline{p})] \psi(p_n).$$

By the induction hypothesis, this is

$$\geqq p_n \psi[h(1_n, \underline{p})] + (1 - p_n) \psi[h(0_n, \underline{p})] + [h(1_n, \underline{p}) - h(0_n, \underline{p})] \psi(p_n).$$

We will be done if we can show

$$p_n \psi[h(1_n, \underline{p})] + (1 - p_n) \psi[h(0_n, \underline{p})] + [h(1_n, \underline{p}) - h(0_n, p)] \psi(p_n) \geqq \psi[h(\underline{p})].$$

Let $r = p_n$, $h_1 = h(1_n, \underline{p})$ and $h(0_n, \underline{p}) = h_0$. We must show that

$$r\psi(h_1) + (1 - r)\psi(h_0) + [h_1 - h_0]\psi(r) \geqq \psi[rh_1 + (1 - r)h_0],$$

i.e.

$$\psi[rh_1 + (1 - r)h_0] - \psi(h_0) \leqq r\psi(h_1) - r\psi(h_0) + (h_1 - h_0)\psi(r).$$

To show (i). Let $\psi(u) = -u \log u$. We claim

$$r\psi(h_1) - r\psi(h_0) + (h_1 - h_0)\psi(r) = \psi(rh_1) - \psi(rh_0).$$

Substituting in for $\psi(u)$ it is obvious. Hence we need only show

$$\psi[rh_1 + (1 - r)h_0] - \psi[h_0] \leqq \psi[rh_1] - \psi[rh_0].$$

This is geometrically obvious from the concavity of $\psi(u) = -u \log u$.

The proof of (ii) is similar.

To show (iii). Let $\psi(u) = u(1 - u)$. We need only show

$$rh_1(1 - h_1) + (1 - r)h_0(1 - h_0) + (h_1 - h_0)r(1 - r) \geqq [rh_1 + (1 - r)h_0][1 - rh_1 - (1 - r)h_0]$$

or

$$-rh_1^2 - (1 - r)h_0^2 + (h_1 - h_0)r(1 - r) \geqq -[rh_1 + (1 - r)h_0]^2$$

or

$$r^2h_1^2 + 2r(1 - r)h_0h_1 + (1 - r)^2h_0^2 - rh_1^2 - (1 - r)h_0^2 + (h_1 - h_0)r(1 - r) \geqq 0$$

or

$$-h_1^2 + 2h_0h_1 - h_0^2 + h_1 - h_0 \geqq 0$$

or

$$-(h_1 - h_0)^2 + (h_1 - h_0) \geqq 0,$$

which is obvious.‖

Bounds on failure distributions.

Classes of failure distributions. The IFRA Failure distributions mentioned earlier are theoretically attractive because of their closure property with respect to coherent structures. They also possess an interesting graphical property which is useful in theoretical investigations. Let

$$G(x) = 1 - e^{-x} \quad \text{for } x \geqq 0,$$
$$= 0 \quad \text{otherwise.}$$

Then the graph of

$$-\log[1 - F(x)] = G^{-1}F(x)$$

is starshaped with respect to the origin for $x \geqq 0$, i.e. $G^{-1}F(x)/x \uparrow$ in $x \geqq 0$ implies that the "upper side" of every point on the graph of $G^{-1}F(x)$ is "visible" from the origin. Figure 2 is an illustration of a starshaped function with respect to the origin which is *not* convex. Note that $G^{-1}F(x)$ is convex for $x \geqq 0$ if and only if it is starshaped with respect to every point on its graph.

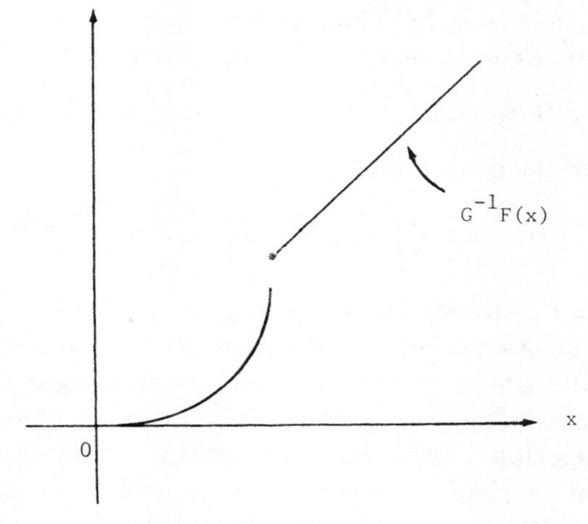

FIGURE 2.

In replacement policy problems especially, one is often concerned with the conditional failure distribution given survival to time t, i.e.

$$\overline{F}_t(x) = P\{X > t + x \mid X > t\} = \overline{F}(t + x)/\overline{F}(t).$$

It is mathematically convenient, and also intuitively plausible in some situations, that the conditional failure distribution of an assembly should also exhibit some "wearout" characteristic if the original failure distribution exhibits such a characteristic. It is, however, easy to provide examples of IFRA distributions which are not only *not* conditionally IFRA but are in fact conditionally DFRA for some $t > 0$. Hence we may ask the question: *What is the largest class of distributions in the IFRA class which remain IFRA upon conditioning on the left?* That is, we want

$$\frac{-\log \overline{F}_t(x)}{x} = \frac{-\log \overline{F}(t+x) - [-\log \overline{F}(t)]}{(t+x) - t}$$

to be nondecreasing in $x \geqq 0$ for every $t \geqq 0$. It follows that $-\log \overline{F}(x)$ is convex for $x \geqq 0$. If F has a density f, then the failure rate function

$$r(x) = f(x)/(1 - F(x))$$

must be nondecreasing in $x \geqq 0$. We call this class of distributions the IFR class for increasing failure rate.

As we have seen, the exponential distribution provides the basis for an interesting hierarchy of failure distributions. The representation in Figure 3 suggested by James Esary, emphasizes the central role of the exponential distribution. Special classes noted in the figure are the Weibull class with densities

$$f(t) = \alpha\lambda t^{\alpha-1}\exp(-\lambda t^{\alpha}) \quad \text{for } \alpha, \lambda > 0,\ t \geqq 0$$

and the gamma class with densities

$$f(t) = \lambda \frac{(\lambda t)^{\alpha-1}}{\Gamma(\alpha)} e^{-\lambda t} \quad \text{for } \alpha, \lambda > 0,\ t \geqq 0.$$

The no-data problem. In the aerospace and electronics industries one of a kind assemblies are common. For such assemblies no life testing results are available. All that is available in many cases is the engineers' past experience with similar components. Contractual obligations, however, often require a reliability statement—also warranties require a reliability assessment, e.g. it may be required that the assembly operate properly for 1000 hours with probability .99.

The mean life of an assembly is a concept with which all laymen are familiar, and engineers will make statements in terms of mean

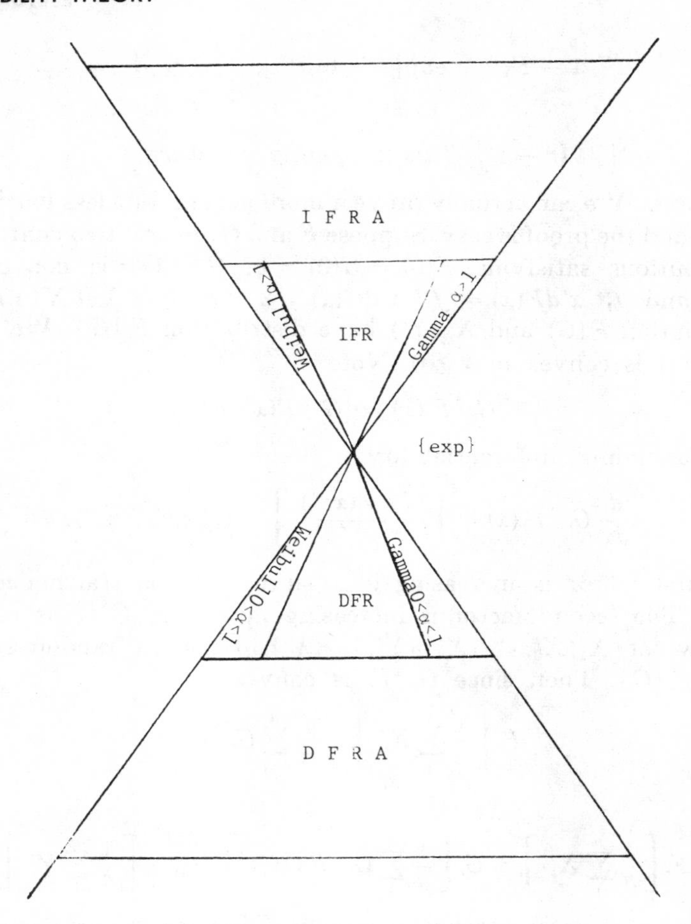

FIGURE 3. Representation for Families of Life Distributions

life far sooner than they will make a probability statement. Hence, even in the absence of data, engineers will often estimate the mean life for a new piece of equipment based on past experience. We can translate this mean life statement into a conservative probability statement using bounds based only on intuitively reasonable assumptions. More generally, given an rth moment we can state the following result (see Barlow and Marshall [1]):

THEOREM 3. *If F is IFR, $F(0) = 0$, $r \geq 1$ and $\mu_r = \int_0^\infty x^r dF(x)$, then*

$$1 - F(t) \geqq \exp[-t/(\lambda_r)^{1/r}], \qquad t \leqq \mu_r^{1/r}, \tag{1}$$
$$\geqq 0, \qquad t > \mu_r^{1/r},$$

where $\lambda_r = \mu_r/\Gamma(r+1)$. *This inequality is sharp.*

Proof. We can actually prove a more general but less motivated result and the proof is easy. Suppose F and G are any two continuous distributions satisfying $F(0) = G(0) = 0$, $G^{-1}F(x)$ is convex for $x \geqq 0$ and $\int_0^\infty x^r dF(x) = \int_0^\infty x^r dG(x) = \mu_r$ for $r \geqq 1$. Let X (Y) have distribution F (G) and X^r (Y^r) have distribution F_r (G_r). We claim $G_r^{-1}F_r(x)$ is convex in $x \geqq 0$. Note

$$G_r^{-1}F_r(x) = [G^{-1}F(x^{1/r})]^r$$

and, assuming differentiability,

$$\frac{d}{dx} G_r^{-1}F_r(x) = \left[\frac{G^{-1}F(x^{1/r})}{x^{1/r}}\right]^{r-1} (G^{-1}F)'(x^{1/r}).$$

The first factor is increasing in x since $G^{-1}F$ is starshaped and $r \geqq 1$. The second factor is increasing in x since $G^{-1}F$ is convex.

Now let $X_1^r, X_2^r, \cdots, X_n^r$ $(Y_1^r, \cdots, Y_n^r)$ denote a random sample from F_r (G_r). Then since $G_r^{-1}F_r$ is convex

$$G_r^{-1}F_r\left[\frac{1}{n}\sum_1^n X_i^r\right] \leqq \frac{1}{n}\sum_{i=1}^n G_r^{-1}F_r(X_i^r)$$

and

$$F_r\left[\frac{1}{n}\sum_1^n X_i^r\right] \leqq G_r\left[\frac{1}{n}\sum_1^n G_r^{-1}F_r(X_i^r)\right] =_{\mathrm{st}} G_r\left[\frac{1}{n}\sum_1^n Y_i^r\right]$$

where $=_{\mathrm{st}}$ denotes stochastic equality. Letting $n \to \infty$, we have by the strong law of large numbers

$$F_r(\mu_r) \leqq G_r(\mu_r)$$

or

$$F(\mu_r)^{1/r} \leqq G(\mu_r)^{1/r}.$$

Since F crosses G at most once and from below if at all, we have

$$F(t) \leqq G(t) \quad \text{for } t \leqq \mu_r^{1/r}.$$

Letting $G(t) = 1 - \exp[-t/\lambda_r^{1/r}]$ for $t \geqq 0$, we easily see that $\int_0^\infty t^r dG(t) = \mu_r$ and (1) is immediate. Since F IFR allows the possibility of a jump at the right end of its interval of support, the proof is completed using limiting arguments.

The bound for $t > \mu_r^{1/r}$ is attained by the distribution degenerate at $\mu_r^{1/r}$, which is a limit of IFR distributions.‖

Since it is well known that $(\mu_r)^{1/r}$ is always nondecreasing in $r > 0$ for distributions on the positive axis, we see that higher moments enable us to obtain nontrivial bounds over a greater range. To prove that $(\mu_r)^{1/r}$ is nondecreasing in $r > 0$ for distributions on the positive axis, let $\phi(x) = x^{r'/r}$ where $r \leqq r'$. Then ϕ is convex for $x \geqq 0$ and

$$\phi\left[\frac{1}{n}\sum_1^n X_i^r\right] = \left[\frac{1}{n}\sum_1^n X_i^r\right]^{r'/r} \leqq \frac{1}{n}\sum_1^n \phi(X_i^r) = \frac{1}{n}\sum_1^n X_i^{r'}$$

or

$$\left[\frac{1}{n}\sum_1^n X_i^r\right]^{1/r} \leqq \left[\frac{1}{n}\sum_1^n X_i^{r'}\right]^{1/r'},$$

where $X_1, X_2, \cdots, X_n$ is a random sample from a distribution F such that $F(0^-) = 0$ and $\mu_r = \int_0^\infty x^r dF(x)$. Applying the strong law of large numbers we have

$$(\mu_r)^{1/r} \leqq (\mu_{r'})^{1/r'}$$

for $r \leqq r'$.

Unfortunately the nontrivial part of the bound in (1) is decreasing in $r \geqq 1$. This follows from the fact that for IFRA distributions $\lambda_r^{1/r} = [\mu_r/\Gamma(r+1)]^{1/r}$ is decreasing in $r > 0$, or equivalently, $-\log \lambda_r$ is starshaped for $r > 0$. It is interesting that for IFRA and IFR distributions the geometrical properties of $\overline{F}(x)$ are inherited by the "normalized" moments $\lambda_r = \mu_r/\Gamma(r+1)$, i.e. if $-\log \overline{F}(x)$ is starshaped (convex) in $x \geqq 0$, then $-\log \lambda_r$ is starshaped (convex) in $r > 0$.

THEOREM 4. *If F is IFRA and $\mu_r = \int_0^\infty x^r dF(x)$, then $\lambda_r^{1/r} = (\mu_r/\Gamma(r+1))^{1/r}$ is nonincreasing in $r > 0$.*

PROOF. Let $s \leqq t$ and note

$$\mu_s = s\int_0^\infty x^{s-1}\overline{F}(x)\,dx$$

$$= s\int_0^\infty x^{s-1}\exp\left(-\frac{x}{\lambda_s^{1/s}}\right)dx.$$

Since F is IFRA, $x^{s-1}\overline{F}(x)$ crosses $x^{s-1}\exp(-x/\lambda_s^{1/s})$ exactly once and from above, say at x_0. Hence if ϕ is an increasing function

$$\int_0^\infty \phi(x) x^{s-1}\overline{F}(x)\,dx - \int_0^\infty \phi(x) x^{s-1}\exp\left(-\frac{x}{\lambda_s^{1/s}}\right) dx$$

$$= \int_0^\infty [\phi(x) - \phi(x_0)]\left[x^{s-1}\overline{F}(x) - x^{s-1}\exp\left(-\frac{x}{\lambda_s^{1/s}}\right)\right] dx \leqq 0.$$

Let $\phi(x) = x^{t-s}$. Then

$$\frac{s}{t}\mu_t = \int_0^\infty x^{t-s}[sx^{s-1}\overline{F}(x)]\,dx \leqq \int_0^\infty x^{t-s}\left[sx^{s-1}\exp\left(-\frac{x}{\lambda_s^{1/s}}\right)\right] dx$$

$$= \frac{s}{t}\Gamma(t+1)(\lambda_s^{1/s})^t$$

or $(\lambda_t)^{1/t} \leqq (\lambda_s)^{1/s}$, which was to be proved.‖

Additional probability bounds may be found in Barlow and Marshall [1], [2], [4]. A generalization of Theorem 4 may be found in Marshall, Olkin, and Proschan [7].

References

1. R. E. Barlow and A. W. Marshall, *Bounds for distributions with monotone hazard rates*. I, Ann. Math. Statist. **35** (1964), 1234-1257.

2. ———, *Tables of bounds for distributions with monotone hazard rate*, J. Amer. Statist. Assoc. **60** (1965), 872-890.

3. R. E. Barlow and F. Proschan, *Mathematical theory of reliability*, Wiley, New York, 1965.

4. R. E. Barlow and A. W. Marshall, *Bounds on interval probabilities for restricted families of distributions*, Proc. of the Fifth Berkeley Symposium, Berkeley, Calif., 1967.

5. Z. W. Birnbaum and J. D. Esary, *Modules of coherent binary systems*, J. Soc. Indust. Appl. Math. **13** (1965), 444-462.

6. Z. W. Birnbaum, J. D. Esary and A. W. Marshall, *A stochastic characterization of wear-out for components and systems*, Ann. Math. Statist. **37** (1966), 816-825.

7. A.W. Marshall, I. Olkin and F. Proschan, *Monotonicity of means and other applications of majorization*, Boeing Sci. Res. Labs. Doc. DI-82-0490, Seattle, Wash., 1965.

8. E. F. Moore and C. E. Shannon, *Reliable circuits using less reliable relays*, J. Franklin Inst. Part I, **262** (1956), 191-208.

UNIVERSITY OF CALIFORNIA, BERKELEY

H. Robbins
D. Siegmund

Iterated Logarithm Inequalities and Related Statistical Procedures

1. **Introduction and summary.** Let $x, x_1, x_2, \cdots$ be a sequence of independent and identically distributed (i.i.d.) random variables with $Ex = 0$, $Ex^2 = 1$, and put $S_n = x_1 + \cdots + x_n$, $\bar{x}_n = S_n/n$. For any sequence of positive constants a_n, $n = 1, 2, \cdots$, tending to 0 as $n \to \infty$, let

$$P_m^+ = P\{\bar{x}_n \geqq a_n, \text{ some } n \geqq m\},$$
$$P_m = P\{|\bar{x}_n| \geqq a_n, \text{ some } n \geqq m\}.$$

The law of the iterated logarithm gives conditions on the sequence (a_n) sufficient to insure that $\lim_{m\to\infty} P_m = 0$—e.g. it suffices that for some $A > 1$, $a_n \geqq (2A \log_2 n/n)^{1/2}$ (we write $\log_2 = \log\log$). In this paper we give explicit upper bounds tending to 0 as $m \to \infty$ for P_m^+ and P_m for various sequences (a_n). Our attention naturally focuses on the case

$$a_n = O((\log_2 n/n)^{1/2}).$$

Such bounds have statistical applications based on the following considerations.

Suppose that $y, y_1, y_2, \cdots$ are i.i.d. with distribution of known form depending on an unknown mean μ and with known variance σ^2. Put $x_n = (y_n - \mu)/\sigma$.

EXAMPLE 1. *Confidence sequences.* Define the intervals

$$J_n = (\bar{y}_n - \sigma a_n, \bar{y}_n + \sigma a_n)$$

and put $I_n = \bigcap_{k=m}^{n} J_k$, $n = m, m+1, \cdots$. Then, knowing bounds on P_m, we can for given $\epsilon > 0$ choose m so large that

$$P\{\mu \in I_n, \text{ all } n \geqq m\} = P\{|\bar{x}_n| < a_n, \text{ all } n \geqq m\}$$
$$= 1 - P_m \geqq 1 - \epsilon.$$

EXAMPLE 2. *Tests with uniformly small error probabilities.* To test $H_0: \mu < \mu_0$ against $H_1: \mu > \mu_0$, stop sampling with $T = \text{first } n \geqq m$ such that $|\bar{y}_n - \mu_0| \geqq \sigma a_n$ and accept H_0 or H_1 according as $\bar{y}_T - \mu_0 \leqq -\sigma a_T$ or $\bar{y}_T - \mu_0 \geqq \sigma a_T$. The error probabilities are then uniformly $\leqq P_m$. Moreover, $ET < \infty$ for all $\mu \neq \mu_0$ (see §4).

EXAMPLE 3. *Tests with* 0 *type* II *error probability.* To test $H_0: \mu \leqq \mu_0$ against $H_1: \mu > \mu_0$, stop sampling and reject H_0 with

$$T = \text{first } n \geqq m \text{ such that } \bar{y}_n - \mu_0 \geqq \sigma a_n. \tag{1}$$

If $\mu \leqq \mu_0$, $P(\text{reject } H_0) = P(T < \infty) \leqq P_m^+$, i.e. the type I error probability is $\leqq P_m^+$, which may be made less than any preassigned ϵ by taking m sufficiently large. If $\mu > \mu_0$, $P(\text{reject } H_0) = 1$, i.e. the type II error probability is 0. Of all possible choices for (a_n) which satisfy the above requirements, the best is presumably that for which ET (as a function of μ) is smallest when $\mu > \mu_0$.

In §2 we mention Chow's generalization to submartingales of the Hájek-Rényi inequality and apply it to the present problem. Next (§3) we show that if $Ee^{tx} < \infty$ for t in some neighborhood of the origin, it is possible to find sequences (a_n) such that $a_n = O((2\log_2 n/n)^{1/2})$ for which $P_m^+ = O(1/\log m)$. In §4 we show how Wald's lemma may be used to find bounds on ET for the stopping rules T of Examples 2 and 3 above. §5 suggests application of these results to the problems of ranking and selection; and in §6 we use a truncation argument to deduce results similar to those of §3 without the assumption of the finiteness of Ee^{tx}.

2. **An application of the Chow-Hájek-Rényi inequality.** Generalizing a result of [8], Chow [2] has shown that if $\{z_n, F_n, n \geqq 1\}$ is a nonnegative submartingale, then for any nondecreasing sequence (b_n) of positive constants

$$P\{z_n \geqq b_n, \text{ some } n \geqq m\} \leqq b_m^{-1} E z_m + \sum_{n=m+1}^{\infty} b_n^{-1} E(z_n - z_{n-1}). \tag{2}$$

(See [6] for the definition and basic properties of submartingales.)

Putting $z_n = S_n^2$, $b_n = A(n+k)[\log(n+k)]^{1+\delta}$, $n = 1, 2, \cdots$, with $A > 0$, $k > 0$, we see after a direct substitution that

$$P\{|\bar{x}_n| \geqq [A(1+k/n)(\log(n+k))^{1+\delta}/n]^{1/2}, \text{ some } n \geqq m\}$$
$$\leqq \frac{1}{A\delta[\log(k+m)]^{\delta}}\left(1 + \frac{\delta}{\log(k+m)}\right), \quad m = 1, 2, \cdots. \tag{3}$$

This continues to hold for $k = 0$ provided that $m \geqq 2$.

3. **A sharper result.** If $\{z_n, F_n, n \geqq 1\}$ is a nonnegative martingale with $Ez_1 = 1$, then (2) specializes to

$$P\{z_n \geqq b, \text{ some } n \geqq 1\} \leqq b^{-1}, \quad (b > 0). \tag{4}$$

Assume now that $\phi(t) \equiv Ee^{tx} < \infty$ for t in some neighborhood of the origin. Putting $z_n = [\phi(t)]^{-n}\exp(tS_n)$, $b = \exp(mt^2/2)$ for any fixed $m \geqq 1$, we see from (4) that

$$P\{\bar{x}_n \geqq t(m/2n + \log\phi(t)/t^2) \text{ some } n \geqq 1\} \leqq \exp(-mt^2/2). \tag{5}$$

Define the function

$$g(t) = 2\log\phi(t)/t^2, \quad \text{for } t > 0,$$
$$= 1, \quad \text{for } t = 0.$$

Since $Ex = 0$, $Ex^2 = 1$ it follows that $g(t)$ is continuous at $t = 0$, and from (5) we have

$$P\{\bar{x}_n \geqq (m + ng(t))t/2n, \text{ some } n \geqq 1\} \leqq \exp(-mt^2/2), \tag{6}$$

valid for all m and $t \geqq 0$. Now let $t_0, t_1, \cdots$ and $m = m_0, m_1, \cdots$ be any two sequences of nonnegative constants such that $m_i \uparrow \infty$ as $i \to \infty$, and define for all integers $n \geqq m_0$ the sequence (a_n) by setting

$$a_n = (m_i + ng(t_i))t_i/2n$$

for all integers n in the range $m_i \leqq n < m_{i+1}$, where $i = 0, 1, \cdots$. Then from (6) we have a fortiori the inequality

$$P\{\bar{x}_n \geqq a_n, \text{ some } n \geqq m\} \leqq \sum_{i=0}^{\infty} \exp(-m_i t_i^2/2). \tag{7}$$

It remains to make definite choices of the sequences (m_i), (t_i).

For example, let $A > 1$, $B \geqq 1$, $\alpha > 1$, $m \geqq 1$, $k \geqq 0$ be five given constants such that either $m > 1$ or $k > 1$, and define for $i = 0, 1, \cdots$

$$
\begin{aligned}
&m_i = m\alpha^i, \\
(8)\qquad &t_i = \{[2\log B + 2A\log(i + k + \log m/\log\alpha)]/m_i\}^{1/2}.
\end{aligned}
$$

Then, the RHS of (7) is

$$
\begin{aligned}
\sum_{i=0}^{\infty}\exp(-m_i t_i^2/2) &= \frac{1}{B}\sum_{i=0}^{\infty}\frac{1}{(i+k+\log m/\log\alpha)^A} \\
&\leqq \frac{1}{B(A-1)(k+\log m/\log\alpha - 1/2)^{A-1}}.
\end{aligned}
$$

We shall now find an upper bound on the sequence (a_n) in (7). In the range of integers n such that

$$
m\alpha^i = m_i \leqq n < m_{i+1} = m\alpha^{i+1}
$$

for any fixed $i = 0, 1, \cdots,$ we have $\log m + i\log\alpha \leqq \log n$ so $\log(i + k + \log m/\log\alpha) \leqq \log_2 n\alpha^k - \log_2\alpha$ and $m_i > n/\alpha$ so

$$
0 \leqq t_i < \alpha^{1/2}[(2A\log_2 n\alpha^k - 2A\log_2\alpha + 2\log B)/n]^{1/2}.
$$

Define

$$
\begin{aligned}
&v_n = \max g(t) \\
(9)\qquad &\qquad \text{for all } 0 \leqq t < \alpha^{1/2}[(2A\log_2 n\alpha^k - 2A\log_2\alpha + 2\log B)/n]^{1/2},
\end{aligned}
$$

so that

$$
g(t_i) \leqq v_n \geqq 1;
$$

then

$$
\frac{m_i + ng(t_i)}{2(m_i n)^{1/2}} \leqq \frac{m_i + nv_n}{2(m_i n)^{1/2}} \leqq \frac{m_i + m_{i+1}v_n}{2(m_i m_{i+1})^{1/2}} = \frac{1+\alpha v_n}{2\alpha^{1/2}}
$$

and hence

$$
\begin{aligned}
a_n &= \frac{m_i + ng(t_i)}{2(m_i n)^{1/2}}\left(\frac{m_i}{n}\right)^{1/2} t_i \\
&< \frac{(1+\alpha v_n)}{2\alpha^{1/2}}\left(\frac{2A\log_2 n\alpha^k - 2A\log_2\alpha + 2\log B}{n}\right)^{1/2}.
\end{aligned}
$$

Thus from (7) we have the inequality

$$
\begin{aligned}
&P\left\{\bar{x}_n \geqq \frac{(1+\alpha v_n)}{2\alpha^{1/2}}\left(\frac{2A\log_2 n\alpha^k - 2A\log_2\alpha + 2\log B}{n}\right)^{1/2}, \text{ some } n \geqq m\right\} \\
(10)\qquad &\qquad \leqq 1/[B(A-1)(k+\log m/\log\alpha - 1/2)^{A-1}],
\end{aligned}
$$

where v_n satisfies (9). In particular, if the distribution of x is "subnormal" in the sense that for all $t \geqq 0$

$$\phi(t) \leqq \exp(t^2/2),$$

then v_n can be replaced by 1 in (10).

4. **Bounds and approximations for** ET. In terms of $x_n = (y_n - \mu)/\sigma$ the stopping rule T defined by (1) with $m = 1$ is for any $\mu > \mu_0$ given by

$$T = \text{first } n \geqq 1 \text{ such that } S_n + n\delta \geqq na_n,$$

where we have put $\delta = (\mu - \mu_0)/\sigma$. Suppose that a_t is defined for all $t \geqq 1$ with $t^{1/2}a_t$ nondecreasing and concave. For any $q > 1$ such that $\nu_q = E(x^{+q}) < \infty$ we have by Wald's lemma

$$E(S_T) = ETEx = 0,$$

and

$$E^q x_T^+ \leqq E\left(\sum_1^T x_n^{+q}\right) = ET\nu_q, \tag{11}$$

so from $S_T + \delta T \leqq Ta_T + x_T^+ + \delta$ we have

$$\begin{aligned} \delta ET &\leqq E(Ta_T) + (ET)^{1/q}(\nu_q)^{1/q} + \delta \\ &\leqq ETa_{ET} + (ET)^{1/q}(\nu_q)^{1/q} + \delta. \end{aligned}$$

This may be written as

$$\delta < a_{ET} + \nu_q^{1/q}/(ET)^{1-1/q} + \delta/ET. \tag{12}$$

(In order to apply Wald's lemma in our derivation of (12) we have tacitly assumed that $ET < \infty$. That this is so follows from a similar line of reasoning with T replaced by $T_n = \min(T, n)$. After deriving (12) for T_n, we let $n \to \infty$ so that $ET_n \to ET$. Since $a_n \to 0$, (12) shows that $ET < \infty$.)

The inequality (12) may be used to find upper bounds on ET. For any $q \geqq 2$, if $t^{1/2}a_t \to 0$ as $t \to \infty$, we can express (12) asymptotically as $\delta \to 0^+$ and $ET \to \infty$ (provided ν_q remains bounded) by $\delta \leqq a_{ET}(1 + o(1))$. This asymptotic inequality can easily be inverted for the sequences (a_n) of §§2 and 3. E.g. with the stopping boundary (a_n) of §3, we have

$$ET \leqq [(1+\alpha)^2/2\alpha\delta^2]A \log_2\delta^{-1}(1 + o(1)). \tag{13}$$

Referring again to the stopping boundary (a_n) of §3, we see that P_1^+ can be made arbitrarily small (less than any preassigned ϵ, say) by taking any of the constants A, B, or k sufficiently large. This suggests studying the behavior of T as B, say, tends to infinity. In this case $\lim_{B\to\infty} T = \infty$ a.s. and it is possible to find an asymptotic expression for ET. It turns out that this result states that asymptotically the "time" that $S_n + n\delta$ crosses the boundary na_n is on the average the same as the "time" that the line $n\delta$ crosses it.

From (12), with

$$a_n = \lambda_n[(2A \log_2 n\alpha^k - 2A \log_2 \alpha + 2\log B)/n]^{1/2},$$

where (λ_n) is any sequence of positive constants such that $\lambda_n \to \lambda$, a finite, nonzero constant, we have

$$\limsup_{B\to\infty} \frac{\delta^2 ET}{2\lambda^2 \log B} \leqq 1. \tag{14}$$

From the definition of T we see that

$$S_{T-1} + (T-1)\delta \leqq Ta_T \leqq S_T + T\delta,$$

and since $(S_T)/T \to 0$ a.s. by the strong law of large numbers, we have

$$\lambda_T[(2A \log_2 T\alpha^k - 2A \log_2 \alpha + 2\log B)/T]^{1/2} \to \delta \quad \text{a.s.},$$

and hence by Fatou's lemma,

$$\liminf_{B\to\infty} \frac{\delta^2 ET}{2\lambda^2 \log B} \geqq 1. \tag{15}$$

REMARK. Although the inequality (11) is quite crude, it is completely general and sufficient to obtain the asymptotic results (13) and (14). In particular cases other methods of estimating the "excess over the boundary" may be more appropriate. For example, if

$$b = \sup_{r>0} E(x + \delta - r \mid x + \delta \geqq r) < \infty,$$

we may use

$$\delta ET = E(S_T + \delta T) \leqq E(Ta_T) + b \leqq ET a_{ET} + b. \tag{12'}$$

5. **Selection and ranking problems.** The results of the preceding sections may easily be applied to so-called selection and ranking problems. To be specific, suppose that we have $J+1$ normal populations $\pi_0, \pi_1, \cdots, \pi_J$, the jth population having unknown

mean μ_j and variance 1. We would like a procedure for selecting the population with the largest mean (assuming that a unique one exists), i.e. given $0 < \beta < 1$ we want a sampling rule and a terminal decision procedure with the property that the probability of correctly selecting the population with the largest mean is at least β. One such procedure is the following. Let $\bar{y}_{j,n}$ denote the mean of the first n observations on the jth population, $j = 0, 1, \cdots, J$, $n = 1, 2, \cdots$. Let (a_n) be as in §3 with B so large that

$$P\{ |\bar{y}_{j,n} - \mu_j| < a_n, \text{ all } n \geqq 1\} \geqq \beta^{1/(J+1)}, \quad j = 0, 1, \cdots, J.$$

For $j = 0, 1, \cdots, J$, $n = 1, 2, \cdots$, define the intervals

$$J_{j,n} = (\bar{y}_{j,n} - a_n, \bar{y}_{j,n} + a_n), \qquad I_{j,n} = \bigcap_{n'=1}^{n} J_{j,n'}.$$

Then $P\{\mu_j \in I_{j,n} \text{ all } n = 1, 2, \cdots, j = 0, \cdots, J\} \geqq \beta$. For any $j = 0, \cdots, J$, $n = 1, 2, \cdots$, let $\tau(j) = n$ if n is the smallest positive integer such that for some $j' \neq j$, $I_{j',n}$ lies completely to the right of $I_{j,n}$. We stop sampling from π_j at time $\tau(j)$ and select π_j if for all $j' \neq j$, $I_{j,\tau(j)}$ lies completely to the right of $I_{j',\tau(j')}$.

The above procedure is not completely specified, since it may (with probability $\leqq 1 - \beta$) occur that for some n some or all of the $I_{j,n}$ are empty. If so we may, for example, stop all sampling and select the population with the largest value $\bar{y}_{j,n}$.

Using the results of §4, we may obtain approximations for the expected total number of observations required. For example, set $\delta_j = \mu_0 - \mu_j, j = 1, 2, \cdots, J$ and suppose that for some $\delta > 0$, $0 \leqq \gamma \leqq 1$,

$$\delta_j \geqq \delta, \quad j = 1, 2, \cdots, J, \tag{16}$$

$$\sum_1^J \frac{1}{\delta_j^2} \leqq \frac{J\gamma}{\delta^2} + o(J) \quad \text{as } J \to \infty. \tag{17}$$

Then $\tau(0) \leqq \max(\tau(1), \cdots, \tau(J)) \leqq \tau(1) + \cdots + \tau(J)$. In the case where we arrive at the correct decision, for $j = 1, 2, \cdots, J$, $\tau(j) \leqq T_j$ $=$ first $n \geqq 1$ such that $\bar{y}_{0,n} - \bar{y}_{j,n} > 2a_n$, and from (10) and (14) we see that the expected total number of observations is $\leqq 4((1 + \alpha)/\alpha^{1/2})^2$ $(\gamma J \log J/\delta^2)(1 + o(1))$ as $J \to \infty$.

By way of comparison the classical procedure of Bechhofer [1] requires knowledge of the constant δ appearing in (16); and from the well-known result [7] that if $z_1, z_2, \cdots$ are independent standard normal random variables, then

$$\max(z_1, \cdots, z_J) - (2 \log J)^{1/2} \to 0$$

in probability it follows that the fixed number of observations required by the Bechhofer procedure is asymptotically $(2J\log J)/\delta^2$ as $J \to \infty$.

6. **Some generalizations.** The results of §3 depend on the assumption that $Ee^{tx} < \infty$ for t in some neighborhood of the origin. The law of the iterated logarithm, however, guarantees that for each $\delta > 0$, if $a_n = (1+\delta)(2\log_2 n/n)^{1/2}$, then $P_m \to 0$ as $m \to \infty$ [**9**]. This suggests the following questions:

(i) If we do not assume the existence of moments of order greater than two, can the approach of §3 be adapted to prove the above mentioned part of the law of the iterated logarithm?

(ii) If we do not assume the existence of a moment generating function, is it possible to find explicit upper bounds for the values P_m for sequences (a_n) such that

$$a_n = O((\log_2 n/n)^{1/2})?$$

Affirmative answers to both questions are provided below. We use the following elementary inequality [**10**, p. 255]: if $|x| \leqq c$ a.s., then for any t such that $0 \leqq t \leqq 1/c$, we have

$$Ee^{tx} \leqq \exp[tEx + \tfrac{1}{2}t^2Ex^2(1 + tc/2)]. \tag{18}$$

Let $m \geqq \alpha^2 > 1$, $A > 1$, $\epsilon > 0$ be constants to be specified later. For any sequence (b_j) of positive constants such that $b_j \to \infty$ as $j \to \infty$ define for $j = 1, 2,$

$$\beta_j = x_j I\{|x_j| \leqq b_j\},$$

$$\gamma_j = x_j - \beta_j,$$

$$\phi_j(t) = E\exp(t\beta_j), \quad -\infty < t < \infty.$$

Then for any sequence of positive constants (a_n),

$$\begin{aligned} &P\{\bar{x}_n \geqq (1+\epsilon)a_n, \text{ some } n \geqq m\} \\ &\quad \leqq P\{\bar{\beta}_n \geqq a_n, \text{ some } n \geqq m\} + P\{\bar{\gamma}_n \geqq \epsilon a_n, \text{ some } n \geqq m\}. \end{aligned} \tag{19}$$

It remains to make specific choices of the parameters involved and analyze the RHS of (19) under specific assumptions about the distribution of x.

Initially we assume only that $Ex = 0$, $Ex^2 = 1$. Suppose for simplicity that $\alpha < e$ (and hence $\log_2\alpha < 0$) and let

$$0 < \delta < \{2(A\alpha)^{1/2}(1 - \log_2\alpha/\log_2 3)\}^{-1}. \tag{20}$$

Set $b_j = \delta(j/\log_2(j+2))^{1/2}$, $j = 1, 2, \cdots$ and $c_j = (j \log_2 j)^{-1/2}$, $j = 3, 4, \cdots$. We see that

$$\begin{aligned}\sum_{j=4}^{\infty} c_j E|\gamma_j| &= \sum_{j=4}^{\infty} c_j \sum_{l=j}^{\infty} \int_{\{b_l \leqq |x| < b_{l+1}\}} |x| \\ &= \sum_{l=4}^{\infty} \sum_{j=4}^{l} c_j \int_{\{b_l \leqq |x| < b_{l+1}\}} |x| \\ &\leqq \text{const.} \sum_{l=4}^{\infty} b_l \int_{\{b_l \leqq |x| < b_{l+1}\}} |x| \leqq \text{const.}\, Ex^2 < \infty.\end{aligned}$$

Hence by Fatou's lemma

$$\sum_{4}^{\infty} \frac{|\gamma_j|}{(j \log_2 j)^{1/2}} < \infty \text{ a.s.},$$

and thus

$$P\{\bar{\gamma}_n \geqq \epsilon(\log_2 n/n)^{1/2}, \text{ some } n \geqq m\} \to 0. \tag{21}$$

Likewise

$$\left|\sum_{1}^{n} E\beta_j\right| = \left|\sum_{1}^{n} E\gamma_j\right| = o((n \log_2 n)^{1/2}). \tag{22}$$

Let (m_i), (t_i) be as in (8) with $B = 1$, $k = 0$. Put

$$a_n = t_i \left(\frac{m_i}{2n} + \frac{1}{nt_i^2} \sum_{j=1}^{n} \log \phi_j(t_i)\right) \tag{23}$$

for all integers n in the range $m_i \leqq n < m_{i+1}$, $i = 0, 1, \cdots$. Then by the results of §3

$$P\{\bar{\beta}_n \geqq a_n, \text{ some } n \geqq m\} \leqq 1/[(A-1)(\log m/\log \alpha - \tfrac{1}{2})^{A-1}]. \tag{24}$$

For $i = 0, 1, \cdots$, and each $m_i \leqq n < m_{i+1}$,

$$\begin{aligned}\left(\frac{2A \log[\log n/\log \alpha - 1]}{n}\right)^{1/2} &\leqq t_i \\ &\leqq \alpha^{1/2} \left(\frac{2A \log_2 n - 2A \log_2 \alpha}{n}\right)^{1/2}.\end{aligned} \tag{25}$$

By (20) we may apply (18), and we find that

$$\left[\frac{m_i}{2n} + \frac{1}{nt_i^2}\sum_{j=1}^{n} \log \phi_j(t_i)\right] \leqq \left[1 + \frac{1}{nt_i}\sum_{1}^{n} E\beta_j + \frac{t_i}{4n}\sum_{1}^{n} b_j\right].$$

Moreover, we can find a C not depending on δ such that

$$\sum_{j=1}^{n} b_j \leqq \frac{\delta C n^{3/2}}{(\log_2 n)^{1/2}}, \quad n = 3, 4, \cdots;$$

and since (22) and (25) show that

$$\lim_{n\to\infty;\, m_i \leqq n < m_{i+1}} \frac{1}{nt_i}\sum_{1}^{n} E\beta_j = 0,$$

we have from (25)

$$\limsup_{n\to\infty;\, m_i \leqq n < m_{i+1}} \left[\frac{m_i}{2n} + \frac{1}{nt_i^2}\sum_{j=1}^{n} \log \phi_j(t_i)\right] \leqq 1 + \frac{\delta C}{4}\,(2\alpha A)^{1/2}.$$

Thus for any $\lambda > 1$, we have after a judicious choice of α, A, ϵ, and δ,

$$\lim_{m\to\infty} P\{\bar{x}_n \geqq \lambda(2\log_2 n/n)^{1/2}, \text{ some } n \geqq m\} = 0.$$

If we assume that $E|x|^{2+\lambda} < \infty$ for some $\lambda > 0$, then a relatively straightforward substitution in (2) provides an explicit upper bound for the second term on the RHS of (19), which together with obvious modifications of the above discussion yields explicit upper bounds for P_m^+.

Instead of carrying out these calculations in detail, we shall sketch another method which is slightly easier to apply in the important special case that x is *symmetrically distributed* and

$$E|x|^3 = D < \infty.$$

We use the well-known inequality [**6**, p. 106]

$$(26) \quad P\{S_k \geqq a, \text{ some } k \leqq n\} \leqq 2P\{S_n \geqq a\}, \; 0 < a < \infty, \; n = 2, 3, \cdots,$$

the Berry-Esseen Theorem (e.g. [**10**, p. 288]), and the fact that

$$1 - \Phi(x) \equiv \int_x^\infty \frac{1}{(2\pi)^{1/2}} \exp(-\tfrac{1}{2}y^2)\,dy \leqq \frac{1}{(2\pi)^{1/2}x} \exp(-\tfrac{1}{2}x^2),$$

$$0 < x < \infty.$$

Then for all $\lambda > \alpha^{1/2} > 1$, $m > \max(3, \alpha^{1/2})$, putting $m_i = m\alpha^i$, $i = 0, 1, \cdots$, we have

$$
\begin{aligned}
P\{S_n &\geqq \lambda(2n\log_2 n)^{1/2}, \text{ some } n \geqq m\} \\
&\leqq \sum_{i=0}^{\infty} P\{S_n \geqq \lambda(2m_i\log_2 m_i)^{1/2}, \text{ some } n \leqq m_{i+1}\} \\
&\leqq 2\sum_{i=0}^{\infty} P\left\{\frac{S_{m_{i+1}}}{m_{i+1}^{1/2}} \geqq \frac{\lambda}{\alpha^{1/2}}(2\log_2 m_i)^{1/2}\right\} \\
&\leqq 2\sum_{i=0}^{\infty}\left(1-\Phi\left(\frac{\lambda}{\alpha^{1/2}}(2\log_2 m_i)^{1/2}\right)+\frac{22D/5}{m_{i+1}^{1/2}}\right) \\
&\leqq \left\{\left(\frac{\pi}{\alpha}\right)^{1/2}\lambda(\log\alpha)\left(\frac{\lambda^2}{\alpha}-1\right)(\log m-\tfrac{1}{2}\,\alpha)^{\lambda^2/\alpha-1}\right. \\
&\qquad\qquad \left.\times\left[\log(\log m-\tfrac{1}{2}\log\alpha)\right]^{1/2}\right\}^{-1} \\
&\quad +\frac{88D}{5(\log\alpha)m^{1/2}}.
\end{aligned}
$$

7. **Remarks.** (a) The introductory comments as well as the results of §§2 and 3 are taken from Darling and Robbins [**3**], [**4**]. In [**4**] a slight generalization of (2) is used to obtain a large class of results analogous to (3) under various assumptions about the finiteness of $E|x|^p$ for different values of p. In addition to the results of §3, in [**3**] Darling and Robbins exhibit a sequence (a_n) such that $a_n \sim (2\log_2 n/n)^{1/2}$ for which $P_m = O(1/\log_2 m)$.

(b) The inequality (4) is easily proved directly. It is by no means necessary to regard it as a special case of (2) (cf. [**6**, p. 314]).

(c) The methods of §4 are well known and are scattered throughout the literature of probability and statistics. See [**11**] for more general results along the lines of the present context.

(d) Condition (17) provides a measure of the average disparity between the best population π_0 and the remaining π_j's. Since $((1+\alpha)/\alpha^{1/2})^2$ may be made as close to 4 as desired, it would seem even from our crude analysis that when J is large the procedure of §5 is superior to that of Bechhofer for a wide range of values of $\mu_0, \mu_1, \cdots, \mu_J$. In addition, our procedure does not require a knowledge of the δ appearing in (16) or for that matter even of σ (see Remark (e)).

(e) Further applications of the results of §3 may be found in [**5**]. In particular Darling and Robbins show that in problems of statistical inference about the mean of a normal random variable their procedures can be easily adapted to take care of the case in which the variance is also unknown.

(f) The first person to use the martingale inequality (4) in studying the law of the iterated logarithm appears to have been Ville [12], who in the case of Bernoulli variables applied (4) to

$$z_n = \int_{-\infty}^{\infty} \frac{\exp tS_n}{\phi^n(t)} dG(t), \tag{27}$$

where G is a probability measure on the line.

More generally, let $x_1, \cdots$ be any sequence of random variables on a probability space, F_n the Borel field generated by $x_1, \cdots, x_n$, and F the Borel field generated by $x_1, \cdots$. Let P, Q be any two probability measures on F, P_n, Q_n their restrictions to F_n, μ_n any σ-finite measure on F_n which dominates P_n and Q_n, and $p_n = dP_n/d\mu_n$, $q_n = dQ_n/d\mu_n$. For any positive b and positive integer m, define $N =$ first $n \geqq m$ such that $bp_n \leqq q_n$. Then

$$\begin{aligned} &P(bp_n \leqq q_n \text{ for some } n \geqq m) \\ &\quad = P(m \leqq N < \infty) \leqq P(N = m) + b^{-1}Q(N \neq m) \\ &\quad = P(bp_m \leqq q_m) + b^{-1}Q(bp_m > q_m) \leqq b^{-1}. \end{aligned} \tag{28}$$

(If P and Q are orthogonal, then $P(bp_m \leqq q_m)$ and $Q(bp_m > q_m) \to 0$ as $m \to \infty$, so that $P(bp_n \leqq q_n \text{ i.o.}) = 0$.) The inequality (4) applied to (27) is the special case of (28) for $m = 1$ when

$$p_n = \prod_1^n p(x_i), \qquad q_t(x) = e^{tx}p(x)\phi^{-1}(t), \qquad q_n = \int_{-\infty}^{\infty} \prod_1^n q_t(x_i)\, dG(t).$$

For example, if $x_1, \cdots$ are i.i.d. $N(0, 1)$ under P, with p.d.f. $p(x) = (2\pi)^{-1/2}\exp(-x^2/2)$ and m.g.f. $\phi(t) = \exp(t^2/2)$, we have

$$(bp_n \leqq q_n) = \left(\int_{-\infty}^{\infty} \exp(tS_n - nt^2/2)\, dG(t) \geqq b \right).$$

Letting G put unit mass at some positive θ and $b = \exp(m\theta^2/2)$, (28) gives the inequality (cf. (6))

$$\begin{aligned} &P(S_n \geqq (m + n)\theta/2 \text{ for some } n \geqq m) \\ &\qquad \leqq \Phi(-\theta\sqrt{m}) + \tfrac{1}{2}\exp(-m\theta^2/2), \end{aligned} \tag{29}$$

where Φ is the normal $(0, 1)$ distribution function. Putting $G = \Phi$ and taking $m = 1$, (28) gives the inequality

$$P(|S_n| \geqq ((n+1)(\log(n+1) + 2\log b))^{1/2} \text{ for some } n \geqq 1) \leqq b^{-1}. \tag{30}$$

Other choices for G with more mass near 0 give iterated logarithm inequalities.

References

1. R. E. Bechhofer, *A single sample multiple decision procedure for ranking means of normal populations with known variances,* Ann. Math. Statist. **25** (1954), 16-39.
2. Y. S. Chow, *A martingale inequality and the law of large numbers,* Proc. Amer. Math. Soc. **11** (1960), 107-111.
3. D. A. Darling and H. Robbins, *Iterated logarithm inequalities,* Proc. Nat. Acad. Sci. **57** (1967), 1188-1193.
4. ———, *Inequalities for the sequence of sample means,* Proc. Nat. Acad. Sci. **57** (1967), 1577-1580.
5. ———, *Confidence sequences for mean, variance and median,* Proc. Nat. Acad. Sci. **58** (1967), 66-68.
6. J. L. Doob, *Stochastic processes,* Wiley, New York, 1953.
7. B. V. Gnedenko, *Sur la distribution limite du terme maxime d'une série aléatoire,* Ann. Math. **44** (1943), 423-453.
8. J. Hájek and A. Rényi, *Generalization of an inequality of Kolmogorov,* Acta. Math. Acad. Sci. Hungar. **6** (1955), 281-283.
9. P. Hartman and A. Wintner, *On the law of the iterated logarithm,* Amer. J. Math. **63** (1941), 169-176.
10. M. Loève, *Probability theory,* 3rd ed., Van Nostrand, Princeton, N. J., 1963.
11. D. Siegmund, *Some one-sided stopping rules,* Ann. Math. Statist. **38** (1967), 1641-1646.
12. J. Ville, *Étude critique de la notion de collectif,* Gauthier-Villars, Paris, 1939.

Note added to the second printing

The idea sketched in §7(f) has been further exploited in a number of papers to derive exact probabilities for a standard Wiener process $W(t)$ to cross various boundary curves $g(t)$ for some $t > \tau \geqq 0$. By means of an invariance theorem these probabilities have been shown to be limits of certain probabilities for sample sums S_n. For example, corresponding to the inequality (30) for normal sums it has been shown that for any $a > 0$

$$\begin{aligned} &P(|W(t)| \geqq ((t+1)(\log(t+1) + a^2))^{1/2} \text{ for some } t > 0) \\ &= \lim_{m\to\infty} P(|S_n| \geqq ((n+m)(\log(n/m+1) + a^2))^{1/2} \\ &\qquad \text{for some } n \geqq 1) = \exp(-a^2/2) \end{aligned} \tag{31}$$

and

$$
\begin{aligned}
&P(|W(t)| \geqq (t(\log t + a^2))^{1/2} \text{ for some } t \geqq 1) \\
(32) \qquad &= \lim_{m\to\infty} P(|S_n| \geqq (n(\log(n/m) + a^2))^{1/2} \text{ for some } n \geqq m) \\
&= 2[1 - \Phi(a) + a(2\pi)^{-1/2}\exp(-a^2/2)]
\end{aligned}
$$

whenever the assumptions of the first sentence of §1 hold. These and other matters are discussed in the additional references below.

13. D. A. Darling and H. Robbins, *Some further remarks on inequalities for sample sums,* Proc. Nat. Acad. Sci. U.S.A. **60** (1968), 1175-1181.

14. ———, *Some nonparametric sequential tests with power one,* Proc. Nat. Acad. Sci. U.S.A. **61** (1968), 804-809.

15. H. Robbins, D. Siegmund and J. Wendel, *The limiting distribution of the last time* $s_n \geqq n\epsilon$, Proc. Nat. Acad. Sci. U.S.A. **61** (1968), 1228-1230.

16. H. Robbins and D. Siegmund, *Probability distributions related to the law of the iterated logarithm,* Proc. Nat. Acad. Sci. U.S.A. **62** (1969), 11-13.

17. ———, *Boundary crossing probabilities for the Wiener process and sample sums,* Ann. Math. Statist. (to appear).

18. ———, *Confidence sequences and interminable tests,* Proc. Thirty-Seventh Session Bull. Internat. Statist. Inst. (London, 1969) (to appear).

19. H. Robbins, *Statistical methods related to the law of the iterated logarithm,* Ann. Math. Statist. (to appear).

University of Michigan,
Ann Arbor, Michigan

Stanford University,
Stanford, California

X
MATHEMATICAL PSYCHOLOGY AND LINGUISTICS

M. Frank Norman[1]

Mathematical Learning Theory

I. **Introduction.** The purpose of mathematical learning theory is to provide simple quantitative descriptions of those processes that are basic to behavior modification in organisms. The objective of providing simple descriptions of basic learning processes determines the flavor of the field: learning models, as we will see, are, by design, simple gadgets for doing simple things.

Most of the experiments to which learning models have been applied consist of a sequence of *trials* on each of which a configuration of stimuli is presented to the subject. The subject then makes one of several *responses*, and this is followed by an *outcome*, perhaps a reward or punishment, determined by the experimenter. If an outcome does not decrease the probability of a certain response on subsequent trials, it is said to *reinforce* that response. The sequence of responses of a subject in such an experiment is conceived to be a realization of a discrete parameter stochastic process. Mathematical learning theorists try to predict the statistics of this process. Most learning models are intrinsically stochastic. I will now give some examples of learning experiments and models that have been applied to them.

[1] The author's work in this area is supported by the National Science Foundation under grant GP-7335.

A. *Two-choice simple learning experiments.* A *simple learning experiment* is one in which the stimulus situation is the same on every trial. If the stimulus situation varies from trial to trial, we have *discrimination learning*. Consider now a simple learning experiment in which the same two responses, A_1 and A_2, are available to the subject on each trial. Let $A_{i,n}$ be the event "A_i on trial n", $n = 1, 2, \cdots$; let X_n be the indicator random variable for the event $A_{1,n}$ (i.e. $X_n = 1$ or 0 depending on whether or not $A_{1,n}$); and let p_n be the probability of $A_{1,n}$ for an individual subject. Suppose that either response can be reinforced on any trial, regardless of what response is made on that trial. Let O_{ij} be the outcome that follows A_i and reinforces A_j, and let $O_{ij,n}$ be the event "outcome O_{ij} occurred on trial n". Suppose that, on any trial, $O_{ij,n}$ follows $A_{i,n}$ with probability π_{ij} fixed throughout the experiment.

1. *Examples.*

a. *Probability learning.* A human subject is seated before a panel to which are affixed two lamps numbered 1 and 2. The subject's task is to predict on each trial which of the two lamps will flash. After he has made his prediction one of the lights flashes. Under suitable conditions the flashing of a light will reinforce, in some degree, the prediction of that light, regardless of what light was predicted on that trial. Here A_i is the prediction of light i and O_{ij} is the consequent flashing of light j. Monetary payoffs are sometimes used. Experiments of this kind are called *probability learning experiments.*

The special case of *noncontingent reinforcement*, in which the probability of reinforcing A_j does not depend on the response made ($\pi_{1j} = \pi_{2j} = \pi_j$), has been studied most intensively experimentally. It is often found that the asymptotic probability of $A_{1,n}$ is approximately equal to the probability π_1 that A_1 will be reinforced (Estes [7]). This probability matching is somewhat surprising since the frequency of correct predictions is obviously maximized by always choosing A_1 if $\pi_1 > 1/2$ and always choosing A_2 if $\pi_1 < 1/2$.

b. *T-maze learning.* On each trial, an animal, say a rat, is placed in the bottom of a T-shaped alley. Eventually he enters the top of the maze and goes to the end of the left (A_1) or right (A_2) arm. There he may receive food, which reinforces the response just made, or he may simply be detained, which, in some cases, reinforces the other response. This is called a *noncorrection procedure* to distinguish it from *correction procedures* in which, after failing to find food in one arm, the rat is allowed to go to the end on the other arm where he receives food. This difference has profound behavioral consequences. Correction procedures sometimes lead to probability matching under noncontingent reinforcement, while noncorrection invariably leads to overwhelming preference for the more favorable side (see Norman [**19**]).

c. *Avoidance and punishment.* Suppose that, within a specified time after the beginning of a trial, a dog may (A_1) or may not (A_2) jump over a barrier, and that shock may or may not follow. We consider experiments in which the avoidance of shock reinforces the response that preceded it, while shock reinforces the other response. When the animal is shocked if and only if he fails to jump ($\pi_{11} = \pi_{21} = 1$), he usually learns to avoid shock altogether. Thus this paradigm is called *avoidance conditioning*. If, on the other hand, he is shocked only when he jumps, i.e. punished for jumping, it is clear that he will learn not to jump. For a comparative evaluation of a number of learning models for an avoidance conditioning experiment, see Bush and Mosteller [**4**].

2. *Models.* The following models have been proposed to account for data obtained in two-choice simple learning experiments.

i. *Stimulus-sampling theory.* (See Atkinson and Estes [**1**]). The subject's experimental environment is represented by N "stimulus elements," each of which is "conditioned" to A_1 or A_2. On each trial the subject "samples" s of these and makes A_1 with probability k/s if k elements of the sample are conditioned to A_1. If O_{ij} occurs, conditioning is "effective" with probability c_{ij}, in which case everything in the sample becomes conditioned to the reinforced response A_j. If conditioning is not effective, then nothing in the sample changes its state of conditioning. (Symmetries in the experimental situation may, of course, reduce the number of distinct c_{ij} that need be considered. More generally, theoretical or empirical considerations may suggest the relative or absolute orders of magnitude of these parameters. Similar comments apply to all the models discussed below.) Variants of this model may be distinguished on the basis of the composition of the sample. For definiteness we will restrict our attention to the *fixed sample size model* in which s is constant and any sample of size s has the same probability. However, the methods to be discussed are equally applicable to other stimulus-sampling models.

In this model we identify p_n, the probability of $A_{1,n}$, with the proportion of the N stimulus elements that are conditioned to A_1 at beginning of trial n. Given p_n, the probability that the sample contains j elements conditioned to A_2, A_1 occurs and is reinforced, and conditioning is effective, so that $\Delta p_n = p_{n+1} - p_n = j/N$, is

$$\phi(p_n) = \frac{\binom{Np_n}{s-j}\binom{N(1-p_n)}{j}}{\binom{N}{s}}\left(1 - \frac{j}{s}\right)\pi_{11}c_{11}.$$

The entire conditional probability distribution of Δp_n given p_n is determined similarly (see Norman [18, Appendix]).

ii. *The four-operator linear model.* (See Sternberg [24].) Here it is assumed that if $A_{i,n}$ and $O_{i1,n}$ then the increment Δp_n is a fixed proportion θ_{i1} of the maximum possible increment $1 - p_n$, while if $A_{i,n}$ and $O_{i2,n}$ then the decrement Δp_n is a proportion θ_{i2} of the maximum possible decrement $-p_n$, $0 \leqq \theta_{ij} \leqq 1$. Thus p_n satisfies the stochastic difference equation

$$\begin{aligned}\Delta p_n &= \theta_{11}(1 - p_n) && \text{if } A_{1,n}O_{11,n},\\ &= \theta_{21}(1 - p_n) && \text{if } A_{2,n}O_{21,n},\\ &= -\theta_{12}p_n && \text{if } A_{1,n}O_{12,n},\\ &= -\theta_{22}p_n && \text{if } A_{2,n}O_{22,n},\end{aligned}$$

where, denoting $P(A_{i,n}O_{ij,n} \mid p_n = p)$ by $\phi_{ij}(p)$,

$$\begin{aligned}\phi_{11}(p) &= p\pi_{11},\\ \phi_{12}(p) &= p\pi_{12},\\ \phi_{21}(p) &= (1 - p)\pi_{21},\end{aligned}$$

and

$$\phi_{22}(p) = (1 - p)\pi_{22}.$$

iii. *The beta model.* (See Sternberg [24].) Let v_n, the "strength" of the response A_1 on trial n, be related to p_n by the equation $p_n = v_n/(1 + v_n)$, or, equivalently $v_n = p_n/(1 - p_n)$. We suppose that to every outcome O_{ij} there is a constant β_{ij} such that $v_{n+1} = \beta_{ij}v_n$ if $A_{i,n}O_{ij,n}$. At the level of p_n we have the transition equation

$$p_{n+1} = \frac{\beta_{ij}p_n}{(1 - p_n) + \beta_{ij}p_n} \quad \text{if } A_{i,n}O_{ij,n}.$$

The quantities $\phi_{ij}(p) = P(A_{i,n}O_{ij,n} \mid p_n = p)$ are the same as in the four-operator linear model. Since O_{ij} reinforces A_j, we must have $\beta_{11} \geqq 1$, $\beta_{21} \geqq 1$, $\beta_{12} \leqq 1$, and $\beta_{22} \leqq 1$.

B. *A discrimination learning experiment.* Consider a T-maze with two lights S_1 and S_2 beside each other at the choice point. On each trial one or the other is lit, and food is available on the left (A_1) or right (A_2) depending on which. $S_{1,n}$ and $S_{2,n}$ each occur on a random 50% of the trials. Clearly, most rats will eventually learn to make the appropriate response on every trial of such an experiment.

The model that will now be presented is Bush's linearization of a theory proposed by L. B. Wyckoff, Jr. (see Bush [**3**, Section 1.3]). We will refer to it as the *Wyckoff model.* Let u_n be the rat's probability of noting which light is lit on trial n (event T_n). Let $x_{1,n}$ and $x_{2,n}$ be, respectively, the probabilities of getting food (event F_n) on S_1 and S_2 trials if the rat attends to the lights. The probability of getting food is clearly 1/2 if he does not attend. If the rat attends to the lights on an S_i trial, it is assumed that either a rewarded A_i response or a nonrewarded alternative response increases the probability x_i of making the A_i response on subsequent trials of this sort. These are the only circumstances in which x_i changes. The probability u of attending to the lights is assumed to increase if attention is followed by reward or nonattention is followed by nonreward, and to decrease in complementary cases. Finally, it is assumed that all of the increments and decrements are effected by linear functions. These assumptions are summarized in Table 1,

TABLE 1

Event	Probability	Δu	Δx_1	Δx_2
S_1TF	$(1/2)ux_1$	+	+	0
$S_1T\tilde{F}$	$(1/2)u(1-x_1)$	−	+	0
S_2TF	$(1/2)ux_2$	+	0	+
$S_2T\tilde{F}$	$(1/2)u(1-x_2)$	−	0	+
$\tilde{T}F$	$(1-u)/2$	−	0	0
$\tilde{T}\tilde{F}$	$(1-u)/2$	+	0	0

in which + (−, 0) indicates a transformation of the form $\Delta y = \theta(1-y)$ ($\Delta y = -\theta y$, $\Delta y = 0$) where $\theta > 0$. The θ's associated with different events may be different.

C. *An experiment with a continuum of responses.* The subject is seated before a large disc. On each trial a spot of light appears somewhere on the rim R. The subject's task is to predict where.

Let Y_n be the subject's nth prediction, and let Z_n be the position of the reinforcement spot. It is assumed that the probability distribution $\pi(Y_n, \cdot)$ of Z_n depends at most on Y_n.

Let μ_n be the probability distribution of Y_n for a single subject. In the model proposed by Suppes [25], μ_n is a random measure satisfying the stochastic difference equation

$$\mu_{n+1} = (1-\theta)\mu_n + \theta\tau(Z_n, \cdot),$$

where θ is a constant, and, for every z, $\tau(z, \cdot)$ is a probability distribution on R with mode at z. Thus the occurrence of a reinforcement at z moves the subject in the direction of a response distribution centered at z as, intuitively, it should.

D. *A general theoretical framework.* All of these examples have the following structure. The behavior of the subject on trial n is determined by his *state* S_n at the beginning of the trial. S_n is a random variable taking on values in a *state space* S. On trial n an *event* E_n occurs that results in a change of state. The random variable E_n takes on values in an *event space* E. That is, to each event e there corresponds a function f_e from S into S such that

$$S_{n+1} = f_{E_n}(S_n), \quad n = 1, 2, \cdots.$$

The function f_e will be called the *operator for the event* e or simply an event operator. The probability distribution $\phi\,.\,(S_n)$ of E_n depends only on the state S_n at the beginning of the trial. Henceforth by a *learning model* I will mean a structure (S, E, f, ϕ).

In the stimulus-sampling model described above, $S_n = p_n$, $S = \{k/N : k = 0, 1, \cdots, N\}$, and each specification of number of elements in the trial sample conditioned to A_2, response, outcome, and effectiveness or ineffectiveness of conditioning determines an event. The operators f_e are translations when restricted to $\{s : \phi_e(s) > 0\}$. Outside of this set the value of f_e is inconsequential. In the four-operator linear model, $S_n = p_n$, $S = [0, 1]$, E is the set of possible response-outcome pairs, and $f_{ij}(p) = (1-\theta_{ij})p + \theta_{ij}\delta_{1j}$. For the beta model we may take $S_n = p_n$, v_n, or $\ln v_n$; S, f and ϕ vary accordingly. As in the linear model, E is the set of response-outcome pairs. In the Wyckoff model, it is natural to take $S_n = (u_n, x_{1,n}, x_{2,n})$ and S the unit cube. The events, their conditional probabilities, and their corresponding event operators are given, respectively, by the first, second, and last three columns of Table 1. Finally, in

Suppes' model, we take $S_n = \mu_n$ and $E_n = Z_n$, so that E is the rim R of the disc and S is the set of Borel probability measures on R.

For any learning model within the framework presented above, it is clear that both the stochastic processes $\{S_n\}$ and $\{W_n\}$, where $W_n = (S_n, E_n)$, are Markov with stationary transition probabilities. This observation is basic to much of what follows.

II. Compact Markov processes and distance diminishing models.

A. *Models with finite state and event spaces.* Learning models like the stimulus-sampling models for which both S and E have only a finite number of elements will be called *finite state models*. In this case the stochastic processes $\{S_n\}$ and $\{W_n\}$ $(W_n = (S_n, E_n))$ are finite Markov chains. The general theory of such chains is exceedingly well developed (see Chung [**5**]) and provides a powerful tool for analyzing these models. Whereas the fact that $\{S_n\}$ is a Markov chain has been used extensively by learning theorists, they have seldom applied Markov chain theory to the process $\{W_n\}$.

As an example of what is available, suppose that the chain $\{W_n\}$ has only one recurrent class and that this class is aperiodic. (For this it suffices that the same be true for $\{S_n\}$.) Let ψ be an arbitrary real valued function on $S \times E$. Then, whatever the initial value s of S_1, the limits

$$\lim_{n\to\infty} E_s[\psi(W_n)] = \mu$$

and

$$\lim_{n\to\infty} nE_s\left[\left(\frac{1}{n}\sum_{j=1}^{n}\psi(W_j) - \mu\right)^2\right] = \sigma^2$$

exist and do not depend on s. The average $(1/n)\sum_{j=1}^{n}\psi(W_j)$ converges with probability 1 to μ and, if $\sigma^2 > 0$, is asymptotically normally distributed with mean μ and variance σ^2/n.

In the stimulus-sampling model, the event that occurs on trial n includes a description of the subject's response on that trial, hence $X_n = \psi(W_n)$ for some ψ, where X_n is the indicator random variable of the event $A_{1,n}$. Thus the above theorems give us precise information about the asymptotic behavior of the proportion of A_1 responses over the first n trials for a single subject. Similar information about the frequency of compound events such as $A_{1,n}O_{11,n}A_{1,n+1} \cup A_{2,n}O_{21,n}A_{1,n+1}$ (reinforcement of A_1 followed by A_1 on the next trial) can be obtained by considering the corresponding compound finite Markov chain, in this case $\{(W_n, W_{n+1})\}$.

B. *Distance diminishing models*. The four-operator linear model, the Wyckoff model, and, more generally, the class of distance diminishing models to be defined shortly, can be treated by a Markov process theory completely analogous to the theory of finite Markov chains. Later in this section, I will introduce this theory; In Section C, I will outline it; and in Section D, I will indicate how it can be applied to learning models.

Suppose that d is a metric on the state space S. For mappings ψ and g of S into the complex numbers and into S, respectively, their maximum "difference quotients" $m(\psi)$ and $\mu(g)$ are defined by

$$m(\psi) = \sup_{s \neq s'} \frac{|\psi(s) - \psi(s')|}{d(s, s')}$$

and

$$\mu(g) = \sup_{s \neq s'} \frac{d(g(s), g(s'))}{d(s, s')}.$$

If $m(\psi) < \infty$, then ψ is said to satisfy a Lipschitz condition. The mapping g is *distance diminishing* if $\mu(g) \leqq 1$ and *strictly distance diminishing* if $\mu(g) < 1$. A learning model is distance diminishing (with respect to d) if (S, d) is compact; the event space E is a finite set; for each event e, $\phi_e(\cdot)$ satisfies a Lipschitz condition and $f_e(\cdot)$ is distance diminishing; and, finally, for any state s there is a positive integer k and there are k events $e_1, \cdots, e_k$ such that this sequence of events succeeds s with positive probability, and the composite mapping of S corresponding to it is strictly distance diminishing:

$$\phi_{e_1 \cdots e_k}(s) > 0 \quad \text{and} \quad \mu(f_{e_1 \cdots e_k}) < 1,$$

where

$$\phi_{e_1 \cdots e_k}(s) = P_s(E_j = e_j, j = 1, \cdots, k)$$

and

$$f_{e_1 \cdots e_k} = f_{e_k} \circ f_{e_{k-1}} \circ \cdots \circ f_{e_1}.$$

The four-operator linear model is distance diminishing if and only if for every i there is a j_i such that $\theta_{ij_i} > 0$ and $\pi_{ij_i} > 0$ (Norman [18, Lemma 3.2]). The Wyckoff model can be shown to be distance diminishing under the restrictions on its parameters given when it was introduced. In both cases the metric considered is Euclidean.

Let $K(\cdot, \cdot)$ be the transition function for a Markov process $\{S_n\}_{n=1}^{\infty}$ in a compact metric space (S, d), i.e.

$$K(s, A) = P(S_{n+1} \in A \mid S_n = s).$$

It is assumed that

(i) $K(s, \cdot)$ is a Borel probability measure for each $s \in S$, and

(ii) $K(\cdot, A)$ is Borel measurable for each Borel set A.

Let U be the linear operator on bounded measurable complex valued functions on S defined by

$$U\psi(s) = \int K(s, dt)\psi(t) = E[\psi(S_{n+1}) \mid S_n = s].$$

Let $C(S)$ be the Banach space of continuous complex valued functions on S under the norm

$$|\psi| = \max_{s \in S} |\psi(s)|,$$

and let CL be the subspace of functions that satisfy a Lipschitz condition. CL is a Banach space with respect to the norm

$$\|\psi\| = m(\psi) + |\psi|.$$

I have shown (Norman [18, Proof of Theorem 5.1]) that, if the Markov process $\{S_n\}$ is associated with a distance diminishing learning model, then the following conditions are satisfied:

(iii) U maps CL into CL and is bounded with respect to the norm $\|\cdot\|$, i.e.

$$\|U\| = \sup_{\psi \in CL; \psi \neq 0} \frac{\|U\psi\|}{\|\psi\|} < \infty,$$

and

(iv) there is a positive integer k and there are two real numbers $r < 1$ and R such that

$$m(U^k\psi) \leq rm(\psi) + R|\psi| \quad \text{for all } \psi \in CL.$$

Any Markov process in a compact metric space that satisfies (i)—(iv) will be called a *compact Markov process*.

Let δ be any metric on the event space E of a distance diminishing model $((S, d), E, f, \phi)$ (e.g. $\delta(e, e') = 0$ or 1 depending on whether or not $e = e'$). Since E is finite, (E, δ) is compact. It follows that $S \times E$ is compact with respect to the metric $D((s, e), (s', e')) = d(s, s') + \delta(e, e')$. Using the fact that the process $\{S_n\}$ for a distance diminishing model is compact, it can be shown that the process $\{W_n\}$ $(W_n = (S_n, E_n))$ is also a compact Markov process with respect to the metric D. Thus compact Markov processes stand in the same relation to distance diminishing models that finite Markov chains do to finite state models.

C. *Compact Markov processes.* If $\{S_n\}$ is any finite Markov chain and d is any metric on its state space S, then it is clear that $C(S)$ and CL consist of all complex valued functions on S. For any such function ψ, $m(\psi) \leqq \nu|\psi|$ where $\nu = 2/\min_{s \neq s'} d(s,s')$, so

$$m(U\psi) \leqq \nu|U\psi| \leqq \nu|\psi|,$$

and (iv) is satisfied. Thus any finite Markov chain is a compact Markov process. On the other hand, it is easy to show that, if $\{S_n\}$ is a compact Markov process, then U maps $C(S)$ into $C(S)$, and, for any $\psi \in C(S)$, the sequence $\{U^n\psi\}$ is equicontinuous. An extensive theory of Markov processes having this property is emerging (Jamison [**11**], [**12**]; Rosenblatt [**22**], [**23**]), and this theory has been helpful in the work discussed below. Thus compact Markov processes stand between finite Markov chains and processes for which the sequence $\{U^n\psi\}$ is equicontinuous. Their theory is actually much closer to that of the former than to that of the latter.

The following basic theorem is a specialization of a uniform ergodic theorem of Ionescu Tulcea and Marinescu ([**9**, Section 9]) along lines suggested by these authors.

THEOREM 1. *Let the operator U correspond to a compact Markov process. Then*

(a) *there are at most a finite number of eigenvalues $\lambda_1, \lambda_2, \cdots, \lambda_p$ of U for which $|\lambda_i| = 1$;*

(b) *for all positive integers n*

$$U^n = \sum_{i=1}^{p} \lambda_i^n U_i + V^n,$$

where V and U_i are linear operators on CL, bounded with respect to $\|\cdot\|$;

(c) $U_i^2 = U_i$, $U_i U_j = 0$ *for* $i \neq j$, $U_i V = VU_i = 0$;

(d) *if* $D(\lambda_i) = \{\psi \in CL\colon U\psi = \lambda_i\psi\}$ *then* $D(\lambda_i) = U_i(CL)$ *is finite dimensional,* $i = 1, \cdots, p$; *and*

(e) *for some* $M < \infty$ *and* $h > 0$,

$$\|V^n\| \leqq M/(1+h)^n$$

for all positive integers n.

From the fact that the norms $\|\cdot\|$ and $|\cdot|$ are equivalent on the finite dimensional linear space $U_i(CL)$, it follows that the condition (iv) is necessary for (a)—(e). It is also immediate from (a)—(e) that, for n sufficiently large, the strongly completely continuous operator $\sum_{i=1}^{p} \lambda_i^n U_i$ on CL satisfies

$$\left\| U^n - \sum_{i=1}^{p} \lambda_i^n U_i \right\| < 1.$$

Thus the operator U on CL is quasi-strongly completely continuous (see Yosida and Kakutani [27]).

If the higher transition kernels $K^{(n)}$ are defined in the usual way:

$$\begin{aligned} K^{(0)}(s, A) &= 1 \qquad \text{if } s \in A, \\ &= 0 \qquad \text{if } s \in' A, \end{aligned}$$

and

$$K^{(n+1)}(s, A) = \int K(s, dt)\, K^{(n)}(t, A);$$

then

$$K^{(n)}(s, A) = P_s(S_{n+1} \in A)$$

and

$$U^n \psi(s) = \int K^{(n)}(s, dt)\, \psi(t) = E_s[\psi(S_{n+1})].$$

In view of the latter equations, Theorem 1 gives us a firm grip on the asymptotic behavior of the quantities $K^{(n)}(s, A)$ and $E_s[\psi(S_{n+1})]$, $\psi \in CL$.

Clearly 1 is an eigenvalue of U and all of the nonzero complex constants are corresponding eigenfunctions. Without loss of generality we suppose that $\lambda_1 = 1$ and $\lambda_i \neq 1$ for $i \neq 1$. Theorem 1 implies that $|\overline{U}_n \psi - U_1 \psi| \to 0$ as $n \to \infty$ for any $\psi \in C(S)$, where $\overline{U}_n = (1/n) \sum_{j=0}^{n-1} U^j$ and U_1 has been extended (uniquely) to a bounded linear operator on $C(S)$. It follows, in turn, that U_1 has the representation

$$U_1 \psi(s) = \int K^{\infty}(s, dt)\, \psi(t)$$

where K^{∞} satisfies (i) and (ii), and that $\overline{K}_n(s, \cdot)$ converges weakly to $K^{\infty}(s, \cdot)$ for every $s \in S$, where $\overline{K}_n = (1/n) \sum_{j=0}^{n-1} K^{(j)}$. In fact, the convergence is, in a certain sense, uniform over s (see Norman

[18, Definition 2.1, and the proof of Lemma 5.3]). If 1 is the only eigenvalue of modulus 1 of U, the Cesaro averaging in the above statements is unnecessary.

For any $s \in S$, $K^\infty(s, \cdot)$ is a *stationary probability distribution* in the sense that S_{n+1} has this distribution if S_n has. Let $M(S)$ be the space of complex valued Borel measures on S, and let T be the operator on $M(S)$ defined by

$$T\nu(A) = \int \nu(ds)K(s, A).$$

Then T is the adjoint of the operator U on continuous functions, and, more intuitively, T takes the distribution of S_n into that of S_{n+1}. Thus the stationary probability distributions are those that are fixed under T.

A Borel set B is *stochastically closed* if it is nonempty and if $K(s, B) = 1$ for all $s \in B$. B is an *ergodic kernel* if it is stochastically and topologically closed and if it has no proper subset with these properties. It is easy to show that distinct ergodic kernels are disjoint. Whatever the initial state of the process, it is attracted to its ergodic kernels as the following theorem shows.

THEOREM 2. *A compact Markov process $\{S_n\}$ has l ergodic kernels, where l is the dimension of $D(1)$. Denote these $E_1, \cdots, E_l$, and let*

$$\gamma_i(s) = P_s\left(\lim_{n\to\infty} d(S_n, E_i) = 0\right).$$

Then $\sum_{i=1}^{l} \gamma_i(s) \equiv 1$ and $\{\gamma_1, \cdots, \gamma_l\}$ is a basis for $D(1)$. There is a unique stationary probability measure μ_i with support E_i, and $\{\mu_1, \cdots, \mu_l\}$ is a basis for $\{\mu \in M(S) : T\mu = \mu\}$. Finally

$$K^\infty(s, B) = \sum_{i=1}^{l} \gamma_i(s)\,\mu_i(B).$$

It follows that a process has a unique stationary probability distribution if and only if there is only one ergodic kernel. In this case we denote this distribution K^∞. Thus $U_1\psi(s) \equiv \int \psi(t)K^\infty(dt)$.

Any subprocess of a compact Markov process obtained by restricting the process to a stochastically and topologically closed subset of the state space is a compact Markov process. Just as in the theory of finite Markov chains, the subprocesses on the ergodic kernels may have a cyclic character. The periods of these cycles determine the eigenvalues of modulus 1 of the original process.

THEOREM 3. *A complex number of modulus* 1 *is an eigenvalue of a compact Markov process if and only if it is an eigenvalue of the subprocess on some ergodic kernel. For any ergodic kernel E there is a positive integer k* (*called the period of E*) *such that the eigenvalues of modulus* 1 *of the corresponding subprocess are* $\exp[2\pi ij/k]$, $j = 1, \cdots, k$. *There are k topologically closed pairwise disjoint sets* $E^1, \cdots, E^k$ *with union E, such that* $K(s, E^{j+1}) = 1$ *for all* $s \in E^j$, $j = 1, \cdots, k$ $(E^{k+1} = E^1)$.

If U has no eigenvalues of modulus 1 other than 1, we say that the process is *aperiodic*.

Let us now consider the asymptotic behavior of the sums $(1/n)\sum_{j=1}^n \psi(S_j)$ for a compact Markov process $\{S_n\}$ and a real valued continuous function ψ. The following strong law of large numbers is obtained by combining a theorem of Jamison [**12**, Theorem 3.2] with Theorem 2 above: Let M_i be the indicator random variable for the event "$\lim_{n\to\infty} d(S_n, E_i) = 0$". Then for any initial state s the probability is 1 that

$$\lim_{n\to\infty} \frac{1}{n} \sum_{j=1}^{n} \psi(S_j) = \sum_{i=1}^{l} M_i \int \psi(t)\, \mu_i(dt)$$

for all continuous real valued functions ψ. Thus, if there is a unique stationary probability distribution K^∞,

$$\frac{1}{n} \sum_{j=1}^{n} \psi(S_j) \to \int \psi(t) K^\infty(dt) \quad \text{as } n \to \infty .$$

Suppose now that ψ is real valued and satisfies a Lipschitz condition, and that the process is aperiodic and has only one ergodic kernel. Assume, without loss of generality, that $\int \psi(t) K^\infty(dt) = 0$. Then $R(m-n) = E[\psi(S_m)\psi(S_n)]$ converges geometrically to 0 as $|m-n| \to \infty$, so the series $\sigma^2 = \sum_{j=-\infty}^{\infty} R(j)$ converges absolutely. (In these expressions and those that follow, expectations without subscripts are taken with respect to the stationary process that arises when S_1 has distribution K^∞.) Moreover, for any $s \in S$

$$E_s\left[\left(\frac{1}{\sqrt{n}} \sum_{j=1}^{n} \psi(S_j)\right)^2\right] = \sigma^2 + O\left(\frac{1}{n}\right).$$

If $\sigma^2 > 0$, then $(1/n)\sum_{j=1}^n \psi(S_j)$ is asymptotically normally distributed with mean 0 and variance σ^2/n, for any initial state. As in the case of the analogous central limit theorem for Markov processes satisfying Doeblin's condition (Doob [**6**, Theorem 7.5]), this can be proved by reduction to the case of sums of independent random variables. Since $\int \psi(t) K^{\infty}(dt) = 0$, the series

$$\bar{\psi} = \sum_{j=0}^{\infty} U^j \psi$$

converges absolutely in CL. From the representation

$$\sigma^2 = E[(\bar{\psi}(S_{n+1}) - U\bar{\psi}(S_n))^2],$$

which follows from the series representation given above, and the fact that $\bar{\psi} - U\bar{\psi} = \psi$, it is clear that if $\sigma^2 = 0$, then $\psi(S_n) = \bar{\psi}(S_n) - \bar{\psi}(S_{n+1})$ with probability 1 for the stationary process. Using this condition, Norma Graham and I have shown that if ψ has only two values, say a and b ($a \neq b$) and if both have positive probability asymptotically ($K^{\infty}(\psi^{-1}\{a\}) > 0$ and $K^{\infty}(\psi^{-1}\{b\}) > 0$), then $\sigma^2 > 0$.

Since $R(j)$ converges geometrically to zero as $|j| \to \infty$, the process $\{\psi(S_n)\}$ ($\{S_n\}$ stationary) has a spectral density function

$$f(t) = \frac{1}{2\pi} \sum_{j=-\infty}^{\infty} e^{-itj} R(j).$$

The quantity σ^2 is $2\pi f(0)$. Let

$$\begin{aligned} p_{a,b,c} = {} & E[\psi(S_n)\psi(S_{n+a})\psi(S_{n+b})\psi(S_{n+c})] \\ & - R(a)R(c-b) - R(b)R(c-a) - R(c)R(b-a). \end{aligned}$$

Under our assumptions (ψ real valued, $\psi \in CL$, $\{S_n\}$ aperiodic with only one ergodic kernel) it can be shown that

$$\sum_{a,b,c=-\infty}^{\infty} |p_{a,b,c}| < \infty.$$

Thus the spectral estimation techniques for processes whose covariances decrease exponentially described by Parzen [**21**] are applicable. Briefly, there are estimators of *exponential type* for which various measures of the error of estimation are of the order of magnitude of $(\ln n)/n$, where n is the number of successive observations of which the estimator is a function.

One final fact: just as the compound process $\{(S_n, S_{n+1}, \cdots, S_{n+k})\}_{n=1}^{\infty}$ is a finite Markov chain if $\{S_n\}$ is, processes compounded out of any compact Markov process are also compact Markov processes. Furthermore, if $\{S_n\}$ is aperiodic or has only one ergodic kernel the same will be true of $\{(S_n, S_{n+1}, \cdots, S_{n+k})\}_{n=1}^{\infty}$.

D. *Application of compact Markov process theory to distance diminishing models.* I stated earlier that the sequence $\{W_n\}$ of state-events pairs from a distance diminishing learning model is a compact Markov process. I now add that if the sequence $\{S_n\}$ of states is aperiodic or has only one ergodic kernel then the same is true of $\{W_n\}$. These facts are important since they permit us to establish regularities of $\{W_n\}$ by considering the simpler process $\{S_n\}$.

Returning to our examples: I have shown that if a four-operator linear model has no absorbing states (stochastically closed unit sets), then it is distance diminishing and $\{p_n\}$ has only one ergodic kernel E_1. If $\theta_{ij} = 1$ and $\pi_{ij} = 1$ for $i \neq j$ then $E_1 = \{0, 1\}$ and E_1 has period 2. Otherwise the process is aperiodic. If, on the other hand, either 0 or 1 is absorbing for a distance diminishing four-operator linear model then the process is aperiodic and these absorbing states are the ergodic kernels (Norman [**18**, Theorems 3.1, 3.2]). For the Wyckoff model it can be shown that, for any state s, the distance from the support $T_n(s)$ of the distribution $K^{(n)}(s, \cdot)$ of S_{n+1} to the absorbing state $(1, 1, 1)$ converges to 0 as $n \to \infty$. From this it follows that $E_1 = \{(1, 1, 1)\}$ is the only ergodic kernel. Clearly the process is aperiodic.

For a critical discussion of earlier mathematical work on distance diminishing models see Norman [**18**, Section 4]. That paper also contains an alternative treatment of the event process $\{E_n\}$, via the theory of random systems with complete connections (see Iosifescu [**10**]).

III. Some special results for the four-operator linear model.

A. *No absorbing barriers.* Consider first the special case where all θ_{ij}'s are equal to some constant $\theta > 0$ and reinforcement is noncontingent ($\pi_{1j} = \pi_{2j} = \pi_j$) with $0 < \pi_1 < 1$. The difference equation for a subject's probability p_n of making response A_1 on trial n reduces in this case to

$$\Delta p_n = \theta(k_n - p_n),$$

where k_n is 1 if A_1 is reinforced on trial n and 0 if A_2 is reinforced on trial n. The random variables $\{k_n\}$ are mutually independent with $P(k_n = 1) = \pi_1$ and $P(k_n = 0) = 1 - \pi_1$. The solution to this equation is

$$p_{n+1} = (1-\theta)^n p_1 + \theta \sum_{m=1}^{n} (1-\theta)^{n-m} k_m.$$

We are going to consider what happens to p_n as n becomes large and θ becomes small. To this end let $\{\theta_n\}$ be a sequence in $(0,1)$ converging to 0. Assuming $\mathrm{var}(p_1) = 0$,

$$p_{n+1} - \xi_{n+1} = \sum_{m=1}^{n} X_{n,m}$$

where $\xi_{n+1} = E_{p_1}[p_{n+1}]$ and

$$X_{n,m} = \theta_n(1-\theta_n)^{n-m}(k_m - \pi_1).$$

Clearly, $X_{n,1}, \cdots, X_{n,n}$ are independent and $E[X_{n,m}] = 0$. Furthermore, a routine computation shows that the triangular array $\{X_{n,m}\}$ of random variables satisfies the Lindeberg condition

$$\lim_{n\to\infty} \frac{1}{\sigma_n^2} \sum_{m=1}^{n} \int_{|x| \geqq \lambda\sigma_n} x^2 dG_{n,m}(x) = 0$$

for all $\lambda > 0$, where $\sigma_n^2 = \mathrm{var}_{p_1}(p_{n+1})$ and $G_{n,m}$ is the probability distribution of $X_{n,m}$. It follows that $(p_n - \xi_n)/\sigma_n$ is asymptotically normally distributed with mean 0 and variance 1. Thus

$$\lim_{n\to\infty;\,\theta\to 0} F_{n,\theta,p_1}(x) = \Phi(x)$$

for all x, where

$$F_{n,\theta,p_1}(x) = P_{p_1}((p_n - E_{p_1}[p_n])/\mathrm{var}_{p_1}^{1/2}[p_n] < x),$$

Φ is the standard normal cumulative distribution function, and the limit is taken as $n \to \infty$ and $\theta \to 0$ simultaneously. *After long experience in a situation in which learning occurs slowly, p_n is approximately normally distributed according to this model.*

Now

$$J_n(x) = \lim_{\theta\to 0} F_{n,\theta,p_1}(x) \quad \text{and} \quad H_\theta(x) = \lim_{n\to\infty} F_{n,\theta,p_1}(x)$$

both exist. As $\theta \to 0$, $(p_n - \xi_n)/\sigma_n$ converges to the normalized binomial random variable

$$\sum_{m=1}^{n-1} \frac{k_m - \pi_1}{[(n-1)\pi_1(1-\pi_1)]^{1/2}},$$

so J_n is a normalized binomial distribution. The existence of the limit of F_{n,θ,p_1} as $n \to \infty$ and its independence of p_1 follow from the fact that the process $\{p_n\}$ has no absorbing states and is aperiodic.

The asymptotic normality of F_{n,θ,p_1} as $n \to \infty$ and $\theta \to 0$ simultaneously then implies that

$$\lim_{n\to\infty} J_n(x) = \Phi(x),$$

and

$$\lim_{\theta\to 0} H_\theta(x) = \Phi(x).$$

The former result is, of course, nothing but DeMoivre's central limit theorem. Since $\sigma_n^2 \to \pi_1(1-\pi_1)\theta/(2-\theta)$ and $\xi_n \to \pi_1$ as $n \to \infty$, the latter result can be rewritten

$$\lim_{\theta\to 0}\lim_{n\to\infty} P_{p_1}\left(\frac{p_n - \pi_1}{[\pi_1(1-\pi_1)\theta/2]^{1/2}} < x\right) = \Phi(x).$$

I will now show how the latter result can be generalized to four-operator linear models for which the process $\{p_n\}$ is aperiodic and has no absorbing states. The method that I will describe leads to an asymptotic expansion of the characteristic function of the distribution

$$\lim_{n\to\infty} P_{p_1}\left(\frac{p_n - \rho}{\theta^{1/2}} < x\right)$$

in powers of $\theta^{1/2}$, where all θ_{ij} are assumed to be proportional to θ, and ρ does not depend on θ.

Consider then a family $\{p_n^{(\theta)}\}_{n=1}^{\infty}$ $(0 < \theta < 1)$ of Markov processes in the unit interval satisfying stochastic difference equations of the form

$$\Delta p_n^{(\theta)} = \theta u_j(p_n^{(\theta)}) \quad \text{with prob. } \phi_j(p_n^{(\theta)}),$$

$1 \leqq j \leqq N$, where $-p \leqq u_j(p) \leqq 1-p$ and $\sum_j \phi_j(p) = 1$ for $0 \leqq p \leqq 1$, and u_j and ϕ_j are infinitely differentiable on $[0,1]$. Let

$$h_m(p) = \sum_j u_j^m(p)\phi_j(p)$$

$$= E[(\Delta p_n^{(\theta)}/\theta)^m \mid p_n^{(\theta)} = p],$$

and suppose that $h_1(0) > 0$, $h_1(1) < 0$, and that there is a unique $\rho \in (0,1)$ such that $h_1(\rho) = 0$. Necessarily $h_1'(\rho) \leqq 0$ and $h_2(\rho) \geqq 0$. We assume that $h_1'(\rho) < 0$ and $h_2(\rho) > 0$. Finally we suppose that, for every $0 < \theta < 1$, F_θ is a stationary cumulative distribution function for the Markov process $\{p_n^{(\theta)}\}$.

For the four-operator linear model, for instance, let b_{ij} $(1 \leq i, j \leq 2)$ be a constant in $[0,1]$ and let $\theta_{ij} = \theta b_{ij}$. In terms of this convenient double index notation we then have

$$u_{i1}(p) = b_{i1}(1-p), \qquad u_{i2}(p) = -b_{i2}p,$$

and

$$\phi_{1j}(p) = p\pi_{1j}, \qquad \phi_{2j}(p) = (1-p)\pi_{2j}.$$

The assumption that the process $\{p_n^{(\theta)}\}$ is aperiodic with no absorbing states is equivalent to the restrictions $b_{ij} > 0$ and $\pi_{ij} > 0$ if $i \neq j$. For such a model the distribution of $p_n^{(\theta)}$ converges as $n \to \infty$ to a stationary distribution F_θ that is, in fact, independent of p_1. That $h_1(p)$ has a unique zero ρ in $(0,1)$ and that $h_1'(\rho) < 0$ follow from the fact that h_1 is quadratic with $h_1(0) > 0$ and $h_1(1) < 0$.

We return now to the general case and henceforth omit θ superscripts. Let $W(\cdot, p)$ be the conditional characteristic function of the normalized increment $\Delta p_n/\theta$ given p_n:

$$W(s,p) = E[\exp(is\Delta p_n/\theta) \mid p_n = p].$$

Notice that W is independent of θ. Let

$$W(s,p) = \sum_{j,k=0}^{\infty} a_{j,k}(is)^j(p-\rho)^k \tag{1}$$

be the formal Taylor expansion of W in powers of $i \cdot s$ and $p - \rho$. Then

$$\sum_{k=0}^{\infty} a_{j,k}(p-\rho)^k = \frac{1}{j!}E\left[\left(\frac{\Delta p_n}{\theta}\right)^j \,\middle|\, p_n = p\right].$$

These equations give

$$a_{0,k} = \delta_{0,k}, \quad a_{1,0} = h_1(\rho) = 0, \quad a_{1,1} = h_1'(\rho) < 0,$$

and

$$a_{2l,0} = h_{2l}(\rho)/(2l)! > 0, \qquad l = 0, 1, 2, \cdots.$$

For any distribution of p_1 we have

$$E[\exp(it(p_{n+1}-\rho)/\theta^{1/2})] = E[W(\theta^{1/2}t, p_n)\exp(it(p_n-\rho)/\theta^{1/2})].$$

Letting p_1 have distribution F_θ and introducing $G_\theta(x) = F_\theta(\theta^{1/2}x + \rho)$, we obtain

$$(2) \qquad \int_{-\infty}^{\infty} e^{itx} dG_\theta(x) = \int_{-\infty}^{\infty} e^{itx} W(\theta^{1/2}t, \rho + \theta^{1/2}x)\, dG_\theta(x).$$

Let

$$g(t,\theta) = \int_{-\infty}^{\infty} e^{itx} dG_\theta(x)$$

be the characteristic function of G_θ. The result toward which we are working is that g has an asymptotic expansion of the form

$$(3) \qquad g(t,\theta) = \sum_{j=0}^{\infty} g_j(t)\, \theta^{j/2}$$

in powers of $\theta^{1/2}$. The heuristic argument that follows will suggest the form of the g_j. That (3) (and (5)) actually gives an asymptotic expansion can then be established by a supplementary argument that will not be presented here.

Substituting (1) into (2) and using

$$\int_{-\infty}^{\infty} x^k e^{itx} dG_\theta(x) = \frac{1}{i^k} \frac{\partial^k}{\partial t^k} g(t,\theta)$$

we obtain

$$(4) \qquad g(t,\theta) = \sum_{j,k=0}^{\infty} a_{j,k} \theta^{(j+k)/2} (it)^j \frac{1}{i^k} \frac{\partial^k}{\partial t^k} g(t,\theta).$$

Formal differentiation of (3) k times with respect to t yields

$$(5) \qquad \frac{\partial^k}{\partial t^k} g(t,\theta) = \sum_{l=0}^{\infty} g_l^{(k)}(t)\, \theta^{l/2}.$$

Substitution of (3) on the left and (5) on the right in (4), and equation of coefficients of like powers of $\theta^{1/2}$ on the two sides of the resulting equation leads to the system

$$g_m(t) = \sum_{\substack{j+k+l=m \\ j,k,l \geqq 0}} a_{j,k} (it)^j (-i)^k g_l^{(k)}(t), \qquad m \geqq 0,$$

of difference-differential equations for $g_m(t)$. In terms of the quantities

$$Q_m(t) = \sum_{\substack{j+k+l=m+2 \\ 1 \leqq j; l \leqq m-1}} \frac{a_{j,k}}{-a_{1,1}} \, i^{j+k}(-1)^k t^{j-1} g_l^{(k)}(t),$$

$m \geqq 1$, and

$$\sigma^2 = a_{2,0}/-a_{1,1} = h_2(\rho)/-2h_1'(\rho),$$

this can be rewritten

$$g_0'(t) + \sigma^2 t g_0(t) = 0,$$
$$g_m'(t) + \sigma^2 t g_m(t) = Q_m(t), \qquad m \geqq 1.$$

Since

$$1 = g(0, \theta) = \sum_{m=0}^{\infty} g_m(0) \theta^{m/2},$$

the initial conditions $g_m(0) = \delta_{0,m}$ are plausible. Assuming them we obtain

$$g_0(t) = \exp[-\sigma^2 t^2/2]$$

and

$$g_m(t) = \exp\left[\frac{-\sigma^2 t^2}{2}\right] \int_0^t \exp\left[\frac{\sigma^2 s^2}{2}\right] Q_m(s)\, ds$$

for $m \geqq 1$. Since Q_m depends only on $g_0, \cdots, g_{m-1}$, these equations permit the successive calculation of $g_1, g_2, \cdots$. It can also be shown inductively from them that g_m is of the form $g_m(t) = P_m(t) \exp(-\sigma^2 t^2/2)$ where P_m is a polynomial, $m \geqq 1$.

In particular $g(t, \theta) \to \exp(-\sigma^2 t^2/2)$ as $\theta \to 0$, so $G_\theta(x) \to \Phi(x/\sigma)$ as $\theta \to 0$. The asymptotic expansion derived above has the same relation to this central limit theorem that the Edgeworth expansion has to the standard central limit theorem for sums of independent random variables.

A proof that $G_\theta(x) \to \Phi(x/\sigma)$ as $\theta \to 0$ in a slightly more general setting and some additional examples are given by Norman and Graham [**20**].

B. *One absorbing barrier*. If A_1 is reinforced regardless of what response occurs ($\pi_{11} = \pi_{21} = 1$) (e.g. A_1 is avoidance in an avoidance conditioning experiment), then the probability q_n of $A_{2,n}$ (an error or failure to avoid on trial n) satisfies

$$q_{n+1} = \alpha_1 q_n \quad \text{if } A_{1,n},$$
$$= \alpha_2 q_n \quad \text{if } A_{2,n},$$

where $\alpha_1 = 1 - \theta_{11}$ and $\alpha_2 = 1 - \theta_{21}$. Clearly,

$$q_n \leqq q_1 \max{}^{n-1}(\alpha_1, \alpha_2),$$

so, if $\alpha_1, \alpha_2 < 1$, $q_n \to 0$ with probability 1, i.e. the animal learns. To test the model in a given experimental situation, one wants to to know such things as the predicted distribution of the total number of errors, the mean and variance of the trial of last error, the mean number of runs of errors of length j, the probability of an error on trial n, etc. All of these quantities depend on the initial error probability q_1. Explicit series expansions of all of these quantities in powers of q_1 have been worked out by a number of investigators using a variety of methods (see Bush [**2**], Tatsuoka and Mosteller [**26**], and Sternberg [**24**, Equation 80, p. 85]). For example, Bush reduced the computation of many of these quantities to that of the quantities

$$S_k(q_1) = \sum_{n=1}^{\infty} E_{q_1}[q_n^k]$$

and obtained the expansion

$$S_k(q_1) = \frac{q_1^k}{1 - \alpha_1^k} + \sum_{i=k+1}^{\infty} \frac{q_1^i}{1 - \alpha_1^i} \prod_{j=k}^{i-1} \frac{\alpha_2^j - \alpha_1^j}{1 - \alpha_1^j}$$

by an interesting direct computation (see Bush [**2**, Equation 9, p. 219]).

C. *Two absorbing barriers*. The four-operator linear model will have both 0 and 1 absorbing if and only if $\pi_{12} = 0$ or $\theta_{12} = 0$, and $\pi_{21} = 0$ or $\theta_{21} = 0$. This means that the only response that is ever effectively reinforced is the one just made. Generalizing slightly, we consider an experimental situation in which A_i has N_i possible outcomes (e.g. various amounts of food) all of which reinforce A_i and the jth one of which has probability π_{ij}. In the corresponding linear model, p_n satisfies

$$\Delta p_n = \theta_{1j}(1 - p_n) \quad \text{with prob. } p_n \pi_{1j},$$
$$= -\theta_{2k} p_n \quad \text{with prob. } (1 - p_n)\pi_{2k}$$

where $1 \leqq j \leqq N_1$, $1 \leqq k \leqq N_2$, $0 \leqq \theta$, $\pi_{il} \leqq 1$, and $\sum_j \pi_{1j} = \sum_k \pi_{2k} = 1$. We assume further that, for some j, $\theta_{1j} > 0$ and $\pi_{1j} > 0$ and, for some k, $\theta_{2k} > 0$ and $\pi_{2k} > 0$. The model is then distance diminishing, and the absorbing states $\{0\}$ and $\{1\}$ are the only ergodic kernels.

In contrast to the one absorbing barrier linear model discussed above, few explicit expressions for properties of two absorbing barrier linear models are available at this time. For example, the absorption probability

$$\gamma(p) = P_p\left(\lim_{n \to \infty} p_n = 1\right)$$

is basic to the theory and application of the models. It can be shown that γ can be extended to a function meromorphic in the entire complex plane (see Norman [19, Section 5] for the four-operator case). However, except for the few relatively uninteresting cases when γ is a polynomial, no one has yet been able to compute the parameters of any of the standard function theoretic representations of γ.

Some information is available. Let $b_i = \sum_l \theta_{il}\pi_{il}$, $r = b_2/b_1$, $t = \min\{\theta_{il} : \theta_{il} > 0\}$, $T = \max\{\theta_{il}\}$, and

$$V(y) = (e^y - 1)/y.$$

The function V is obviously strictly increasing, hence has an inverse V^{-1}. The quantity r is a measure of the efficacy of reinforcement of A_2 relative to reinforcement of A_1. Thus $r > 1$ means that A_2 is the "better" response. The following theorem treats only this case. The comparable result for the case $r < 1$ is obtained by reversing the roles of A_1 and A_2.

THEOREM 1. *Suppose $r > 1$ and define x and x^* by $x = -V^{-1}(1/r)/t$ and $x^* = V^{-1}(r)/T$. Then*

$$\frac{e^{xp} - 1}{e^x - 1} \leqq \gamma(p) \leqq \frac{e^{x^*p} - 1}{e^{x^*} - 1}$$

for all $0 \leqq p \leqq 1$.

This theorem generalizes Theorem 4 of Norman [**19**]. A slight modification of the proof of the latter theorem will yield a proof of Theorem 1.

It is easy to show that $\gamma(p) = p$ if $r = 1$. Hence we expect $\gamma(p)$

$\doteq p$ if $r \doteq 1$ and $\gamma(p) \leqq p$ when $r > 1$. If $r - 1$ is "large" $\gamma(p)$ is the probability of being absorbed on a very unfavorable response. Thus, we expect $\gamma(p) \doteq 0$. The following theorem on the behavior of γ as the π's and θ's vary follows from Theorem 1 and shows that departures of r from 1 should be measured relative to the θ's.

THEOREM 2. *Suppose* $r > 1$.

(a) *If* $(r-1)/T \to \infty$, *then* $\gamma(p) \to 0$ *for all* $0 < p < 1$. *If* $\gamma(p) \to 0$ *for some* $0 < p < 1$, *then* $(r-1)/t \to \infty$.

(b) *If* $(r-1)/t \to 0$, *then* $\gamma(p) \to p$ *for all* $0 < p < 1$. *Furthermore*

$$\limsup \frac{p - \gamma(p)}{p(1-p)(r-1)/t} \leqq 1 \quad \text{and} \quad \liminf \frac{p - \gamma(p)}{p(1-p)(r-1)/T} \geqq 1$$

for all $0 < p < 1$. *If* $\gamma(p) \to p$ *for some* $0 < p < 1$, *then* $(r-1)/T \to 0$ *and*

$$\liminf \frac{p - \gamma(p)}{p(1-p)(r-1)/T} \geqq 1$$

for all $0 < p < 1$.

In applying a model like this one to experimental data, one might want to predict the mean number $\chi(p)$ of times a response configuration of the type

$$A_{i_1,n+1}A_{i_2,n+2} \cdots A_{i_k,n+k}$$

occurs, when $p_1 = p$. If all the responses in the configuration are the same, then this number will be infinite if the subject is absorbed on that response. Assume then that both responses occur in the configuration. In that case the kth degree (real) polynomial

$$g(p) = P_p(A_{i_1,1} \cdots A_{i_k,k})$$

vanishes on the ergodic kernels $\{0\}$ and $\{1\}$, so $\| U^n g\| \to 0$ geometrically as $n \to \infty$. Thus the series $\sum_{n=0}^{\infty} U^n g$ converges to a *CL* function, and from

$$U^n g(p) = P_p(A_{i_1,n+1} \cdots A_{i_k,n+k})$$

it is clear that this limit is $\chi(p)$. Thus χ satisfies the functional equation

$$(*) \qquad \chi = g + U\chi$$

and the boundary conditions $\chi(0) = \chi(1) = 0$. If Ω is any *CL* solution of $(*)$, then $\Delta = \chi - \Omega$ is a CL solution of the homogeneous equation $U\Delta = \Delta$. It then follows from Theorem 2 of §II that $\Delta = \Delta(1)\gamma + \Delta(0)(1-\gamma)$, i.e.

$$\chi(p) = \Omega(p) - \Omega(1)\gamma(p) - \Omega(0)(1-\gamma(p)).$$

The next theorem gives a sufficient condition for the existence of polynomial solutions of $(*)$ when g is a polynomial that vanishes on the absorbing states as well as a closely related necessary and sufficient condition for γ to be a polynomial. The usefulness of this theorem is enhanced by the fact that, if a polynomial solution of $(*)$ exists, it is very easy to calculate it explicitly. Let

$$\delta_n = \sum_{k=1}^{N_2} (1-\theta_{2k})^n \pi_{2k} - \sum_{j=1}^{N_1} (1-\theta_{1j})^n \pi_{1j}.$$

THEOREM 3. (a) *If $\delta_n \neq 0$ for all $n \geqq 1$, then γ is not a polynomial. Let g be any polynomial of degree at least 2, such that $g(0) = g(1) = 0$. Then there is a unique polynomial ψ such that $\psi = g + U\psi$ and $\psi(0) = 0$. The degree of ψ is one less than that of g.*

(b) *If $\delta_1 = 0$ then $\gamma(p) \equiv p$. Suppose $n \geqq 1$, $\delta_j \neq 0$ for $1 \leqq j \leqq n$ and $\delta_{n+1} = 0$. Then γ is a polynomial of degree $n+1$. If g is a polynomial of degree at least 2 and at most $n+1$ such that $g(0) = g(1) = 0$, there is a unique polynomial ψ such that (degree ψ) $\leqq n$, $\psi = g + U\psi$, and $\psi(0) = 0$. The degree of ψ is one less than that of g.*

IV. Learning models with noncompact state spaces.

A. *The beta model.*

1. *Recurrence criteria.* At the level of the logarithm w_n of the strength v_n of the A_1 response on trial n the stochastic difference equation for the four-operator beta model is

$$\Delta w_n = \ln \beta_{ij} \quad \text{if } A_{i,n} O_{ij,n},$$

where $\beta_{i1} \geqq 1$ and $\beta_{i2} \leqq 1$. In terms of w_n, the probability of $A_{1,n}$ is

$$p_n = e^{w_n}/(1+e^{w_n}),$$

and, as usual, $P(O_{ij,n} | A_{i,n}) = \pi_{ij}$.

To begin to appreciate the magnitude of the difference in the asymptotic behavior of the linear and beta models, consider the special case of noncontingent reinforcement under the additional assumption that outcomes O_{11} and O_{21} have exactly the same effect on A_1 response probability that O_{12} and O_{22} have on A_2 response probability ($\beta_{11} = \beta_{21} = \beta > 1$, $\beta_{12} = \beta_{22} = 1/\beta$). In this case the above difference equation reduces to

$$\begin{aligned} \Delta w_n &= \ln \beta \quad \text{if } A_{1,n}O_{11,n} \text{ or } A_{2,n}O_{21,n} \\ &= -\ln \beta \quad \text{if } A_{1,n}O_{12,n} \text{ or } A_{2,n}O_{22,n}, \end{aligned}$$

so that w_n is just w_1 plus the sum of the $n-1$ independently and identically distributed random variables $\Delta w_1, \cdots, \Delta w_{n-1}$ with mean

$$E[\Delta w_n | w_n] = E[\Delta w_n] = 2(\pi_1 - 1/2) \ln \beta.$$

Since $\ln \beta > 0$ it follows from the strong law of large numbers that $\lim_{n\to\infty} w_n = \infty$ with probability 1 if $E[\Delta w_n] > 0$ (i.e. if $\pi_1 > 1/2$) while $\lim_{n\to\infty} w_n = -\infty$ if $E[\Delta w_n] < 0$ ($\pi_1 < 1/2$). This means that the asymptotic probability of making the more frequently reinforced response (e.g. predicting the light that flashes most often) is 1. By way of comparison,

$$P_{p_1}\left(\liminf_{n\to\infty} p_n = 0\right) = 1,$$

$$P_{p_1}\left(\limsup_{n\to\infty} p_n = 1\right) = 1,$$

and

$$\lim_{n\to\infty} E_{p_1}[p_n] = \pi_1$$

for any $0 < p_1$, $\pi_1 < 1$ in the comparable linear model. Thus the process $\{p_n\}$ for $0 < p_1 < 1$ and $\pi_1 \neq 0, 1/2$, or 1 is recurrent in the linear model but not in the beta model.

In the general case, where the distribution of Δw_n depends on w_n, the qualitative asymptotic behavior of $\{w_n\}$ and $\{p_n\}$ is still determined by the conditional mean $E[\Delta w_n | w_n = w]$, or, more precisely, by its limits at $\pm\infty$. Let

$$\mu_+ = \lim_{w\to\infty} E[\Delta w_n | w_n = w] = \pi_{11} \ln \beta_{11} + \pi_{12} \ln \beta_{12}$$

and

$$\mu_- = \lim_{w\to-\infty} E[\Delta w_n | w_n = w] = \pi_{21} \ln \beta_{21} + \pi_{22} \ln \beta_{22}.$$

THEOREM 1. *For the four-operator beta model with* $\beta_{i1} > 1$ *and* $\beta_{i2} < 1$, $i = 1, 2$, *and for all* $0 < p_1 < 1$:

(a) *If* $\mu_+ < 0$ *and* $\mu_- > 0$, *then*

$$P_{p_1}\left(\liminf_{n\to\infty} p_n = 0\right) = 1 \tag{1}$$

and

$$P_{p_1}\left(\limsup_{n\to\infty} p_n = 1\right) = 1. \tag{2}$$

(b) *If* $\mu_+ < 0$ (> 0) *and* $\mu_- < 0$ (> 0), *then*

$$P_{p_1}\left(\lim_{n\to\infty} p_n = 0 \ \ (1)\right) = 1. \tag{3}$$

(c) *If* $\mu_+ > 0$ *and* $\mu_- < 0$, *then*

$$P_{p_1}\left(\lim_{n\to\infty} p_n = 1\right) = \delta(p_1) \tag{4}$$

and

$$P_{p_1}\left(\lim_{n\to\infty} p_n = 0\right) = 1 - \delta(p_1) \tag{5}$$

for some $0 < \delta(p_1) < 1$.

This theorem is due to Lamperti and Suppes [**15**].

In comparison, for a distance diminishing four-operator linear model, (1) and (2) hold if neither 0 nor 1 is absorbing, (3) holds if 0 (1) is absorbing and 1 (0) is not, while (4) and (5) (with $0 < \delta(p_1) < 1$) hold if both 0 and 1 are absorbing. These results follow from the theory presented in §II above.

2. *Other results*. Far fewer analytic formulas for predictions are available for beta models than for linear models. The best results to date are those of L. Kanal. He has shown [**13**] that, for the two-operator model,

$$\begin{aligned} \Delta w_n &= \ln \beta_{12} \quad \text{if } A_{1,n}, \\ &= \ln \beta_{22} \quad \text{if } A_{2,n} \end{aligned}$$

$(\pi_{12} = \pi_{22} = 1,\ \beta_{12} = \beta_{22} < 1)$, many functions of interest in testing the model empirically are solutions of the functional equation $\chi = g + U\chi$ that vanish at $w = -\infty$ $(v = 0,\ p = 0)$. His general solution of this equation is closely related to the standard expression $\sum_{n=0}^{\infty} U^n g$. Concerning the model

$$\begin{aligned} \Delta w_n &= \ \ \ln \beta \quad \text{if } A_{1,n}, \\ &= -\ln \beta \quad \text{if } A_{2,n} \end{aligned}$$

$(\pi_{11} = \pi_{22} = 1,\ \beta_{11} = 1/\beta_{22} = \beta)$, he has shown (Kanal [**14**]) that the probability $\delta(w_1)$ that $w_n \to \infty$ $(p_n \to 1)$ as $n \to \infty$ is given by

$$\delta(w_1) = \frac{\sum_{k=0}^{\infty} \exp\left\{-\frac{1}{2b}\left[w_1 - \left(k+\frac{1}{2}\right)b\right]^2\right\}}{\sum_{k=-\infty}^{\infty} \exp\left\{-\frac{1}{2b}\left[w_1 - \left(k+\frac{1}{2}\right)b\right]^2\right\}},$$

where $b = \ln \beta$.

3. *Commutativity*. A central feature conceptually of the beta model is the fact that the operators

$$p \to F(p,\beta) = \beta p/((1-p) + \beta p)$$

corresponding to different β's (different events) commute. This is obvious at the level of $v = p/(1-p)$ or $w = \ln v$ where the comparable transformations are $v \to \beta v$ and $w \to w + \ln \beta$. Luce [16] has studied abstract one parameter families of functions $F(p,\beta)$ corresponding intuitively to learning operators, with special emphasis on the commutative case. His article contains a discussion (see his Section 6, p. 396) of the extent to which beta model results like those discussed above generalize to models generated by other commutative families of operators.

B. *Models with distance diminishing event operators*. Generalizing Suppes' continuous linear model (see IC above) we consider the following set-up: (i) The state space (S,d) is a bounded metric space (that is, $d(s,s') \leqq \nu$ for some $\nu < \infty$ and all $s, s' \in S$). (ii) The event space (E,Γ) is a measurable space. As before, the state S_{n+1} on trial $n+1$ is related to the state S_n and event E_n on trial n by $S_{n+1} = f_{E_n}(S_n)$, $n \geqq 1$. We impose the very strong condition that the event operators be uniformly strictly distance diminishing: (iii) there is some $r < 1$ such that $\mu(f_e) \leqq r$ for all $e \in E$. Also (iv) $f.(s)$ is a measurable transformation for every $s \in S$. The conditional probability distribution of E_n given $S_n = s$ is $\phi.(s)$. Thus (v) $\phi.(s)$ is a probability measure on Γ for each $s \in S$. It is assumed that the functions $\phi_G(\cdot)$ for $G \in \Gamma$ satisfy a uniform Lipschitz condition (vi) $m(\phi_G) \leqq \omega$ for some $\omega < \infty$ and all $G \in \Gamma$. Our final condition is (vii) There is a nonnegative integer j, a positive real number λ, and a probability measure ζ on Γ such that

$$\phi_G(f_{e_1 \cdots e_j}(s)) \geqq \lambda \zeta(G)$$

for all $e_1, \cdots, e_j \in E$, $s \in S$, and $G \in \Gamma$. (For $j = 0$ read $\phi_G(s) \geqq \lambda\zeta(G)$ for all s and G.) This condition is quite restrictive. Theorem 2 below shows that, under these hypotheses, the Markov process $\{S_n\}$ behaves like a compact Markov process with only one, aperiodic, ergodic kernel.

In what follows, $C(S)$ is the set of bounded continuous real valued functions on S, $|\psi| = \sup_{s \in S} |\psi(s)|$, CL is the subset of $C(S)$ whose elements satisfy a Lipschitz condition, and $\|\psi\| = m(\psi) + |\psi|$ for $\psi \in CL$. The linear operator U is defined by

$$U\psi(s) = E[\psi(S_{n+1}) \,|\, S_n = s] = \int \psi(f_e(s))\, \phi_{de}(s).$$

It is easy to show that, under the above assumptions, U is a bounded mapping of $C(S)$ into $C(S)$ ($|U| = 1$) and of CL into CL ($\|U\| \leqq 2\omega + 1$).

THEOREM 2. *Under conditions* (i)—(vii) *there are constants* ρ *and* η *such that, for every* $\psi \in CL$, *there is a constant function* $U^\infty\psi$ *for which*

$$|U^{n+j}\psi - U^\infty\psi| \leqq \eta\rho^{n^{1/2}} \|\psi\|$$

for all $n \geqq 1$. *If* k *is* 1 *when* $\omega\nu r/(1 - r) \leqq 1/8$ *and otherwise is the least integer greater than or equal to*

$$\frac{\ln[(1 - r)/8\omega\nu]}{\ln r},$$

and $h = 1 - \lambda^k/4$, *then we can take*

$$\rho = \max(r, h) \quad \text{and} \quad \eta = \max\left(\frac{\nu}{r(1 - h)}, \frac{4\omega\nu}{r(1 - r)(1 - h)} + \frac{2}{h}\right).$$

The case $j = 0$ is a consequence of a theorem of Ionescu Tulcea [8, Theorem 1], and the other cases are consequences of this one.

It follows from results of Iosifescu [10, Chapter 3, Section 3] that, under (i)—(vii),

$$\lim_{n \to \infty} E_s[f(E_n)] = E^\infty(f)$$

and

$$\lim_{n \to \infty} \frac{1}{n} E_s \left\{ \left[\sum_{j=1}^{n} (f(E_n) - E^\infty(f)) \right]^2 \right\} = \sigma^2$$

exist for any real valued bounded measurable function f on E and do not depend on s. If $\sigma^2 > 0$, then $\sum_{j=1}^n f(E_j)$ is asymptotically normally distributed with mean $nE^\infty(f)$ and variance $n\sigma^2$.

To apply these results to Suppes' model we take the state space S to be the Borel probability measures on the rim R of the disc, the distance $d(\mu, \mu')$ between two states to be the total variation of their difference, the event space E to be R, and Γ to be the Borel subsets of R. The operator for the event z is

$$f_z(\mu) = (1-\theta)\mu + \theta\tau(z, \cdot)$$

where $\theta > 0$. Then $\mu(f_e) = (1-\theta) < 1$ for all events e, so (iii) holds. The measurability condition (iv) on $f.(\mu)$ is satisfied if $\tau(\cdot, A)$ is measurable for each Borel set A. Assuming $\pi(\cdot, G)$ measurable for any G, we take

$$\phi_G(\mu) = \int \mu(dy)\pi(y, G).$$

Then $m(\phi_G) \leqq 1$ for all G so (vi) holds. Finally, (vii) is satisfied with $j = 0$ if $\pi(y, \cdot)$ has a nontrivial response independent component: $\pi(y, G) \geqq a\xi(G)$ for some probability measure ξ and $a > 0$, and all y and G. For in that case

$$\phi_G(\mu) \geqq \int \mu(dy) a\xi(G) = a\xi(G)$$

for all μ and G.

CL includes functions of the form

$$\psi(\mu) = \int \chi(y)\, \mu\, (dy)$$

where χ is bounded and measurable on R. Hence it includes finite products of such functions. Thus, when Theorem 2 applies, we can conclude that

$$\lim_{n\to\infty} E_{\mu_1}\left[\prod_{j=1}^k \int \chi_j\,(y)\,\mu_n(dy)\right]$$

exists and does not depend on μ_1 for bounded measurable functions $\chi_1, \cdots, \chi_k$. The central limit theorem mentioned above is not very useful in this case, since the subject's response on a trial is not a function of the event that occurred on that trial.

Theorem 2 (with $j = 1$) and Iosifescu's central limit theorem are also applicable to my linear model for operant conditioning (Norman [**17**]).

Bibliography

1. R. C. Atkinson and W. K. Estes, "Stimulus sampling theory" in *Handbook of mathematical psychology*. vol. II edited by R. D. Luce et al, Wiley, New York, 1963, pp. 121-268.

2. R. R. Bush, "Sequential properties of linear models" in *Studies in mathematical learning theory* edited by R. R. Bush and W. K. Estes, Stanford Univ. Press, Stanford, Calif., 1959, pp. 215-227.

3. ———, "Identification learning" in *Handbook of mathematical psychology*. vol. III edited by R. D. Luce et al, Wiley, New York, 1965, pp. 161-203.

4. R. R. Bush and F. Mosteller, "A comparison of eight models" in *Studies in mathematical learning theory* edited by R. R. Bush and W. K. Estes, Stanford Univ. Press, Stanford, Calif., 1949, pp. 293-307.

5. K. L. Chung, *Markov chains*. 2nd ed., Springer-Verlag, New York, 1967.

6. J. L. Doob, *Stochastic processes*, Wiley, New York, 1953.

7. W. K. Estes, "Probability learning" in *Categories of human learning* edited by A. W. Melton, Academic Press, New York, 1964, pp. 89-128.

8. C. T. Ionescu Tulcea, *On a class of operators occurring in the theory of chains of infinite order*, Canad. J. Math. **11** (1959), 112-121.

9. C. T. Ionescu Tulcea and G. Marinescu, *Theorie ergodique pour des classes d'operations noncompletement continues*, Ann. of Math. **52** (1950), 140-147.

10. M. Iosifescu, *Random systems with complete connections with an arbitrary set of states*, Rev. Math. Pures Appl. **8** (1963), 611-645.

11. B. Jamison, *Asymptotic behavior of successive iterates of continuous functions under a Markov operator*, J. Math. Anal. Appl. **9** (1964), 203-214.

12. ———, *Ergodic decompositions induced by certain Markov operators*, Trans. Amer. Math. Soc. **117** (1965), 451-468.

13. L. Kanal, *A functional equation analysis of two learning models*, Psychometrika **27** (1962), 89-104.

14. ———, *The asymptotic distribution for the two-absorbing-barrier beta model*, Psychometrika **27** (1962), 105-109.

15. J. Lamperti and P. Suppes, *Some asymptotic properties of Luce's beta learning model*, Psychometrika **25** (1960), 233-241.

16. R. D. Luce, "Some one-parameter families of commutative learning operators" in *Studies in mathematical psychology* edited by R. C. Atkinson, Stanford Univ. Press, Stanford, Calif., 1964, pp. 380-398.

17. M. F. Norman, *An approach to free-responding on schedules that prescribe reinforcement probability as a function of interresponse time*, J. Math. Psychology **3** (1966), 235-268.

18. ———, *Some convergence theorems for stochastic learning models with distance diminishing operators*, J. Math. Psychology **5** (1968), 61-101.

19. ———, *On the linear model with two absorbing barriers*, J. Math. Psychology **5** (1968), 225-241.

20. M. F. Norman and N. V. Graham, *A central limit theorem for families of stochastic processes indexed by a small average step size parameter, and some applications to learning models*, Psychometrika **33** (1968), 441-449.

21. E. Parzen, *On asymptotically efficient consistent estimates of the spectral density function of a stationary time series*, J. Roy. Statist. Soc. B. **20** (1958), 303-322.

22. M. Rosenblatt, *Equicontinuous Markov operators*, Theor. Probability Appl. **9** (1964), 205-222.

23. ———, *Almost periodic transition operators acting on the continuous functions on a compact space,* J. Math. Mech. **13** (1964), 837-847.

24. S. Sternberg, "Stochastic learning theory" in *Handbook of mathematical psychology*. vol. II edited by R. D. Luce et al, Wiley, New York, 1963, pp. 1-120.

25. P. Suppes, "A linear model for a continuum of responses" in *Studies in mathematical learning theory* edited by R. R. Bush and W. K. Estes, Stanford Univ. Press, Stanford, Calif., 1959, pp. 400-414.

26. M. Tatsuoka and F. Mosteller, "A commuting-operator model" in *Studies in mathematical learning theory* edited by R. R. Bush and W. K. Estes, Stanford Univ. Press, Stanford, Calif., 1959, pp. 228-247.

27. K. Yosida and S. Kakutani, *Operator-theoretical treatment of Markoff's process and mean ergodic theorem,* Ann. of Math. **42** (1941), 188-228.

Addendum (January, 1970)

Proofs of the theorems in II.C and II.D are given in [**30**]. The theory of slow learning, some aspects of which are considered in III.A, is further developed in [**29**], [**31**, §6], and [**32**]. Some new methods and new results for *additive models,* such as the beta model (IV.A), are presented in [**31**]. §IV.B is superseded by [**33**].

A comprehensive presentation of the theory of random systems with complete connections is given in [**28**].

Supplementary Bibliography

28. M. Iosifescu and R. Theodorescu, *Random processes and learning,* Springer-Verlag, New York, 1969.

29. M. F. Norman, *Slow learning,* British J. Math. Statist. Psychology **21** (1968), 141-159.

30. ———, *Compact Markov processes,* Technical Report No. 2, NSF grant GP-7335, August, 1968.

31. ———, *Limit theorems for additive learning models,* J. Math. Psychology **7** (1970), (to appear).

32. ———, *Slow learning with small drift in two-absorbing-barrier models,* J. Math. Psychology **7** (1970), (to appear).

33. ———, *A uniform ergodic theorem for certain Markov operators on Lipschitz functions on bounded metric spaces,* Z. Wahrscheinlichkeitstheorie und Verw. Gebiete. **15** (1970), 51-56.

UNIVERSITY OF PENNSYLVANIA
PHILADELPHIA, PENNSYLVANIA

David Krantz

A Survey of Measurement Theory[1]

Table of Contents

[1] This work was partially supported by Public Health Service Grant GM-1231 and by National Science Foundation Grant GB 4947.

1. Examples of measurement theory.

1.1. *Definition of measurement.* Measurement, in its broadest sense, consists of the correspondence between mathematical objects, such as real numbers, vectors, or operators, and empirical objects, such as heavy bodies, forces, colors, etc. The correspondence is based on an isomorphism between observable formal properties of the empirical objects and the formal properties characterizing the mathematical objects. For example, in the measurement of mass, positive real numbers are assigned to heavy objects, so that the order of the numbers reflects the order of the objects, as determined by a suitable balance, and the addition of real numbers corresponds to combining of objects.

Many instances of measurement are like the measurement of mass, inasmuch as they involve construction of a real-valued function that preserves the order and additive structure of an empirical system. Such constructions are based ultimately on the theorem, due to Hölder [13], that any Archimedean fully ordered group is isomorphic to a subgroup of the ordered group of additive real numbers. I shall present a formal statement and proof of this theorem in the next section. Following this, I shall present two applications of Hölder's theorem. In the first application, the additive structure in the empirical objects is given directly, similar to combining heavy objects in the same pan of a balance. This is called *extensive measurement*. In the second application, no additive structure is given directly, but nevertheless, an associative binary operation can be defined and Hölder's theorem applied.

1.2. *Holder's Theorem.*

DEFINITION 1.1. Let G be a group, with binary operation $(x, y) \to xy$ and identity e, and let $\geqq$ be a total order on G. The pair $(G, \geqq)$ is called an *ordered group* if for all $x, y, z \in G$, $x \geqq y$ implies both $xz \geqq yz$ and $zx \geqq zy$. The ordered group $(G, \geqq)$ is called *Archimedean* if for all $x, y \in G$, with $x > e$, there exists some positive integer n such that $x^n > y$.

We shall denote the ordered additive group of real numbers by $(\mathrm{Re}, +, \geqq)$.

THEOREM 1.1. *Let $(G, \geqq)$ be an Archimedean ordered group. Then $(G, \geqq)$ is isomorphic to a subgroup of $(\mathrm{Re}, +, \geqq)$. Moreover, the isomorphism is unique up to multiplication by a positive constant.*

PROOF. Let $G^+ = \{x \mid x > e\}$. We can distinguish 2 cases:

(A) G^+ has a lower bound $x_1 > e$.

(B) $\inf G^+ = e$. ($G = \{e\}$ is a trivial case.)

In case (A), for any $y \in G$, there exists a unique integer n (positive, negative, or zero) such that $x_1^n \leqq y < x_1^{n+1}$. If $y \neq x_1^n$, then $x_1^{-n}y$ is in G^+ but is $< x_1$, a contradiction. Thus $y = x_1^n$. Hence, G is cyclic with generator x_1, and the theorem follows; the subgroup of $(\mathrm{Re}, +, \geqq)$ is any discrete subgroup.

For case (B), let $x \in G^+$ and $y \in G$ be arbitrary; then there exists a unique integer $N(x, y)$ such that $x^{N(x,y)} \leqq y < x^{N(x,y)+1}$. Clearly, for $x, x' \in G^+$, $y \in G$, we have

$$(1) \quad [N(x, x') + 1][N(x', y) + 1] > N(x, y) \geqq N(x, x')N(x', y).$$

Let $\{x_k\}$ be any sequence in G^+ such that $x_{k+1}^2 \leqq x_k$. It is easy to show that such $\{x_k\}$ exists and that for any $y \in G^+$, $N(x_k, y) \to +\infty$, while for $y^{-1} \in G^+$, $N(x_k, y) \to -\infty$. For any $y, z \in G$, with $z \neq e$ and for k, l large enough that $N(x_k, z)$, $N(x_l, z)$, and $N(x_k, x_l) > 0$, we have by (1)

$$(2) \qquad \frac{N(x_k, y)}{N(x_k, z)} < \frac{[N(x_k, x_l) + 1][N(x_l, y) + 1]}{N(x_k, x_l)N(x_l, z)}.$$

If we fix l and let $k \to \infty$, taking the lim sup on the left and right in (2), we obtain

$$(3) \qquad \limsup_{k \to \infty} \frac{N(x_k, y)}{N(x_k, z)} \leqq \frac{N(x_l, y) + 1}{N(x_l, z)}.$$

Now taking the lim inf on the right in (3) as $l \to \infty$, we find that $\lim N(x_k, y)/N(x_k, z)$ exists (and is finite, as is easily seen). For fixed $y_1 \in G^+$, define

$$\phi(y) = \lim_{k \to \infty} \frac{N(x_k, y)}{N(x_k, y_1)}.$$

It is easily shown that ϕ is an isomorphism of $(G, \geqq)$ onto a subgroup of $(\mathrm{Re}, +, \geqq)$. To this end, one can use the fact that for any $x \in G^+$, $y, z \in G$,

$$(4) \qquad N(x,y) + N(x,z) + 1 \geqq N(x,yz) \geqq N(x,y) + N(x,z).$$

To show uniqueness of the isomorphism, let ϕ' be any other isomorphism; then clearly, for any k, y,

$$N(x_k, y)\phi'(x_k) \leqq \phi'(y) < [N(x_k, y) + 1]\phi'(x_k).$$

It follows that

$$\frac{\phi'(y)}{\phi'(y_1)} = \lim_{k \to \infty} \frac{N(x_k, y)}{N(x_k, y_1)} = \phi(y),$$

so that $\phi' = \alpha\phi$, where $\alpha = \phi'(y_1) > 0$. This completes the proof of Theorem 1.1.

For a different proof, see Birkhoff [**3**, p. 300]. The proof given above has the advantages of being easily generalized and of constructing the isomorphism ϕ in a manner similar to actual measurement procedures. These points will be made more clearly in sections (1.3) and (2.2).

1.3. *Extensive measurement.* In extensive measurement, one starts with an empirical system that includes an associative binary operation. Placing 2 heavy objects together in the same pan of a balance is one example; others are found in the usual measurement procedures for length, where rods are combined by laying them end to end, and for time, where time intervals are concatenated by using the same event to mark the end of one interval and the beginning of another.

The following set of weak, logically independent axioms is due to Suppes [**33**].

PRIMITIVES. *K, a nonempty set.*
Q, a binary relation on K.
$*$, a binary function on K, $(x, y) \to x * y$.

AXIOMS. For all $x, y, z \in K$:

1. *If* $x Q y$ *and* $y Q z$, *then* $x Q z$.
2. $x * y \in K$.
3. $(x * y) * z Q x * (y * z)$.
4. *If* $x Q y$, *then* $x * z Q z * y$.
5. *If not* $x Q y$, *then there exists* $w \in K$ *such that* $x Q y * w$ *and* $y * w Q x$.
6. *Not* $x * y Q x$.
7. *If* $x Q y$, *then there is a positive integer* n *such that* $y Q nx$ $[1x = x,\ nx = (n-1)x * x]$.

We can prove the following measurement theorem.

THEOREM 1.2. *If* $(K, Q, *)$ *satisfies Axioms* 1—7, *then there exists a real-valued function* ϕ *on* K *such that for all* $x, y \in K$:
(i) $x Q y$ *if and only if* $\phi(x) \leqq \phi(y)$.
(ii) $\phi(x * y) = \phi(x) + \phi(y)$.
Furthermore, ϕ *is unique up to multiplication by a positive constant.*

Theorem 1.2 includes a representation theorem for extensive measurement—a theorem specifying that real-valued assignments can be constructed that preserve the empirically given structure—and a uniqueness theorem, limiting the class of possible isomorphisms. Uniqueness theorems are quite important in measurement, since they determine what sorts of statements about measured values are meaningful. Measurement representations that are unique up to multiplication by a positive constant are called *ratio scales*, because ratios are preserved by permissible changes in representation. Thus, the statement "X is twice as tall as Y" is meaningful, independent of the units chosen for measurement of length.

To prove Theorem 1.2, it is convenient to introduce a relation $\sim$ on K: $x \sim y$ if $x Q y$ and $y Q x$. From the axioms, $\sim$ can be shown to be an equivalence relation. Moreover, it can be shown that $Q, *$ induce a total order, $\geqq$, and a binary operation, $+$, on the set of equivalence classes, $K/\sim$. The system $(K/\sim, +, \geqq)$ satisfies all the properties of G^+ in Hölder's Theorem; in particular, although inverses do not exist, $K/\sim$ is closed under subtraction of smaller elements from larger ones (see Axiom 5). This permits the proof of Hölder's Theorem to be carried through, with no change, for $(K/\sim, +, \geqq)$; this latter is proved isomorphic to a subsemigroup of the additive semigroup of positive real numbers. The isomorphism yields the measurement representation required in Theorem 1.2. Uniqueness follows similarly from the uniqueness argument for Hölder's Theorem.

Finally, I should like to point out the close relation between the construction of the isomorphism ϕ, in Theorem 1.1 or 1.2, and actual procedures for assigning real numbers to objects. Consider the case where K consists of straight rods, and $x * y$ is formed by laying end-to-end one replica of rod x and one of rod y. Laying rods side-by-side permits comparisons, establishing $x\,Q\,y$, etc. Measurement is carried out by forming a *standard sequence* $x, 2x, \cdots, nx$ (we use additive notation) laying $1, 2, \cdots, n$ replicas of x end-to-end. To measure y in feet, we form the ratio of $N(x,y)$ to $N(x,y_1)$, where y_1 is a standard foot-ruler. The generator of the standard sequence, x, is chosen sufficiently small to attain any desired accuracy of measurement. Equation (4) shows that the approximate measures, $N(x,y)/N(x,y_1)$, are approximately additive. The main point of Theorem 1.1 was to show that $N(x,y)/N(x,y_1)$ converges as x is taken arbitrarily small.

1.4. *Conjoint measurement.* In the social sciences, it is rare to find associative binary operations that can be used for extensive measurement. However, it is common to observe an ordering of objects, where position in the ordering depends on the values of 2 or more independently controllable factors. Such a situation is represented formally by a transitive relation $\geqq$ defined over a product set, $A = \prod_{i=1}^{n} A_i$. The simplest law governing the dependence of ordinal position on the different factors is an additive one:

$$\phi(a_1, \cdots, a_n) = \sum_{i=1}^{n} \phi_i(a_i)$$

where ϕ is a real-valued, order-preserving function on A and each ϕ_i is a real-valued function on A_i. If such functions can be constructed, we say that *additive conjoint measurement* is feasible for the system $(A_1, \cdots, A_n, \geqq)$. The functions ϕ_i and ϕ provide measurement scales for the factors A_i and the observed output, relative to which an additive law holds.

Additive conjoint measurement has a complex history. However, it was the publication of a set of sufficient conditions, by Luce and Tukey **[24]** in 1964, that created widespread interest. The most important anticipation of their work was published by Debreu **[9]**. The Luce-Tukey axioms, which apply to the case $A = A_1 \times A_2$, are essentially the following.

Primitives. *A_1, A_2, nonempty sets.*
$\geqq$, a binary relation on $A = A_1 \times A_2$.

Axioms. 1. *$\geqq$ is a weak order; i.e. it is transitive and any 2 elements of A are comparable.*

2. *Any change in one factor can be exactly compensated by a change in the other, i.e. if $a \in A$, $b_1 \in A_1$, then there exists $b_2 \in A_2$, such that $a \sim (b_1, b_2)$ ($\sim$ means $\geqq$ and $\leqq$); and similarly for the other factor.*

Axiom 2 is called the *solvability axiom*, since we "solve" for b_2, given a, b_1.

3. *For any (a_1, a_2), (b_1, b_2), $(c_1, c_2) \in A$, if $(a_1, b_2) \geqq (b_1, c_2)$ and $(b_1, a_2) \geqq (c_1, b_2)$, then $(a_1, a_2) \geqq (c_1, c_2)$.*

Axiom 3 is called the *cancellation axiom*, since, given an additive representation, we can add up the 2 antecedent inequalities and cancel $b_1 + b_2$, yielding the conclusion. This condition is illustrated geometrically in terms of indifference curves in the $A_1 \times A_2$ plane, in Figure 1. In the theory of webs, this is called the Thomsen condition (see Aczél, Pickert, and Rado [1]).

4. *A sequence $\{(a_{1i}, a_{2j}) \mid i, j = 0, \pm 1, \pm 2, \cdots\}$ in A is called a* dual standard sequence *if $(a_{1i}, a_{2j}) \sim (a_{1k}, a_{2l})$ iff $i + j = k + l$. If $\{(a_{1i}, a_{2j})\}$ is a dual standard sequence, then for any $a \in A$, there exist n, m with $(a_{1n}, a_{2n}) \geqq a \geqq (a_{1m}, a_{2m})$.*

Axiom 4 is called the *Archimedean axiom*. It is easily verified that Axioms 1, 3, and 4 are necessary for additive conjoint measurement; solvability is not.

Theorem 1.3. *If $(A_1, A_2, \geqq)$ satisfy Axioms 1—4, then there exist real-valued functions ϕ on A, ϕ_1 on A_1, ϕ_2 on A_2, such that for all (a_1, a_2), $(b_1, b_2) \in A$:*

(i) *$(a_1, a_2) \geqq (b_1, b_2)$ if and only if $\phi(a_1, a_2) \geqq \phi(b_1, b_2)$.*

(ii) *$\phi(a_1, a_2) = \phi_1(a_1) + \phi_2(a_2)$.*

Furthermore, if ϕ', ϕ_1', ϕ_2' are any other such functions, then there are real numbers $\alpha > 0$, β_1, β_2 such that $\phi_i' = \alpha\phi_i + \beta_i$, $\phi' = \alpha\phi + \beta_1 + \beta_2$.

It should be noted that the uniqueness clause of this theorem is the best that could be expected. Such a representation is called *interval scale* measurement; ratios of intervals are invariant under permissible transformations. (A more standard term would be *affine scale*, since the affine ratio is invariant.)

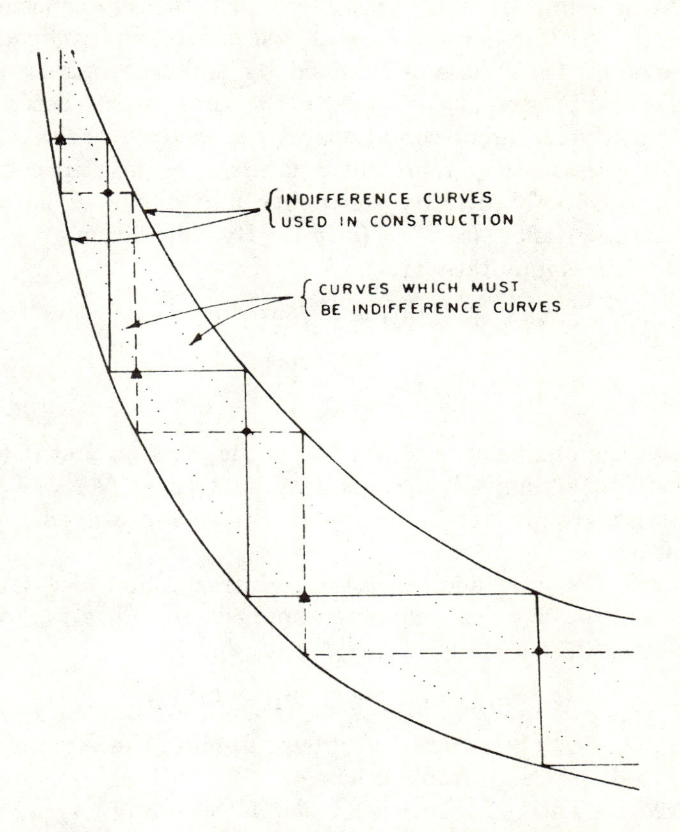

FIGURE 1. The cancellation axiom illustrated for indifference curves. If two "flights of stairs" are inscribed between two indifference curves, as shown, then alternate intersections lie on an indifference curve when the cancellation axiom is true. (Taken from Luce & Tukey [**24**, p. 7].) (The author wishes to thank Academic Press, Inc., for permission to reprint this figure from the Journal of Mathematical Psychology.)

I shall sketch a proof of Theorem 1.3, based on Hölder's Theorem, which was published by Krantz [**17**].

Choose an arbitrary origin $a^0 = (a_1^0, a_2^0)$ in A. By solvability, any equivalence class of A contains elements of forms (b_1, a_2^0) and (a_1^0, b_2). Define an operation, $+$, on $A/\sim$, by

$$(b_1, a_2^0) + (a_1^0, b_2) = (b_1, b_2).$$

I shall show that $+$ is well defined and that $(A/\sim, +, \geqq)$ is an Archimedean ordered group, where $\geqq$ is defined in the natural way on $A/\sim$.

If $(b_1, a_2^0) \sim (a_1^0, c_2)$ and $(a_1^0, b_2) \sim (c_1, a_2^0)$, then by cancellation, $(b_1, b_2) \sim (c_1, c_2)$. It follows that $+$ is *well defined* and *commutative,* since the equivalence class determined by adding arbitrary representations of the form (b_1, a_2^0), (a_1^0, b_2) is the same as that determined by adding arbitrary representations in the reverse order.

To prove *associativity* represent 3 arbitrary equivalence classes as (b_1, a_2^0), (a_1^0, b_2), (c_1, a_2^0). By solvability find $c_2, d_2 \in A_2$ such that $(b_1, b_2) \sim (a_1^0, c_2)$ and $(c_1, b_2) \sim (a_1^0, d_2)$. By definition of $+$, cancellation, and commutativity,

$$\begin{aligned}[(b_1, a_2^0) + (a_1^0, b_2)] + (c_1, a_2^0) &= (c_1, c_2) \\ &= (b_1, d_2) \\ &= (b_1, a_2^0) + [(a_1^0, b_2) + (c_1, a_2^0)].\end{aligned}$$

Obviously, the equivalence class of a^0 is the *identity,* and if $(b_1, b_2) \sim a^0$, then (a_1^0, b_2) and (b_1, a_2^0) are *inverse*. Hence, $(A/\sim, +)$ is a commutative group. Note that the results so far use only properties of $\sim$.

If $(b_1, a_2^0) \gtrsim (c_1, a_2^0)$, and (a_1^0, d_2) is arbitrary, find $d_1 \in A_1$ with $(c_1, a_2^0) \sim (d_1, d_2)$. By cancellation, applied to $(b_1, a_2^0) \gtrsim (d_1, d_2)$, $(d_1, d_2) \sim (c_1, a_2^0)$, we obtain $(b_1, d_2) \gtrsim (c_1, d_2)$, or

$$(b_1, a_2^0) + (a_1^0, d_2) \gtrsim (c_1, a_2^0) + (a_1^0, d_2).$$

Thus $(A/\sim, +, \gtrsim)$ is an ordered group. Finally, the Archimedean property follows easily from Axiom 4.

By Hölder's Theorem, there is an isomorphism ϕ of $(A/\sim, +, \gtrsim)$ onto a subgroup of $(\mathrm{Re}, +, \geqq)$. Let $\phi_1(b_1) = \phi(b_1, a_2^0)$, $\phi_2(b_2) = \phi(a_1^0, b_2)$. Then $\phi(b_1, b_2) = \phi_1(b_1) + \phi_2(b_2)$ as specified by Theorem 1.3. The uniqueness clause follows from the fact that, if ϕ', ϕ_1', ϕ_2' are any functions satisfying (i) and (ii) of Theorem 1.3, then

$$(b_1, b_2) \to \phi_1'(b_1) - \phi_1'(a_1^0) + \phi_2'(b_2) - \phi_2'(a_2^0)$$

is an isomorphism of $(A/\sim, +, \gtrsim)$ into $(\mathrm{Re}, +, \geqq)$, and so, by Hölder's Theorem, must differ from ϕ by multiplication by a positive constant. This completes the proof.

Construction of the isomorphism ϕ, and thus of the measurement scales ϕ, ϕ_1, ϕ_2, depends on the construction of a standard sequence in $(A/\sim, +, \gtrsim)$, as in the proofs of Theorems 1.1 and 1.2. This amounts to measuring the deviation of any (b_1, b_2) from (a_1^0, a_2^0) in terms of multiples of a small unit deviation from (a_1^0, a_2^0).

2. Some new problems generated by applying theories of measurement in the social sciences.

2.1. *The role of measurement theory in social science.* For the physical scientist, measurement theory is properly a branch of philosophy. The axioms for extensive measurement of mass, length, or time provide a foundational analysis of long-established procedures. However, these axioms are too trivial to claim the status of laws of physics; rather, they are obvious properties of measurement operations, and are taken for granted in the actual practice of measurement.

Furthermore, in testing nontrivial laws that specify rules of combination for 2 or more variables, the physicist need not rely on an axiomatic analysis of the sort provided by conjoint measurement theory. For example, the equation of state for an ideal gas, $pV/T = \text{constant}$, and the second law of motion, $F = ma$, are stated in terms of numerical scales obtained by extensive measurement, and are directly testable by numerical calculations.

In the social sciences, there are no measurement procedures comparable to the ones used for measurement of mass, length, and time. Therefore, when an axiomatic theory of measurement is applied in a social science context, the axioms are not obvious properties of long-established procedures; rather, they are a set of proposed laws, which are not at all trivial. Some of the laws may be qualitative, i.e. directly testable by observations involving order or class membership. The axioms of additive conjoint measurement are of this sort. Other laws may be numerical, for example, the assertion that 2 variables combine additively. These numerical laws cannot be tested as simply as in physics, since no numerical scales are specified for the variables. Rather, they must be tested by searching for numerical scales that satisfy the laws, or by testing other laws that imply or are implied by the given numerical laws. In short, at the present stage of development of quantitative theory in social science, it is impossible to separate the search for interesting empirical laws from the discovery and refinement of measurement procedures.

As a consequence of the above situation, measurement theory is of more than philosophical interest for social science. By providing an axiomatic theory for various numerical laws, one proposes qualitative experiments that distinguish among laws, and techniques of measurement where none existed previously. One result of the more dynamic and integral role of measurement theory in social science is that the discoveries or the difficulties encountered in empirical studies constitute an important source of new mathematical problems.

The next 5 sections are devoted to an overview of 5 areas in which new mathematical problems have emerged from the requirements of social science quantification: foundations of geometry, ordered rings, theory of models, semiorders, and error theory. The problems in foundations of geometry and in ordered rings were generated by the attempt to axiomatize laws other than the simple additive combination of variables: geometric laws, and polynomial combination laws, respectively. These topics will be pursued in more depth in §§3 and 4. The problems in theory of models, semiorders, and error theory derive from difficulties in realizing idealized primitives of measurement theory, such as total orderings, amid the doubts and errors of real data.

2.2. *Foundations of geometry*. Geometrical models are heavily used in social science as a basis for quantitative treatments of similarity or correlation. For example, suppose that one has a set of objects, A, and obtains some measure of the dissimilarity of any 2 objects in A. This measure gives rise to an order relation on $A \times A$. To represent the dissimilarity ordering by a geometric model, one tries to map the objects of A into a metric space, where the ordering of metric distances corresponds to the observed ordering of dissimilarities. In 1962, Shepard [**31**] published a practical method of computing a representation for a finite set A, in low-dimensional Euclidean space, which yields the best approximation (for a given dimension) to the dissimilarity ordering on $A \times A$. Since then, this sort of measurement has been widely practiced, with little concern over appropriate foundations.

In terms of measurement theory, the problem of foundations may be stated as follows: given a set A, an observable ordering $\geqq$ of the pairs of elements of A, and a class $\mathscr{C}$ of metric spaces (the desired geometric representation), what axioms (empirical laws) must be satisfied, in order for there to be a metric d on A, such that (A,d) is in class $\mathscr{C}$, and such that $(x,y) \geqq (z,w)$ if and only if $d(x,y) \geqq d(z,w)$? From the viewpoint of the classical field of foundations of geometry, we are asking for an axiomatization of geometries of class $\mathscr{C}$, in terms of the undefined (primitive) notions of a set of points and a *quaternary relation* on the points.

The classical axiom systems for foundations of Euclidean geometry (see Blumenthal [**4**]) generally involve undefined notions of *point, congruence of point pairs* (a quaternary relation), and *collinear betweenness of point triples* (a ternary relation). Sometimes, *lines*, and *incidence* of points and lines are also taken as primitive. In the study of empirical similarity, the required primitives (incidence of a point on a line, or collinear betweenness) do not seem to arise in any natural way. Thus, the problem of developing geometric measurement theories for similarity generates new problems in foundations of geometry, that is, axiomatizing different forms of metric geometry in terms of a single quaternary relation. One such axiomatization will be presented in detail in §3, and some further possibilities will be mentioned briefly in §4.

2.3. *Ordered rings*. Another source of problems is found in the general theory of conjoint measurement. Given a set of factors, $A_1, \cdots, A_n$, and an order relation $\geqq$ on $A = \prod_{i=1}^n A_i$, various laws of combination for the different factors can be considered, besides the additive law discussed in §1. The following definition is quite general.

DEFINITION 2.1. Let $A_1, \cdots, A_n$ be nonempty sets, with $\geqq$ a binary relation on $A = \prod_{i=1}^n A_i$. Let f be a real-valued function of n real variables. We say that $(A_1, \cdots, A_n, \geqq)$ is *decomposable relative to* f if there exist real-valued functions $\phi, \phi_1, \cdots, \phi_n$, with ϕ defined on A and ϕ_i on A_i, such that for $a = (a_1, \cdots, a_n)$, $b = (b_1, \cdots, b_n) \in A$:

(i) $a \geqq b$ if and only if $\phi(a) \geqq \phi(b)$.

(ii) $\phi(a) = f[\phi_1(a_1), \cdots, \phi_n(a_n)]$.

The function f gives the rule of combination for the variables; ϕ, $\phi_1, \cdots, \phi_n$ give appropriate measurement scales. The problem of measurement theory is to specify axioms (empirical laws) that are necessary and/or sufficient for decomposability relative to a specified rule f. This problem becomes fairly tractable when the function f is a polynomial in n variables. Moreover, quite a few miniature theories have been proposed, which explicitly posit polynomial rules of combination for a set of factors.

For one simple illustration of polynomial combination rules, consider the relationship of the evaluative (moral) connotation of combinations of quantitative adverbs with adjectives, as a function of the adverb and of the adjective. The overall moral connotation of a combination such as "slightly evil" is better described as a multiplicative, rather than an additive combination of "slightly" and "evil". To see this, note that "slightly evil" would be rated better than "very evil", while "slightly pleasant" would be rated worse than "very pleasant". These opposite orderings of "slightly" and "very" correspond to multiplying numerical scale values of the modifiers by moral values of opposite sign for "evil" and "pleasant". Studies of moral connotation that include the above examples, and use a multiplicative combination rule, were carried out by Cliff [**8**]. Many other miniature theories involve mixtures of additive and multiplicative combinations, i.e. more general polynomials.

The basic tool for polynomial conjoint measurement is the ring analog of Hölder's Theorem: an Archimedean ordered ring (with nontrivial multiplication) is isomorphic to a unique subring of the ordered ring of real numbers. (See Birkhoff [**3**, p. 398].) This tool can be used in at least 2 ways. One procedure is analogous to the proof of Theorem 1.3 on additive conjoint measurement: one introduces ring operations, $+$, $\cdot$ directly into the set of equivalence classes, $A/\sim$. The definitions of $+$ and $\cdot$ depend on the hypothesized polynomial; the required axioms are those for which the system $(A/\sim, +, \cdot, \geqq)$ becomes an Archimedean ordered ring. A different strategy is to let each relation statement of form $a \geqq b$ correspond to a suitable polynomial inequality. Obtaining the required functions ϕ_i is equivalent to solving a set of simultaneous polynomial inequalities. This leads to the study of partial orders on polynomial rings. In particular, the following question seems to be unsolved and of interest: for what classes of partially ordered rings is an extension possible to an Archimedean total order? This problem is discussed by Tversky [**35**]; for some results on extensions of partial orders, see Fuchs [**11**].

2.4. *Factorial designs*; *theory of models*. Given a binary relation $\geqq$ on $\prod_{i=1}^{n} A_i$, a common experimental procedure is to sample a finite subset B_i of A_i, $i = 1, \cdots, n$, and to observe the ordering $\geqq$ only on $\prod_{j=1}^{n} B_i$. This is called a factorial design. Certain axioms of polynomial conjoint measurement theories, such as solvability and Archimedean axioms, are untestable in such an experiment. But even if the untested axioms are valid in the entire empirical system $(A_1, \cdots, A_n, \geqq)$, while the testable axioms are verified in $(B_1, \cdots, B_n, \geqq)$, it may still be false that $(B_1, \cdots, B_n, \geqq)$ is decomposable relative to the polynomial combination rule in question. Testing this decomposability amounts to searching for a simultaneous solution to a finite set of polynomial inequalities, a problem which is computationally demanding and for which there seems to be no general algorithm.

Thus, the problem arises of axiomatizing polynomial combination rules for finite systems. Here, the theory of models, developed by Tarski [**34**], is relevant. Using results from this theory, Scott and Suppes [**30**] proved a theorem that implies that there is no finite axiomatization for finite systems of additive conjoint measurement, by universal sentences in the first order functional calculus. One may conjecture that there is no finite axiomatization, in first-order functional calculus, for any system of polynomial conjoint measurement.

2.5. *Semiorders*. The binary or quaternary relations of extensive, conjoint, or metric space measurement are generally assumed to be transitive. In practice, 2 types of intransitivity are observed:

(i) $x \sim y,\ y \sim z,$ but $x > z.$

(ii) $x > y,\ y > z,$ but $z > x.$

The first type may occur because differences between x and y and between y and z are too small to be detected, but add up to a detectable difference between x and z.

Type (i) intransitivities occur in a formal system called a *semiorder*, introduced by Luce [**21**]. This involves 2 binary relations, P (strict preference) and I (intransitive indifference).

Let juxtaposition denote the usual relation product, i.e. $xPIz$ if there exists y such that xPy and yIz; let P^* be the reflection of P in the diagonal, i.e. xP^*y if yPx. We can define a semiorder as follows.

DEFINITION 2.2. (X, P, I) is a *semiorder* if X is a set, and P, I are binary relations on X, such that:

1. $\{P, P^*, I\}$ is a partition of $X \times X$.
2. $PIP \subset P$.
3. $P^2 \cap I^2$ is empty.

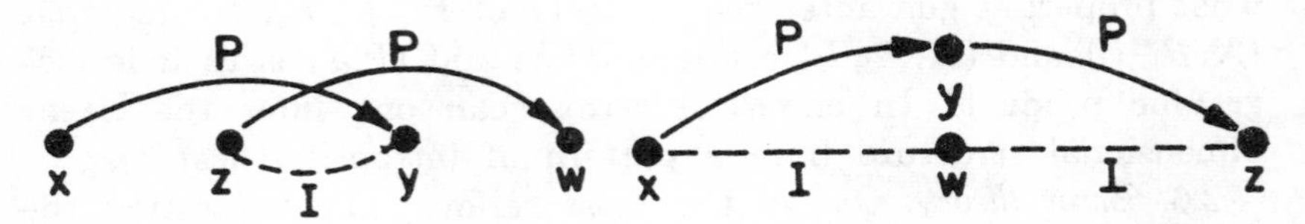

FIGURE 2. Illustration of Axioms 2 (left) and 3 (right) for a semiorder. The configuration on the left implies xPw. The configuration on the right is asserted to be impossible.

The content of conditions 2 and 3 is depicted in Figure 2. In the left diagram, xPy, yIz, zPw, and the conclusion is xPw. In the right-hand diagram, xP^2z (via y) and xI^2z (via w); the assertion is that no such configuration exists.

One of the main results on semiorders is that any semiorder defines a natural complete order.

THEOREM 2.1. *Let (X, P, I) be a semiorder. Define $x \sim y$ if for all $z \in X$, xIz iff yIz, and $x \geqq y$ if neither $yPIx$ nor $yIPx$. Then $\sim$ is an equivalence relation and $\geqq$ induces a total order on the equivalence classes $X/\sim$.*

The proof of this theorem is a useful exercise.

Several problems arise in connection with semiorders. One problem is to axiomatize various forms of measurement, replacing the usual order relation by a semiorder. This can be done in a trivial way using the defined total order of Theorem 2.1, but the real point is to show that, in a semiordered system, one can attain any desired accuracy of measurement from appropriate finite sets of P, I observations. This has been done for extensive measurement by Krantz [18].

A second type of problem is to deal with type (ii) intransitivities. One way to account for these is by assuming shifts in dimensions that determine the decision. For example, in purchasing a new car, each additional accessory may seem worth the added cost, but the total cost of several may drive one back to the basic model. Below some threshold, the cost dimension is ignored; above, it is decisive. One might capture this by assuming 2 semiorders (X, P_1, I_1), (X, P_2, I_2) over the same base set X, and defining the "lexicographic product", $P = P_1 \cup (I_1 \cap P_2)$, $I = I_1 \cap I_2$. That is, xPy if xP_1y (the first dimension is decisive) or if $x(I_1 \cap P_2)y$. Obviously, P need not be transitive. The interesting question is to *characterize* lexicographic products of semiorders: given a pair of relations, P, I, what properties guarantee the existence of P_1, I_1, P_2, I_2 such that (X, P_1, I_1) and (X, P_2, I_2) are semiorders and (P, I) is their lexicographic product? In empirical terms, can one infer the latent dimensional structure from a pattern of intransitivities?

2.6. *Error theory.* One of the most serious bars to testing the axioms of various measurement theories is the presence of "random" error. One way to deal with this difficulty is to superimpose a probability model on the algebraic one. For example, in conjoint measurement, one might assume that a pair (a_1, a_2) corresponds to a Gaussian random variable with expectation $M(a_1, a_2)$; one might interpret $(a_1, a_2) \geqq (b_1, b_2)$ to mean that $M(a_1, a_2) \geqq M(b_1, b_2)$. Axioms such as transitivity or cancellation (Axioms 1, 3 of §1.4) are testable statistical hypotheses.

A criticism of the conventional statistical approach is that, if the measurement axioms are satisfied, then the construction of measurement scales induces transformations of the random variables. It is at least as reasonable to assume that the transformed random variables, rather than the original ones, satisfy a tractable probabilistic model, but this greatly complicates the statistical treatment.

More generally, one may wish to deal with random error in a manner that is less arbitrary than assumption of a special probabilistic model. It may be desirable to incorporate error processes more directly into the system of primitives and axioms.

An extreme version of the incorporation of error processes into a measurement axiomatization is to base the measurement entirely on error. For example, given a family of real-valued random variables, one may seek a transformation of the real numbers such that the transformed random variables are identically distributed except for translations. If this can be done, the transformation in question provides a measurement scale that regularizes the error theory. Levine [**20**] explored this problem quite deeply; among other results, he showed that if such a transformation exists, for a family of 3 or more random variables, then it is unique up to changes of origin and unit, i.e. we have interval scale measurement. Thus, the error theory has run away with the measurement procedure—there is no longer any room for basing measurement on an extensive operation, a geometric model, or a polynomial combination rule. Some intermediate manner of incorporating random error in the measurement process would seem desirable.

3. **Metrics with additive segments.**

3.1. *Preview.* A metric space (X, d) is a *metric with additive segments* if for any $x, z \in X$, there is an isometry f of the real interval $[0, d(x,z)]$ into X, such that $f(0) = x$, $f(d(x,z)) = z$. Most metric spaces studied in geometry are of this type: e.g., Riemannian spaces, or G-spaces (Busemann [**7**]). In this section, I examine the foundations of metrics with additive segments, starting with an ordering of pairs. More precisely, given a set A, and an ordering $\geqq$ on $A \times A$ (or a mapping $(x,y) \rightarrow xy$ of $A \times A$ onto a totally ordered set, $(P, \geqq)$), what axioms guarantee the existence of an order-preserving real-valued function ϕ on P such that, for $d(x,y) = \phi(xy)$, (A,d) is a metric with additive segments? The source of this problem is the demand for a geometric model of dissimilarities, discussed in §2.2.

The key to analyzing foundations of metrics with additive segments is the ternary relation $\langle xyz \rangle$, which, in terms of a metric, can be defined as $d(x,y) + d(y,z) = d(x,z)$. We must define this relation, and establish its main properties, using the ordering alone. Once this is done, we define a binary operation in P as follows: $xy + x'y' = uw$ if $xy = uv$, $x'y' = vw$, and $\langle uvw \rangle$. This operation, however, cannot necessarily be defined for all pairs $(xy, x'y')$ (there may not exist additive segments of arbitrary length). Thus, in order to apply Hölder's Theorem to the system $(P, +, \geqq)$, we must establish a version of it that applies when the binary operation is defined only for sufficiently small elements. This sort of local theorem has other important applications. In the next section, I shall state it, sketch its proof, and indicate the applications to extensive and conjoint measurement. In succeeding sections, I shall return to the question of metrics with additive segments.

3.2. *A bounded version of Hölder's Theorem.*

DEFINITION 3.1. Let G be a set, with binary relations $\geqq$, B on G, and a binary operation $(x,y) \rightarrow x + y$ from B to G. The quadruple $(G, B, +, \geqq)$ will be called a *positive ordered local semigroup* if the following are true for all $x, y, z, x', y' \in G$:

1. $\geqq$ is a total order.
2. If $(x,y) \in B$, $x \geqq x'$, $y \geqq y'$, then $(y', x') \in B$.
3. If (x,y), $(x+y, z) \in B$, then (y,z), $(x, y+z) \in B$ and $(x+y)+z = x+(y+z)$.
4. If $x \geqq y$ and $(x,z) \in B$, then $x+z \geqq y+z$ and $z+x \geqq z+y$.
5. If $(x,y) \in B$, then $x+y > x$.
6. If $z > x$, then there exists $y \in G$ with $(x,y) \in B$ and $z \geqq x+y$.

A positive ordered local semigroup is *Archimedean* if for all $x, y \in G$, $\{n \mid nx \text{ defined}, y \geqq nx\}$ is finite.

Note that by property 2, $(x,y) \in B$ iff $(y,x) \in B$; from this, we know that $y+z$, $z+x$, $z+y$ are defined in property 4 of Definition 3.1.

THEOREM 3.1. *Let $(G, B, +, \geqq)$ be an Archimedean positive ordered local semigroup. Let $G' = \{x \mid \exists y, (x,y) \in B\}$. Then there is a real-valued function ϕ on G' such that for all $x, y \in G'$:*

(i) *$x \geqq y$ iff $\phi(x) \geqq \phi(y)$.*

(ii) *If $(x,y) \in B$, and $x + y \in G'$, then $\phi(x+y) = \phi(x) + \phi(y)$. Moreover, if ϕ, ϕ' are any two such functions, then $\phi' = \alpha\phi$ for some $\alpha > 0$.*

The proof of Theorem 3.1 is like that of Theorem 1.1 in all essential details. We note only two slight differences. First, for $x, y \in G'$, $y \geqq x$, define $N(x,y)$ to be the largest n for which nx is defined and $y \geqq nx$. If $(x,y) \in B$, then $y \geqq nx$ implies $(n+1)x$ is defined; hence, for $(x,y) \in B$, we have $[N(x,y)+1]x > y \geqq N(x,y)x$, as in Theorem 1.1.

Second, the use of inverses in Theorem 1.1 is solely to provide elements of form $y^{-1}x$, where $y < x$. The same effect is achieved here by finding y' such that $y + y' \leqq x$, using property 6 of Definition 3.1.

Theorem 3.1 is clearly applicable to extensive measurement, for the case where there is a practical upper bound on the size of elements that can be compared. This has been discussed by Luce and Marley [**23**]. Less obvious is the application of the theorem to a more realistic version of conjoint measurement. The solvability axiom (Axiom 2 of §1.4) essentially forces the set $A/\sim$ to be a subgroup of real numbers, whereas in practice, one would like to restrict attention to a bounded subset of such a subgroup. This corresponds to the fact that one cannot in practice always solve equations of the form $a \sim (b_1, b_2)$ for b_2, given a, b_1. The change on the first factor may be so large as to be unmatchable within A_2. To deal with this case, solvability has been replaced by a much more realistic assumption (Debreu [**9**]; Luce [**22**]) called *restricted solvability*:

AXIOM 2′. *For all $a \in A$, $b_1 \in A_1$, if there exist $\underline{c}_2, \bar{c}_2$ such that $(b_1, \bar{c}_2) \geqq a \geqq (b_1, \underline{c}_2)$, then there exists c_2 such that $a \sim (b_1, c_2)$; and a similar assumption with the roles of A_1, A_2 interchanged.*

In this more restricted situation, one can order "positive differences" between elements of A_1 by comparison with a "difference" in A_2: namely, define $a_1 - b_1 \geqq a_1' - b_1'$ if there exist $a_2, b_2 \in A_2$ such that

$$(a_1, b_2) \geqq (b_1, a_2), \qquad (b_1', a_2) \geqq (a_1', b_2).$$

Certain positive differences can then be "added" by laying off equivalent differences end-to-end. An additional axiom, similar to the cancellation axiom of §1.4, is required, and minor modifications of the Archimedean axiom (Axiom 4 of §1.4) are needed, but given these, the bounded Hölder's Theorem can be applied to the system of positive differences on each factor, A_1 and A_2. Ultimately, this leads to the same conclusion as that of Theorem 1.3, based on much weaker assumptions.[2]

3.3. *The ternary relation* $\langle xyz \rangle$. We return now to consideration of a set A, and a mapping $(x,y) \to xy$ of $A \times A$ onto a total order $(P, \geqq)$. If there exists a function ϕ from P to the reals, which preserves order, such that $d(x,y) = \phi(xy)$ is a metric with additive segments, then the following 4 axioms are easily seen to be necessary.

AXIOMS. 1. *For* $x \neq y$, $xx = yy < xy$.

2. $xy = yx$.

3. *If* $xy \leqq uw$, *then there exists* v *such that* $xy = uv$ *and* $\langle uvw \rangle$.

4. *If* $x \neq y$, *then for any* u, w *there exist* $x_0, \cdots, x_n$ *such that* $x_0 = u$, $x_n = w$, *and for* $i = 1, \cdots, n$, $x_{i-1}x_i \leqq xy$.

Axioms 1, 2, and 4 are stated entirely in terms of the ordering $\geqq$, but Axiom 3 involves the ternary relation $\langle uvw \rangle$. In terms of the desired metric, this means $d(u,v) + d(v,w) = d(u,w)$. However, we shall define $\langle \quad \rangle$ in terms of the ordering alone. We do this by noticing that, if $d(x,y) + d(y,z) = d(x,z)$, then the distance $d(y',z)$ achieves a minimum at $y' = y$, for all points y' on or inside the sphere with center x and radius $d(x,y)$. This characterization of y uses only ordinal relations.

DEFINITION 3.2. $\langle xyz \rangle_L$ if for all x', y', z' such that $x'y' \leqq xy$ and $xz \leqq x'z'$, both of the following hold:

(i) $yz \leqq y'z'$.

(ii) If $yz = y'z'$, then $xy = x'y'$ and $xz = x'z'$.

Define $\langle xyz \rangle$ if both $\langle xyz \rangle_L$ and $\langle zyx \rangle_L$.

---

[2] This use of positive differences on each component in a system of additive conjoint measurement is unpublished; it draws on material from a book in preparation, tentatively titled *Foundations of measurement*, by D. H. Krantz, R. D. Luce, P. Suppes, and A. Tversky. For a slightly different treatment, also based on Axiom 2′, see Luce [**22**].

Roughly speaking, $\langle xyz \rangle_L$ holds if yz is minimal among all $y'z'$ such that y' is inside a sphere of radius xy and x' is outside a concentric sphere with radius xz. The relation $\langle xyz \rangle$ is simply the symmetric form: clearly, $\langle xyz \rangle$ iff $\langle zyx \rangle$. Henceforth, in Axiom 3 above, the relation $\langle \quad \rangle$ will be understood to be the one defined by Definition 3.2. From this definition, we obtain the following useful lemma (only Axiom 2 is used in the proof).

LEMMA 3.1. *If* $\langle xyz \rangle$, $x'y' \leqq xy$, $y'z' \leqq yz$, *and* $xz \leqq x'z'$, *then* $\langle x'y'z' \rangle$.

This follows because if any inequality were strict, then (ii) of the definition would yield a contradiction of one of the other inequalities. Hence $x'y' = xy$, $y'z' = yz$, and $xz = x'z'$, and $\langle x'y'z' \rangle$ follows. We also note that from Axioms 1 and 2 and Definition 3.2, if $\langle xyz \rangle$, then $xy, yz \leqq xz$. Given these preliminary results, we can prove the following fundamental theorem.

THEOREM 3.2. *If Axioms* 1—3 *hold, and if* $\langle xyz \rangle$ *and* $\langle xzw \rangle$, *then* $\langle yzw \rangle$ *and* $\langle xyw \rangle$.

PROOF. First we prove $\langle yzw \rangle$. From $xz \leqq xw$ and $\langle xyz \rangle_L$, we have $yz \leqq yw$. By Axiom 3, choose z' such that $yz = yz'$ and $\langle yz'w \rangle$. By Lemma 3.1, it suffices to show $zw \leqq z'w$. Since $\langle xyz \rangle_L$ and $yz' \leqq yz$, we have $xz' \leqq xz$; but then, by $\langle xzw \rangle_L$, $zw \leqq z'w$ follows as required.

Next we show $\langle xyw \rangle$. Note that $wy \leqq wx$; otherwise, if $wx < wy$, then $\langle wzx \rangle_L$ implies $zx < zy$, contradicting $\langle zyx \rangle$. By Axiom 3, choose y' with $wy = wy'$ and $\langle wy'x \rangle$. By Lemma 3.1, it will suffice to show that $xy \leqq xy'$.

From $\langle yzw \rangle$, $wz \leqq wy'$. Construct z' with $wz = wz'$ and $\langle wz'y' \rangle$. Suppose $xy' < xy$. Then by $\langle xyz \rangle$, either $xz' < xz$, or $yz < y'z'$. The former is false, since $\langle xzw \rangle$ and $xz' < xz$ imply $zw < z'w$; and the latter is wrong, because $\langle wz'y' \rangle$, $wz = wz'$, and $wy = wy'$ imply $y'z' \leqq yz$. Hence, we conclude that $xy \leqq xy'$, as required. This completes the proof of Theorem 3.2.

Theorem 3.2 states the basic property of the ternary relation $\langle \quad \rangle$ which is needed to construct a metric; as will be seen in the next section, it corresponds to associativity of the operation $+$ defined on P.

3.4. *Existence and uniqueness of a metric.*

DEFINITION 3.3. Let A, P, be as above, and let $\langle \quad \rangle$ be given by Definition 3.2. Define a binary operation $+$ on P by $xy + x'y'$

$= uw$ if $xy = uv$, $x'y' = vw$, and $\langle uvw \rangle$.

We note that this is well defined. Let $P_1 = P - \{xx\}$, and let $B = \{(xy, x'y') \mid xy + x'y' \text{ is defined}\}$. Then the following theorem can be proved.

Theorem 3.3.[3] *If Axioms* 1—3 *of* §3.3 *hold, then* $(P_1, B, +, \geqq)$ *is a positive ordered local semigroup; if Axiom* 4 *also holds, then it is Archimedean as well. Hence, if Axioms* 1—4 *hold, then there is a metric* d *on* A *such that*:

(i) $xy \geqq x'y'$ *iff* $d(x,y) \geqq d(x',y')$.

(ii) $\langle xyz \rangle$ *iff* $d(x,y) + d(y,z) = d(x,z)$.

Moreover, d *is unique up to similarity transformations* (*i.e.* d *is a ratio scale*).

Proof. The proof that $(P_1, B, +, \geqq)$ is a positive ordered local semigroup if Axioms 1—4 hold is almost immediate. To illustrate, we prove associativity (in the sense of property 3, Definition 3.1). Suppose $(xy, x'y')$ and $(xy + x'y', x''y'') \in B$. Let $uv = xy + x'y'$ and $vw = x''y''$, with $\langle uvw \rangle$. Let $u'v' = xy$, $v'w' = x'y'$, with $\langle u'v'w' \rangle$. Then $u'w' = uv$. Since $u'v' \leqq uv$, there exists z with $u'v' = uz$ and $\langle uzv \rangle$. By Theorem 3.2, $\langle zvw \rangle$ and $\langle uzw \rangle$. By Definition 3.2, $zv = x'y'$. Thus, $x'y' + x''y''$ is defined and $= zw$; so $xy + (x'y' + x''y'')$ is defined and $= uw = (xy + x'y') + x''y''$.

From Theorem 3.1, there exists a real-valued function ϕ on P_1' such that $xy \geqq x'y'$ iff $\phi(xy) \geqq \phi(x'y')$ and $\phi(xy + x'y') = \phi(xy) + \phi(x'y')$. We note that there is at most one $p \in P_1 - P_1'$, i.e. a maximal element of P, if such exists. We define d on $A \times A$ by

$$\begin{aligned} d(x,y) &= \phi(xy) \quad \text{if } xy \in P_1', \\ &= 0 \quad \text{if } x = y, \\ &= \sup\{\phi(uv) \mid uv \in P_1'\} \quad \text{if } xy \text{ is maximal.} \end{aligned}$$

Obviously, $xy \geqq x'y'$ iff $d(x,y) \geqq d(x',y')$ and $\langle xyz \rangle$ iff $d(x,y) + d(y,z) = d(x,z)$. The triangle inequality follows from the definition of $\langle\ \ \rangle$. Also, by Axiom 4, if xy is maximal in P, then $\sup\{\phi(uv) \mid uv \in P_1'\}$ is finite. Thus, d is a metric satisfying (i)

[3] The results in §§3.3—3.5 are essentially due to Beals and Krantz [**2**]. They considered a somewhat more general situation. The version presented here, particularly Theorem 3.3, draws on unpublished material for the book mentioned in footnote 2.

and (ii). Clearly, any other metric d' with the same properties defines a function ϕ' on P_1' with the same properties as ϕ; $\phi' = \alpha\phi$, hence, $d' = \alpha d$, follow from the uniqueness assertion of Theorem 3.1. This completes the proof of Theorem 3.3.

3.5. *Existence of segments*. Note that Axioms 1—4 of §3.3 can be satisfied by the set $A = \{x,y,z\}$, with $xy = yz < xz$. In fact, $\langle xyz \rangle$ holds, and the only d satisfying Theorem 3.3 is given by $d(x,y) = d(y,z) = D$, $d(x,z) = 2D$, where $D > 0$ is arbitrary. Such finite examples are avoided if we impose the requirement that any $x,z \in A$ be joined by an *additive segment*, as defined in §3.1. One way to guarantee this is to impose two additional conditions, nondiscreteness and completeness:

AXIOMS. 5. *$P - \{xx\}$ has no minimal element.*

6. *If x_i is a sequence in A such that for $u \neq v$, $x_i x_j \leqq uv$ for all but finitely many (i,j), then there exists $y \in A$ such that for $u \neq v$, $x_i y \leqq uv$ for all but finitely many i. That is, any Cauchy sequence converges.*

We call a subset γ of A a *partial segment from x to z* if (i) $x,z \in \gamma$ and (ii) for any $u,v \in \gamma$, $\langle xuv \rangle$ or $\langle xvu \rangle$. The set γ is a *segment from x to z* if it is a maximal partial segment from x to z. By Zorn's lemma, for any x, z, there exists a segment γ from x to z. We wish to show that γ is isometric to $[0, d(x,z)]$. For $y \in \gamma$, let $f(y) = d(x,y)$. Obviously, $d(u,v) = |f(u) - f(v)|$, so f is an isometry of γ into $[0, d(x,z)]$. It remains only to show that f is onto. For this, we use Axioms 5 and 6 above.

Let t be $\in (0, d(x,z))$. Let

$$\gamma_1 = \{y \in \gamma \mid f(y) \leqq t\}, \qquad \gamma_2 = \{y \in \gamma \mid f(y) \geqq t\}.$$

From Axiom 6 it is easily shown that f attains its maximum in γ_1 at some $y_1 \in \gamma_1$ and its minimum in γ_2 at $y_2 \in \gamma_2$. For example, if u_i is a sequence in γ_1 such that $d(x_i, u_i) \to \sup\{f(y) \mid y \in \gamma_1\}$, then by Axiom 6, $u_i \to y_1 \in A$, and it is easy to see that $\gamma \cup \{y_1\}$ is a partial segment. It follows by maximality that $y_1 \in \gamma$; y_2 is treated similarly. If $y_1 = y_2$, then $f(y_1) = f(y_2) = t$, and the required preimage of t in γ has been constructed. But for $y_1 \neq y_2$, we can use Axioms 5 and 3 to choose y with $\langle y_1 y y_2 \rangle$, and $y \neq y_1, y_2$. By construction, $y \notin \gamma$, but by Theorem 3.2, $\gamma \cup \{y\}$ is a partial segment, contradicting the maximality of γ. Thus, $y_1 = y_2$ as required. This completes the proof of the following theorem (whose converse is obviously true also):

THEOREM 3.4. *Let $A, P, \geqq$ satisfy Axioms 1—6. Then there is a metric d on A, unique up to multiplication by a positive constant, such that (A,d) is a complete metric space with additive segments, and such that $xy \geqq x'y'$ iff $d(x,y) \geqq d(x',y')$.*

4. Polynomial measurement theories.

4.1. *Independence and sign-dependence.* If $\geqq$ is a binary relation on $\prod_{i=1}^{n} A_i$, where $n \geqq 2$, it induces other binary relations on products of any m factors, where $m < n$. For example, if b is a fixed element of $\prod_{i=m+1}^{n} A_i$, and a, a' are elements of $\prod_{i=1}^{m} A_i$, we can define

$$a \leqq (b)\, a' \quad \text{if } (a,b) \geqq (a',b).$$

Thus any choice of a fixed $b \in \prod_{i=m+1}^{n} A_i$ induces a relation $\geqq (b)$ on $\prod_{i=1}^{m} A_i$. Similarly, choosing fixed components in any subset of factors induces a binary relation over the product of the remaining factors.

One of the things that makes the study of polynomial combination laws fruitful is that, for binary relations obeying such laws, the induced relation $\geqq (b)$ varies in regular and interesting ways as a function of the vector of fixed components, b. Recall that, according to Definition 2.1, $(A_1, \cdots, A_n, \geqq)$ satisfies a polynomial combination law f provided that there are real-valued functions ϕ_i on A_i and ϕ on $\prod_{i=1}^{n} A_i$, such that ϕ is order-preserving and $\phi = f(\phi_1, \cdots, \phi_n)$. A simple example of regularity of induced relations occurs if $f(x_1, \cdots, x_n) = g(x_1, \cdots, x_m) + h(x_{m+1}, \cdots, x_n)$. In that case, it is obvious that $\geqq (b)$ is independent of b, for $b \in \prod_{i=m+1}^{n} A_i$; the ordering of elements of $\prod_{i=1}^{m} A_i$ depends only on the values of $g(\phi_1, \cdots, \phi_m)$. We say in this case that $\prod_{i=1}^{m} A_i$ is *independent of* $\prod_{i=m+1}^{n} A_i$. In the simplest case, $f(x_1, \cdots, x_n) = \sum_{i=1}^{n} x_i$, and any subset of factors is independent of its complement. We say in this case that the system $(A_1, \cdots, A_n, \geqq)$ is *completely independent*; complete independence is thus a necessary condition for additive conjoint measurement in n factors.

A subset of factors can be independent of a proper subset of its complement, even when it is not independent of the entire complement; this occurs when an induced relation, $\geqq (b,c)$, does not depend on the vector of components represented by b, for fixed c. For example, when $f(x_1, x_2, x_3) = (x_1 + x_2)x_3$, A_1 is independent of A_2, and vice versa, although A_1 need not be independent of $A_2 \times A_3$.

If $f(x_1, \cdots, x_n) = g(x_1, \cdots, x_m) \cdot h(x_{m+1}, \cdots, x_n)$, then the induced ordering of elements in $\prod_{i=1}^{m} A_i$ depends not only on $g(\phi_1, \cdots, \phi_m)$ but on whether $h(\phi_{m+1}, \cdots, \phi_n)$ is positive, negative, or zero. Thus, $\prod_{i=m+1}^{n} A_i$ can be partitioned into at most 3 subsets, S^+, S^0, S^-. All induced orders $\geqq (b)$ are identical for $b \in S^+$, and are the reverse of $\geqq (b)$ for $b \in S^-$, while $\geqq (b)$ is degenerate (the universal relation) for $b \in S^0$. We express this partition property by saying that $\prod_{i=1}^{m} A_i$ is *sign-dependent on* $\prod_{i=m+1}^{n} A_i$. *Independence of* is the special case of *sign-dependence on* in which only S^+ or S^- is nonempty. Thus, for $(x_1 + x_2) x_3$, $A_1 \times A_2$ and A_3 are mutually sign-dependent. It is not true, however, that $A_1 \times A_3$ is sign-dependent on A_2. In fact, for $x_3' > x_3$.

$$(x_1 + y_2) x_3 \geqq (x_1' + y_2) x_3' \quad \text{iff } y_2 \leqq (x_1 x_3 - x_1' x_3') / (x_3' - x_3)$$

so that the value of $y_2 = \phi_2(a_2)$ at which the order of (x_1, x_3) and (x_1', x_3') reverses is not fixed, leading to a partition, but varies with x_1, x_3, x_1', x_3'.

Independence and sign-dependence properties are good examples of qualitative laws. Independence is subject to straightforward experimental testing, wherever order relations can be determined; when satisfied, it strongly indicates that the effects of certain variables can be evaluated apart from consideration of the fixed values of other variables. Sign-dependence is similarly testable; moreover, it has a kind of special "flavor", since a large effect can be produced in 2 completely different ways, by combination of two "positive" values or of two "negative" values. An instance of sign-dependence was discussed in §2.4 where moral evaluations of descriptive adverb-adjective combinations were studied, e.g., "slightly evil", etc. There, the adverb factor is sign-dependent on the adjective factor; "evil" is in the S^- part of A_2. It scarcely needs experimental testing in this case to show that sign-dependence is more appropriate than independence, and hence that multiplication is more appropriate than addition. The detailed accuracy of a multiplicative model is, of course, another question.

In general, particular polynomials exhibit more or less idiosyncratic patterns of independence and sign-dependence; thus, it is possible, on the basis of empirical information concerning these properties, to diagnose appropriate polynomial combination rules, or at least, to narrow the field.

4.2. *Additive conjoint measurement and independent dimensions in geometry*. It was indicated previously that a system of additive conjoint measurement is completely independent. Surprisingly, a partial converse can be proved: if $(A_1, \cdots, A_n, \geqq)$ satisfies complete independence, restricted solvability, and an Archimedean condition, and if $n \geqq 3$, then $(A_1, \cdots, A_n, \geqq)$ is decomposable relative to $\sum_{i=1}^n x_i$. (Recall that restricted solvability means that any shift on $n-1$ factors can be compensated by a shift on the nth factor, given that certain boundedness conditions hold.) Since restricted solvability and Archimedean conditions are usually assumed to be valid empirically, this means that if $n \geqq 3$ (in the sense that there are at least 3 nontrivial factors), complete independence is the empirical equivalent to additivity of the factors.[4] (No such result holds for $n = 2$; in that case, the cancellation axiom (Axiom 3 of §1.4) is required.) The proof relies on the bounded Hölder theorem established in §3.1, but it is quite complicated and I shall not present it here. It can be greatly simplified if unrestricted solvability is assumed.

It was recognized by Tversky and Krantz [**36**] that additive conjoint measurement can be profitably applied to metric representation of similarity orderings. Suppose that we have a dissimilarity ordering $\geqq$ on $A \times A$, as in §3, but that the set A is endowed with product structure, $A = \prod_{i=1}^n A_i$. If the ordering on $A \times A$ is to be represented by a Euclidean metric d, with the sets A_i as a complete set of orthogonal coordinates, then there are functions ϕ_i on A_i such that

$$d(a, b) = \left[\sum_{i=1}^n |\phi_i(a_i) - \phi_i(b_i)|^2 \right]^{1/2}.$$

This equation can be generalized in 2 ways:

$$(1) \qquad d(a, b) = F\left[\sum_{i=1}^n \phi_i(a_i, b_i) \right],$$

$$(2) \qquad d(a, b) = F[\,|\phi_1(a_1) - \phi_1(b_1)|, \cdots, |\phi_n(a_n) - \phi_n(b_n)|\,].$$

In (1), F is a strictly increasing function of 1 variable, while in (2), F is strictly increasing in each of the n variables. Equation (1) specifies that dimensions combine additively to determine dissimilarity, while (2) specifies that the contribution of any one dimension can be represented by absolute differences of scale values.

If d is not required to be a metric, then equation (1) is simply

[4] A topological version of this theorem was published by Debreu [**9**]. The present version is due to R. D. Luce and is taken from material for the book mentioned in footnote 2.

the equation of additive conjoint measurement, in n variables. Similarly, when d need not be a metric, then the absolute difference representation, on any one dimension, can be analyzed by methods very close to those of additive conjoint measurement in two variables. (Absolute difference representations have been extensively studied by Pfanzagl [**28**].)

If we combine the conditions for n-factor additivity (across dimensions), for 2-factor additivity (within each dimension), and for a metric with additive segments (§3 above), then the metric d takes the form

$$d(a,b) = \left[\sum_{i=1}^{n} | \phi_i(a_i) - \phi_i(b_i) |^r \right]^{1/r}, \tag{3}$$

where $1 \leqq r < \infty$, i.e., d is a Minkowski r-metric. For a proof, see Tversky and Krantz [**36**]. The proof uses equation (2), additive segments, and the Krein-Milman theorem to establish that d is homogeneous; then equations (1), (2), and homogeneity yield a functional equation related to the equation

$$F^{-1}\left[\frac{1}{n}\sum_{i=1}^{n} F(t\alpha_i)\right] = tF^{-1}\left[\frac{1}{n}\sum_{i=1}^{n} F(\alpha_i)\right],$$

whose only continuous solutions are $p + q \log \alpha$ and $p + q\alpha^r$. The form given by equation (3) then follows.

If we let $F(\alpha) = p(e^{q\alpha} - 1)^r$, $(p, \alpha > 0, r \geqq 1)$ then

$$d(a,b) = F^{-1}\left[\sum_{i=1}^{n} F(| \phi_i(a_i) - \phi_i(b_i) |)\right]$$

yields a metric, but only points differing in exactly one dimension can be joined by an additive segment. The geometry of this "exponential metric" seems not to have been studied. The Minkowski r-metric is the limit of the exponential metric, letting $q \to 0$ while pq^r is constant.

4.3. *Simple polynomials.* We now return to consideration of polynomial combination rules more general than additivity. The independence or sign-dependence properties that are logically necessary for a given polynomial are usually not sufficient, even if appropriate solvability and Archimedean conditions are assumed. In this respect, the pure additive and pure multiplicative rules are exceptional. For one class of polynomial combination rules, called simple polynomials, there is a general schema for finding a sufficient set of axioms. This class is defined as follows: (i) single variables are simple polynomials; (ii) if f_1 and f_2 are simple polynomials with disjoint variables, then $f_1 + f_2$ and $f_1 f_2$ are simple polynomials;

(iii) no polynomials are simple except by virtue of (i) and (ii). More formally:

DEFINITION 4.1. Let $F[Y]$ be a ring of polynomials in the indeterminates Y. Then $S[Y]$ is the smallest subset of $F[Y]$ such that

(i) $Y \subset S[Y]$.

(ii) If $Y_1, Y_2 \subset Y$, with $Y_1 \cap Y_2$ empty, then for any $f_1 \in S[Y_1]$ and $f_2 \in S[Y_2]$, $f_1 + f_2$ and $f_1 f_2 \in S[Y]$.

The elements of $S[Y]$ are the *simple polynomials* of $F[Y]$.

To axiomatize polynomial conjoint measurement, relative to a simple polynomial f, for $(A_1, \cdots, A_n, \geqq)$, we proceed to introduce a series of addition and multiplication operations in the set $A/\sim$. An addition operation is introduced for each decomposition of a simple component of f as the sum of smaller simple components with nonoverlapping variables, and similarly for multiplications. The manner of introducing these operations is the same as in §1.4, e.g., $(b_1, a_2^0) + (a_1^0, b_2) = (b_1, b_2)$, where $a^0 = (a_1^0, a_2^0)$ is the origin (or unit, for multiplication). However, one must be careful to keep the origin the same for all additions, the unit the same for all multiplications, and to choose the origin as a multiplicative zero. In addition to suitable sign-dependence, solvability and Archimedean conditions, three classes of axioms need to be introduced:

(i) Appropriate cancellation conditions, like Axiom 3 of §1.4, that guarantee that the operations introduced are well defined, commutative, and associative.

(ii) Conditions guaranteeing that all the addition operations have the same effect, as do all the multiplications.

(iii) A condition that is used to prove distributivity of multiplication over addition.

Of course, all these axioms may not be logically independent, so that some of them may be eliminable in any given instance.

Rather than give complete abstract details of this general schema, I shall sketch an illustration for 4-factor conjoint measurement, relative to the polynomial $x_1x_2 + x_3x_4$.

We choose a suitable origin, $a^0 = (a_1^0, a_2^0, a_3^0, a_4^0)$ and a suitable unit, $a^1 = (a_1^1, a_2^1, a_3^0, a_4^0) \sim (a_1^0, a_2^0, a_3^1, a_4^1)$. Addition is defined by

$$(b_1, b_2, a_3^0, a_4^0) + (a_1^0, a_2^0, b_3, b_4) = (b_1, b_2, b_3, b_4).$$

Two different multiplications are defined by

$$(b_1, a_2^1, a_3^0, a_4^0) \cdot (a_1^1, b_2, a_3^0, a_4^0) = (b_1, b_2, a_3^0, a_4^0),$$

$$(a_1^0, a_2^0, b_3, a_4^1) \cdot (a_1^0, a_2^0, a_3^1, b_4) = (a_1^0, a_2^0, b_3, b_4).$$

These definitions are chosen so that any element b of $A/\sim$ satisfies

$$(b_1, b_2, b_3, b_4) = (b_1, a_2^1, a_3^0, a_4^0) \cdot (a_1^1, b_2, a_3^0, a_4^0) + (a_1^0, a_2^0, b_3, a_4^1) \cdot (a_1^0, a_2^0, a_3^1, b_4).$$

Once we have constructed a ring isomorphism ϕ of $(A/\sim, +, \cdot, \geqq)$ into $(\mathrm{Re}, +, \cdot, \geqq)$, we can define $\phi_1(b_1) = \phi(b_1, a_2^1, a_3^0, a_4^0)$, etc., and obtain, from the previous equation and the isomorphism property, the desired relation

$$\phi(b) = \phi_1(b_1)\phi_2(b_2) + \phi_3(b_3)\phi_4(b_4).$$

Unlike the simple additive case, the origin a^0 and unit a^1 must be chosen with some care, and the solvability and cancellation conditions must be formulated to take note of exceptions. In particular, we start by assuming that $A_1 \times A_2$ and $A_3 \times A_4$ are mutually independent, while A_1 and A_2, as well as A_3 and A_4, are mutually sign-dependent. For $i = 1, 2, 3, 4$, a_i^0 must be chosen in the 0-class determined by sign-dependence, while a_i^1 must be chosen outside the 0-class. This amounts to choosing the additive origin to be the multiplicative zero, and the multiplication to be non-trivial. The solvability and cancellation conditions must be formulated with due care for division by zero. Lastly, some delicacy is required in designating + and − signs, relative to sign-dependence. Once all this is done, and suitable conditions of type (ii) and (iii) are imposed, guaranteeing that the multiplications coincide and are distributive over addition, we obtain an ordered ring, as required.

To illustrate the derivation of type (ii) and (iii) axioms, I indicate one that guarantees distributivity of multiplication over addition. Represent three arbitrary equivalence classes by

$$b = (b_1, a_2^1, a_3^0, a_4^0), \quad c = (a_1^0, a_2^0, b_3, a_4^1),$$
$$d = (a_1^1, b_2, a_3^0, a_4^0) \sim (a_1^0, a_2^0, a_3^1, b_4).$$

Then

$$b \cdot d + c \cdot d = (b_1, b_2, b_3, b_4), \qquad b + c = (b_1, a_2^1, b_3, a_4^1).$$

Choose c_1 such that $(b_1, a_2^1, b_3, a_4^1) \sim (c_1, a_2^1, a_3^0, a_4^0)$. Then

$$(b + c) \cdot d = (c_1, b_2, a_3^0, a_4^0).$$

Thus, the required axiom is one that permits us to infer, from the equivalences

$$(a_1^1, b_2, a_3^0, a_4^0) \sim (a_1^0, a_2^0, a_3^1, b_4), \qquad (b_1, a_2^1, b_3, a_4^1) \sim (c_1, a_2^1, a_3^0, a_4^0),$$

the conclusion

$$(b_1, b_2, b_3, b_4) \sim (c_1, b_2, a_3^0, a_4^0),$$

i.e. $b \cdot d + c \cdot d = (b + c) \cdot d$. With some extra trouble, the required axiom can be formulated as a general condition, independent of the choice of the a_i^0, a_i^1, and logically necessary for the desired polynomial combination rule.

Finally, note that all the axioms introduced, except for solvability, turn out to be logically necessary conditions for the given polynomial combination rule. It would be interesting to reformulate the above treatment of polynomial measurement with restricted solvability, obtaining only a subset of an ordered ring, but this remains an open problem, which may involve considerable technical difficulty.

5. The measurement of color.

5.1. *Metameric matches and vectorial representation*. The psychological laws on which color measurement is based were clearly enunciated by H. Grassman in 1853 [**12**], and bear his name. These laws, and the measurement techniques based on them, are of particular interest, because of their simplicity and beauty, and because they involve an unusual blend of physics, physiology, and psychology. The empirical basis of color measurement involves a physically defined binary operation, *additive color mixture*, and a psychological equivalence relation, *metamerism*. Grassman's laws, which relate these primitives, have clear-cut physiological implications (Brindley [**5**, pp. 198-218]). Thus, color measurement has long been the point of departure for physiological and psychological color theories.

The measurement representation for colors involves vectors over the real numbers, rather than real numbers alone. International standards for the vectorial representation of colors were adopted in 1931 by the International Commission on Illumination (ICI) [**15**]. Extensive discussions of these color measurement standards may be found in the works of Wright [**37**] and Stiles [**32**].

In this first section, we assume that the stimulus whose color is to be measured consists of a small, homogeneous patch of light, viewed under standardized conditions. Such a stimulus is specified by a function giving its energy,[5] or energy density, for each wave-

[5] The term "energy" is used in a broad sense; depending on the nature of the stimulus, various measures derived from energy may be more appropriate, e.g.,

length in the visible portion of the electromagnetic spectrum (wavelengths from 4×10^{-7} to 7×10^{-7} meters, approximately). Thus, a color stimulus may be considered to correspond to a non-negative (energy) measure defined on the Borel subsets of a real interval.

Additive color mixture means summation of the energy in the mixed stimuli. For example, if stimuli b and c are produced by illuminating the same portion of a screen with light from two different projectors, then the mixture, $b + c$, is produced by turning both projectors on at once. In fact, the countably additive real-valued set functions on the Borel subsets of the visible spectrum form a vector space over the reals, which we denote B; the non-negative elements of B, which correspond exactly to the possible specifications of color stimuli, form a convex cone in B, denoted C. Additive color mixture corresponds to vector addition in C.

Two distinct elements of C may correspond to color stimuli that look alike in color. We say that such stimuli are a *metameric match*, or more simply, are *metamers*. We denote the relation of metamerism by $\sim$. With suitable experimental methods, for a normal observer, $\sim$ can be considered to be an equivalence relation on C, to a very high degree of approximation. There are many examples of metameric matches: for instance, a stimulus with a "bimodal" energy distribution, with most of the energy in the "red" and "green" parts of the visible spectrum, is metameric to an appropriate "unimodal" stimulus with most of its energy concentrated in the "yellow" part of the spectrum.

Let M be the set of differences of metameric pairs, i.e. $M = \{d \mid d \in B$, and for some $b, c \in C$, with $b \sim c$, $d = b - c\}$. The content of *Grassman's third law* is that M is a linear subspace of B. This experimental finding has been confirmed in modern studies to a high degree of approximation, over a wide range of conditions (see Brindley, [**5**, p. 211]). We can now state succinctly the classical experimental law of color mixture, the *Law of Trichromacy* (*Grassman's first law*):

$$\text{For a normal observer, } \dim B/M = 3.$$

---

power, power per unit area of source, etc. For any change of units the nonnegative measures that correspond to color stimuli need only be altered by a suitable constant factor.

An observer is called dichromatic if $\dim B/M = 2$, and monochromatic (totally color blind) if $\dim B/M = 1$.

Standardized systems of color measurement employ a convenient basis for B/M. For $b \in C$, the coordinates of $b + M$ relative to the basis in B/M are called *tristimulus coordinates* of b. Thus, two stimuli have the same tristimulus coordinates if and only if they are metameric. The ICI standards specify tristimulus coordinates for an average *standard observer*, for approximately monochromatic stimuli (point measures). The tristimulus coordinates for more general stimuli are computed by approximating these as sums of monochromatic stimuli.

Another useful set of coordinates is obtained by regarding the tristimulus coordinates for b as homogeneous (projective) coordinates for the one-dimensional subspace generated by $b + M$. If these homogeneous coordinates are normalized so they sum to 1, they are called *chromaticity coordinates*. Two stimuli have the same chromaticity coordinates if a scalar multiple (change in overall energy level) of one of them is metameric to the other.

5.2. *Photopigments*. The biological effects of light are mediated through absorption by photopigments. The absorbing properties of a photopigment are specified by a spectral absorptance function p. For wavelength λ, $p(\lambda)$ is the fraction of the incident energy at wavelength λ that is absorbed by the pigments and converted into electrochemical energy. For a photopigment in human visual receptors, it is convenient to include in $p(\lambda)$ the wavelength-dependent alterations in the stimulus between the point where it is specified by a measure and the point where it is absorbed by a receptor, e.g. reflection at the cornea, absorption and scattering in the ocular media. If this is done, then the average number of quanta absorbed and converted into electrochemical energy, from a stimulus b, by a photopigment p, is proportional to

$$\int \lambda p(\lambda)\, db(\lambda).$$

The integral is taken over the visible spectrum; the factor λ is introduced because average quanta/unit energy is proportional to wavelength. Thus, a photopigment p defines a linear functional on C, and by the natural extension, on B as well. (Departures from linearity occur insofar as the photopigment is appreciably depleted by the ensuing photochemical reaction.)

Grassman's laws would be explained if we assume that there are three linearly independent photopigments, p_1, p_2, p_3, which mediate color vision. The intersection of the null-spaces of the p_i would be exactly M. The p_i define linear functionals $\bar{p}_i$ on B/M, which span the dual space of B/M. Their dual basis yields a preferred coordinate system in B/M. Relative to this basis, the ith tristimulus coordinate of b is precisely $p_i(b) = \int \lambda p_i(\lambda)\, db(\lambda)$, the effective quantal absorption of stimulus b by p_i. One may hope to account for various other properties of color (for example, color discriminability or perceived hue) in a simple way in terms of those coordinates, which represent the basic physiological responses (the "Grundempfindungen" of Helmholtz). This is the thesis of the Young-Helmholtz theory of color vision.

An important variant of the 3-pigment hypothesis postulates n linearly independent photopigments, $p_1, \cdots, p_n$, $n \geqq 3$, which in turn contribute linearly to 3 independent outputs q_1, q_2, q_3, where

$$q_i = \sum_{j=1}^{n} a_{ij} p_j, \qquad i = 1, 2, 3.$$

The q_i are again linear functionals on B, whose null-spaces intersect in M, and which induce a preferred coordinate system in B/M. This variant allows a stage of "recoding", represented by the linear transformation (a_{ij}), between photopigment absorption and the basic outputs that determine other aspects of color vision, such as discriminability, subjective appearance, etc. This sort of recoding is a feature of many color theories, particularly, the Hering opponent-colors theory as quantified by Hurvich and Jameson [**14**].

There is strong evidence that, if such a recoding does take place, nevertheless, only three independent photopigments are involved, i.e. $n = 3$. The key to the argument is the finding that metameric matches are not broken down by moderate adaptation to colored lights. If $b \sim c$ for normal adaptation, then the appearance of both b and c changes after adaptation to colored light, but except for extreme adaptations, b and c still match in color.

It is generally assumed that the effect of adaptation involves (among other things) bleaching of the photopigments, i.e. the spectral absorptance function p_i is multiplied by a constant t_i, where $1 - t_i$ represents the fraction bleached. Let P be the space of linear functionals on B spanned by $p_1, \cdots, p_n$. Then the equations

$$Tp_i = t_i p_i, \qquad i = 1, \cdots, n,$$

define a linear operator T of P onto itself, with eigenvectors p_i. Let Q be the subspace of P spanned by q_1, q_2, q_3. Dimensionality considerations show that Q consists of all the linear functionals on B which vanish on M. Since adaptation leaves metameric pairs invariant, $b \sim c$ implies that $Tq_i(b) = Tq_i(c)$ for $i = 1, 2, 3$; that is, if $b - c \in M$, then $Tq_i(b - c) = 0$, $i = 1, 2, 3$. Hence, $Tq_i \in Q$, $i = 1, 2, 3$, and it follows that the subspace Q is invariant under all transformations of form T. This implies that Q is spanned by three of the vectors[6] $p_1, \cdots, p_n$. Hence, so far as mediation of color vision is concerned, there are exactly three independent photopigments.

Recently, improved spectrophotometric techniques have yielded direct evidence that there are three photopigments in human retinal cones (Rushton [**29**]; Marks, Dobelle and MacNichol [**26**]). The question of whether significant recoding of the three photopigment outputs takes place is still a key one for color theory, but seems not to be decidable on the basis of color-matching data alone.

5.3. *Color appearance*. Tristimulus coordinates, determined by fixing an arbitrary basis for B/M, convey no information about the phenomenal appearance of color. Much of the literature on psychology and physiology of color is devoted to establishing a preferred (linear or curvilinear) coordinate system in B/M, in terms of which color appearance, including color discriminability and similarity, can be accounted for, and to characterizing changes of color appearance, correlated with changes in viewing conditions, as transformations in these preferred coordinates. As was shown in the previous section, the photopigments provide one system of preferred coordinates, but the possibility of recoding needs to be considered.

The most popular color theory involving recoding is the opponent-colors theory of Hering. This was quantified by Hurvich and Jameson [**14**]. It is based on the observation that the qualities of color appearance can be grouped into three pairs, red-green, yellow-blue, and white-black. Any color partakes of at most one quality from each pair. The pairs are *opponent*, in the sense that additive mixture produces cancellation: if b looks red and c looks green, then, under the same viewing conditions, $b + c$ looks either less red than b or less green than c or neither red nor green. This suggests

[6] If the t_i are distinct, then the minimum polynomial of T, acting in Q, must have form $(x - t_{i_1})(x - t_{i_2})(x - t_{i_3})$. Since the only eigenvectors associated with t_{i_k} are multiples of p_{i_k}, it follows that $p_{i_1}, p_{i_2}, p_{i_3} \in Q$.

that the three recoded outputs, q_1, q_2, q_3 of the previous section, consist of a red-green output [$q_1(b) > 0$ if b looks red, < 0 if b looks green], a yellow-blue output, and a white-black output. These outputs depend on the photopigment absorptions, as indicated in the linear equations of the previous section, but also on other aspects of viewing conditions, particularly the presence of other stimuli in adjacent parts of the visual field.

Quantitative versions of opponent-colors theory give a good account not only of the basic opponent-color qualities, but of many other aspects of color appearance. For example, Euclidean distances calculated in terms of differences in the red-green and yellow-blue coordinates, give a reasonably good account of color similarity for monochromatic stimuli of constant brightness (Krantz [**16**]). This example shows how color measurement makes contact with some of the more general psychological measurement ideas presented in earlier lectures, in particular, with problems of measurement of similarity. Color measurement also makes contact with polynomial measurement; for example, Hurvich and Jameson postulated interrelations among some of the subjective qualities of color that follow polynomial combination rules.

The facts of color appearance, especially the existence of opponent pairs of qualities such as red-green, etc., and the usefulness of opponent pairs as explanations for other color phenomena, make it relatively certain that opponent-color recoding is the proper basis for color theory, although many quantitative details remain to be worked out. Clear-cut physiological evidence for opponent-color recoding has been obtained from microelectrode recording of neural activity in the monkey's visual system, by DeValois and his co-workers [**10**]. The exact physiological mechanisms remain a mystery.

In concluding the study of color theory and color appearance, I should like to touch on the problem of dependence of color appearance on viewing conditions. This dependence is quite dramatic. For instance, a stimulus that looks reddish-yellow under "normal" conditions may look yellow-green after exposing the eye to a red adapting light, or in the presence of a bright red stimulus simultaneously shown in an adjacent part of the visual field. The most precise tool for studying these changes is *cross-context matching*. If we denote different spatio-temporal contexts by σ, τ, etc., and denote by b^σ stimulus b viewed in context σ, then we can define $b^\sigma \sim c^\tau$ to mean that stimulus b, viewed in context σ, has the same color as stimulus c, viewed in context τ. With appropriate experimental methods, $\sim$ can be regarded as an equivalence relation. $b^\sigma \sim c^\sigma$ means that b is metameric to c, and, as indicated earlier, it has been found that $b^\sigma \sim c^\sigma$ implies $b^\tau \sim c^\tau$ for a wide range of contexts τ.

A pair of contexts, σ, τ defines a function: $f_{\sigma,\tau}(c) = b$ iff $b^\sigma \sim c^\tau$. This is defined for every $c \in C$ whose appearance, in context τ, can be matched by the appearance of some stimulus in context σ. Also, $f_{\sigma,\tau}$ is well defined on the corresponding subset of B/M, and its range can be considered a subset of B/M. The study of context effects on color appearance is thus reduced to study of the properties of vector-valued functions of vectors, of form $f_{\sigma,\tau}$.

In the preceding section, it was pointed out that one effect of exposure to an adapting light is likely to be bleaching of photopigments, $p_i \to t_i p_i$. If we postulate that equal appearance corresponds to equal photopigment outputs, then the corresponding functions $f_{\sigma,\tau}$ representing changes in this effect of adaptation must be all representable by diagonal matrices in the coordinates corresponding to the dual basis of the p_i. This is called the von Kries coefficient law [**27**]. During the 1950's, several unsuccessful attempts were made to use empirical determinations of the functions $f_{\sigma,\tau}$ to determine the required coordinate system, and hence, to find the photopigment absorptance curves [**6**], [**25**]. However, the functions $f_{\sigma,\tau}$ seem much better represented by a linear transformation plus a translation than by linear transformations alone. A paper by Krantz [**19**] provides a general theoretical framework for study of functions defined by cross-context matching, and yields the prediction of linear transformation-plus-translation as a special case in B/M.

References

1. J. Aczél, G. Pickert and F. Radó, *Nomogramme Gewebe, und Quasigruppen*, Mathematica (Cluj) (25) **2** (1960), 5-24 (fasc. 1).

2. R. Beals and D. H. Krantz, *Metrics and geodesics induced by order relations*, Math. Z. **101** (1967), 285-298.

3. G. Birkhoff, *Lattice theory*, 3rd ed., Amer. Math. Soc. Colloq. Publ. No. 25, Amer. Math. Soc., Providence, R. I., 1967.

4. L. Blumenthal, *A modern view of geometry*, Freeman, San Francisco, Calif., 1961.

5. G. S. Brindley, *Physiology of the retina and visual pathway*, Edward Arnold Ltd., London, 1960.

6. R. W. Burnham, R. M. Evans and S. M. Newhall, *Prediction of color appearance with different adaptation illuminations,* J. Opt. Soc. Amer. **47** (1957), 35-42.

7. H. Busemann, *The geometry of geodesics,* Academic Press, New York, 1955.

8. N. Cliff, *Adverbs as multipliers,* Psychol. Rev. **66** (1959), 27-44.

9. G. Debreu, "Topological methods in cardinal utility theory" in *Mathematical methods in the social sciences, 1959* edited by K. J. Arrow, S. Karlin and P. Suppes, Stanford Univ. Press, Stanford, Calif., 1960.

10. R. L. DeValois, I. Abramov and G. H. Jacobs, *Analysis of response patterns of LGN cells,* J. Opt. Soc. Amer. **56** (1966), 966-977.

11. L. Fuchs, *Partially ordered algebraic systems,* Addison-Wesley, Reading, Mass., 1963.

12. H. Grassman, *On the theory of compound colours,* Philos. Magazine (London) 7 (4) (1854), 254-264. (Transl. from Pogg. Ann. Physik **89** (1853), 69-84.)

13. O. Hölder, "Die Axiome der Quantität und die Lehre vom Mass" in *Berichte Verhand. König. Sächs. Gesell. Wiss.* (Leipzig), Math.-Phys. Cl. **53** (1901), 1-64.

14. L. M. Hurvich and D. Jameson, *An opponent-process theory of color vision,* Psychol. Rev. **64** (1957), 384-404.

15. International Commission on Illumination, Proc. 8th Session, Cambridge Univ. Press, Cambridge, 1931.

16. D. H. Krantz, *The scaling of small and large color differences,* Ph. D. Dissertation, Univ. of Pennsylvania, 1964.

17. ———, *Conjoint measurement: the Luce-Tukey axiomatization and some extensions,* J. Math. Psychol. **1** (1964), 248-277.

18. ———, *Extensive measurement in semiorders,* Philos. of Science **34** (1967), 348-362.

19. ———, *A theory of context effects based on cross-context matching,* J. Math. Psychol. **5** (1968), 1-48.

20. M. V. Levine, *Transformations which render curves parallel,* J. Math. Psychol. **7** (1970), in press.

21. R. D. Luce, *Semiorders and a theory of utility discrimination,* Econometrica **24** (1956), 178-191.

22. ———, *Two extensions of conjoint measurement,* J. Math. Psychol. **3** (1966), 348-370.

23. R. D. Luce and A. A. J. Marley, "Extensive measurement when concatenation is restricted and maximal elements may exist" in *Essays in honor of Ernest Nagel,* edited by S. Morgenbesser, P. Suppes, and M. G. White, St. Martin's Press, New York, (to appear).

24. R. D. Luce and J. W. Tukey, *Simultaneous conjoint measurement: a new type of fundamental measurement,* J. Math. Psychol. **1** (1964), 1-27.

25. D. L. MacAdam, *Chromatic adaptation,* J. Opt. Soc. Amer. **46** (1956), 500-513.

26. W. B. Marks, W. H. Dobelle and E. F. MacNichol, Jr., *Visual pigments of single primate cones,* Science **143** (1964), 1181-1183.

27. W. Nagel, *Handbuch der Physiologie des Menschen,* Vol. 3, F. Viewig und Sohn, Braunschweig, 1905, pp. 205-221.

28. J. Pfanzagl, *Die axiomatischen Grundlagen einer allgemeinen Theorie des Messens,* Schriftenreihe Statist. Instituts Univ. Wien, Neue Folge, Nr. 1., Physica-Verlag, Würzburg, 1959.

29. W. A. H. Rushton, *A cone pigment in the protanope,* J. Physiol. **168** (1963), 345-359. H. D. Baker and W. A. H. Rushton, *The red-sensitive pigment in normal cones,* J. Physiol. **176** (1965), 56-72.

30. D. Scott and P. Suppes, *Foundational aspects of theories of measurement,* J. Symbolic Logic **23** (1958), 113-128.

31. R. N. Shepard, *The analysis of proximities: multidimensional scaling with an unknown distance function.* I, Psychometrika **27** (1962), 125-140.

32. W. S. Stiles, *The basic data of colour matching,* Phys. Soc. Yearbook, Phys. Soc. (London), 1955, 44-65. Reprinted in R. D. Luce, R. R. Buch and E. Galanter, editors, *Readings in Mathematical Psychology,* vol. 2, Wiley, New York, 1965.

33. P. Suppes, *A set of independent axioms for extensive quantities,* Portugal. Math. **10** (1951), 163-172.

34. A. Tarski, *Contributions to the theory of models,* I, II, III. Indag. Math. **16** (1954), 572-581, 582-588, and **17** (1955), 56-64.

35. A. Tversky, *A general theory of polynomial conjoint measurement,* J. Math. Psychol. **4** (1967), 1-20.

36. A. Tversky and D. H. Krantz, *The dimensional representation and the metric structure of similarity data,* J. Math. Psychol. **7** (1970), in press.

37. W. D. Wright, *The measurement of colour,* 3rd ed., Hilger & Watts, London, 1964.

University of Michigan
Ann Arbor, Michigan

Andrze J. Ehrenfeucht

Two Dimensional Visual Geometry

We shall try to describe a geometry based exclusively on the concept "from my position I see an object A to the left of object B". Two dimensional means, here, that an observer and all observed objects are on the two dimensional Euclidean plane.

For the objects we shall take marked points on the Euclidean plane (intuitively, they are visible and individually distinguishable objects). The observer will be characterized by his angle of vision which can vary from 0 to 360 degrees. The observer's position on the plane is characterized by a point (the point where he is standing) and the angle equal to his visual angle with the vertex in the point (the angle covers the area he is observing). The observation from such a position is the permutation (sequence) of marked points which are within his visual angle listed in order from left to right.

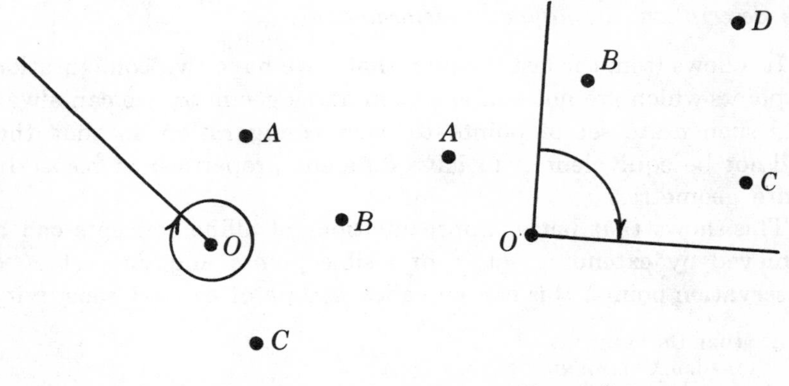

FIGURE 1 FIGURE 2

In the first picture the observer has a visual angle α of 360 degrees. From his position marked by O, he made the observation $[A, B, C]$ (he sees A to the left of B, and B to the left of C). In the second picture the observer's visual angle $\alpha = 90$ degrees and his observation is $[B, D, C]$; A is not in his field of vision.

REMARK. We shall not consider the cases in which the observer's position is in a straight line joining two visible points so that one of them is directly in front of another, but this restriction can be easily avoided.

A set of observational points X and a set of marked points Y from points O in X will be called the description of Y. Such triplet (X, Y, α) will be called a visual geometry. As the concepts of such a geometry we admit only those which can be defined in terms of the description of Y. For example, in the geometry where X is the whole plane and $\alpha \leqq 180$ degrees we can express the fact that point A lies within the triangle B, D, C in the following way: in no observation in which A, B, C, D are visible does A occupy outer-most left or outer-most right position.

In the case in which Y has exactly n elements there are only finitely many possible descriptions of the set Y (the description is a set of permutations with elements in Y). In some cases more exact estimations can be given for the number of descriptions of the set Y. For example, when X is a whole plane and $\alpha = 360$ degrees the number of different descriptions of Y has the magnitude of n^4.

In case when the set Y consists of all points on the plane (or any dense set of points) then the resulting geometry is affine geometry in the following sense:

THEOREM. *If the set X contains at least four noncollinear points, then the only transformations of the plane into itself which preserves the description are affine transformations.*

It follows from the last theorem that if we have two configurations of points which are not equivalent in affine geometry we can always add such finite set of points to each configuration so that they will not be equivalent (will have different properties) in respective finite geometries.

This shows that better approximations of affine concepts can be achieved by extending set Y of visible points and not set X of observation points; this can be called a type of context sensitivity.

STANFORD UNIVERSITY
STANFORD, CALIFORNIA

Stanley Peters

Mathematical Linguistics

I. **Introductory remarks.** The term "mathematical linguistics" is used by various researchers to refer to work in three distinct areas.

(a) One of these areas is sometimes referred to more narrowly as "algebraic linguistics". The term is not really self-explanatory, but it denotes the effort to apply recursive function theory and automata theory in general to the study of grammar.

(b) A second area, sometimes called "computational linguistics", involves application of computing machines to the analysis of linguistic materials. This field of research, which grew out of the now defunct attempt to translate languages mechanically, has drawn heavily on theoretical results supplied by algebraic linguistics for the solution of the practical problems with which it chiefly concerns itself. It has also made use of results provided by

(c) the third area of mathematical linguistics, namely "statistical linguistics". As its name implies, this area involves application of statistics, including information theory, to various problems arising in the study of languages. This is the oldest of the three subfields of mathematical linguistics.

I will discuss only the first of these three areas here. This is the field in which the applications of mathematics to linguistics have been most interesting, I think.

It is difficult to date the beginning of algebraic linguistics exactly. There are papers in the field early in the fifties. But until the very late fifties there was no field, only isolated papers. In contrast, it is easy to date the beginning of the trend which has dominated the field since it became a united field. The crucial paper was published by Noam Chomsky in 1959 with the title "On certain formal properties of grammars" [**3**]. It is this paper that inspired the considerable amount of research which has been done since then. The central idea of the paper had appeared in print in 1956 [**2**], but without the clarity and unaccompanied by the significant results which mark the 1959 paper. We will see in detail what this idea was, but before we can discuss this application of mathematics to linguistics, we must first briefly investigate the area to which mathematics is to be applied.

II. **Preliminary definitions.** Before we can proceed any further, we must know what we mean by "language". So we shall adopt the following definition: A *language* is a set of finite sequences of symbols belonging to a finite vocabulary. Differently put, we must have a finite set V of symbols which can be regarded either as the words or letters which appear in the sentences of the language. This set we call indifferently the *vocabulary* or the *alphabet* of the language. The language itself is then simply the set of sentences, i.e. finite sequences of words or letters, which are "grammatical" or "well-formed" in the language. Obviously, under this definition, a language may be infinite, but only countably so.

This definition includes natural languages (human languages) in both their written and spoken aspects as well as artificial or invented languages such as those used in logic or computer programming. Needless to say, linguists are mainly interested in natural languages, but if we can develop a theory of natural language that says something about artificial languages, so much the better.

One of the most striking features of natural languages is that they are all infinite. The obvious fact that only a finite number of sentences of any natural language will have been used at any given point in time need occasion no confusion. This fact reflects a limitation of the language user rather than a limitation on the language he is using. Each user of a natural language has the capacity to produce completely new sentences of the language on appropriate occasions and to understand totally new sentences. In fact, much of our normal use of language is of exactly this sort. There is no reason to believe that anything in the nature of the LANGUAGE puts an upper bound on the length of these new sentences.

In fact, quite the opposite, there is every reason to believe that there is no upper bound. It follows immediately that the language is infinite.

The limitations of the human as language user result in the fact that sentences that are long or complex enough are difficult or impossible to understand. This means that we must distinguish the performance of a language user from his internalized competence. His *competence* is the knowledge he has which enables him to speak his language. This knowledge allows him to construct and understand totally new sentences. In putting his competence to use in linguistic *performance* other factors such as memory limitations, lapses of attention, etc., enter. Thus a person's actual use of language reflects his competence only indirectly. This is quite analogous to a person knowing an algorithm for multiplying any two integers. We would say he knows how to—i.e., has the competence to—multiply any two integers. But obviously his performance is not going to be infallible. For sufficiently large integers he will make mistakes or simply give up. This does not mean he doesn't know how to multiply the integers, just that memory limitations, etc., circumscribe his ability to use his knowledge.

If natural languages are infinite, at least one other point seems equally clear: a human's knowledge of his language is finitely represented in his brain. Since the brain itself is finite, that part which stores a human's linguistic competence is also. We shall thus define a *grammar* of a natural language to be any explicit finite representation of a native speaker's knowledge of that language. In this sense a grammar is a model of some information stored in the brain; it can be a very abstract model in which case we should not expect correspondence at more than a few points with the structure which is modeled.

III. **The nature of a grammar.** To determine what type of model is appropriate, we must next look at some things that people know about the language they speak. We can use ourselves as informants. Consider the sentence

(1) John looked up the street.

Every speaker of English knows this sentence is ambiguous; either "John looked up the street in the city directory" or "John glanced up the street". This ambiguity is tied to the different structures the sentence can have. If sentence (1) means John used a directory to find the street, then we know that *up* is structurally closer to the verb than the object. On the other hand, if the sentence (1) means John glanced up the street, then *up* is structurally connected to *the street*. We can indicate these structures by using parentheses to group the elements that are connected. Thus

(2)
(i) John (looked up) the street,
(ii) John looked (up the street).

Now look at the sentence

(3) She is a successful businessman's wife.

This sentence is also ambiguous; either "She is the wife of a successful business man" or "She is good at being a businessman's wife". And here also the ambiguity is tied to the structures the sentence may have. The first meaning is associated with the structure

(4) She is a ((successful businessman's) wife)

while the second meaning goes with the structure

(5) She is a (successful (businessman's wife)).

Actually this sentence is not really ambiguous since there is a very slight difference in stress or accent or emphasis which is a direct consequence of the difference in structure. In (4) the stress on *successful* is slightly weaker than that on *wife* and the reverse is true of (5). Thus there is a slight difference of pronunciation tied to a difference in structure.

These examples illustrate the fact that a sentence is not just a sequence of words or letters; each sentence has its elements grouped into units which are called *phrases*. These phrases do not overlap but are arranged in a hierarchy: either two phrases are disjoint or one contains the other. Thus using parentheses to indicate structural groupings is quite appropriate: the possible arrangements of a sequence of elements into phrases are exactly the proper parenthesizations that are possible for that sequence. When a phrase A contains another phrase B, B is called a *constituent* of A. If B is a constituent of A and there is no C (other than A or B) such that C is a constituent of A and B is a constituent of C, then B is an *immediate constituent* of A.

But a sentence does not have just a bracketing into phrases associated with its shape as a sequence of elements. In addition, these phrases are classified into types or grammatical categories.

For example, in sentence (3) the type of the phrase *successful businessman's wife* is the same as that of *she*. These phrases both belong to the category usually called "noun phrase".

This information about the category of a phrase can be represented by labeling the parentheses which delimit the phrases with the name of the category to which the phrase belongs. For example, the sentence (6) would have its phrases and their types marked in (7). I am using square brackets with labels as is customary in linguistic discussions, but no importance is to be attached to the change from parentheses.

(6) Jack built a house

(7) $[_{\text{Sentence}} [_{\text{Noun Phrase}} [_{\text{Noun}} \text{Jack}]_{\text{Noun}}]_{\text{Noun Phrase}} [_{\text{Verb Phrase}} [_{\text{Verb}} \text{built}]_{\text{Verb}}$
$[_{\text{Noun Phrase}} [_{\text{Article}} \text{a}]_{\text{Article}} [_{\text{Noun}} \text{house}]_{\text{Noun}}]_{\text{Noun Phrase}}]_{\text{Verb Phrase}}]_{\text{Sentence}}$

Notice that when a phrase belongs to two different categories—for example *Jack* in (6)—one set of labeled brackets is used to indicate each category.

The information which (7) gives about (6) is just the structure (more specifically, surface structure) of (6). Thus (7) is a (partial) *structural description* of (6). (7) is also sometimes called a *Phrase-marker* of (6).

Thus we see that each sentence of a language has a structure which we will call a *surface structure* associated with it, and this structure can and will be represented by a proper labeled bracketing of the sentence. But there is still more to the structure of a sentence.

Consider the sentences (8i) and (8ii).

(8) (i) John is easy to please.
(ii) John is eager to please.

Although these sentences are identical in surface structure they are understood differently. In (8ii) it is John who is pleasing while in (8i) it is someone else. In (8i) it is John who is pleased but in (8ii) it is someone else. These semantic differences are connected with syntactic differences. For example in (8ii) the direct object of *please*, which is missing, can be added to give a well-formed sentence of English (9ii), but this is not so for (8i) (cf. (9i)).

(9) (i) John is easy to please us.
(ii) John is eager to please us.

The structural descriptions of these sentences must indicate the differences between them, and their surface structures clearly do not.

If we include in the structural description of a sentence another labeled bracketing, which we can call a *deep structure* because it is more abstract than the surface structure, we can indicate that, at some level, sentence (8i) is related to the sentences

(10) (i) It is easy.
(ii) Someone pleases John.

and the *it* of (10i) represents (10ii). This information can be given as the labeled bracketing

(11) $$[_{S} [_{NP} \text{it } [_{S} [_{NP} [_{N} \text{someone}]_{N}]_{NP} [_{VP} [_{V} \text{pleases}]_{V} [_{NP} \text{John}]_{NP}]_{VP}]_{S}]_{NP} [_{VP} \text{is } [_{Adjective} \text{easy}]_{Adjective}]_{VP}]_{S}.$$

A similar deep structure can be written for (8ii). (8ii) is of course related to the sentences

(12) (i) John is eager for it.
(ii) John pleases someone.

Its deep structure is

(13) $$[_{S} [_{NP} [_{N} \text{John}]_{N}]_{NP} [_{VP} \text{is } [_{Predicate} [_{Adjective} \text{eager}]_{Adjective} [_{PP} \text{for } [_{NP} \text{it } [_{S} [_{NP} \text{John}]_{NP} [_{VP} [_{V} \text{pleases}]_{V} [_{NP} [_{N} \text{someone}]_{N}]_{NP}]_{VP}]_{S}]_{NP}]_{PP}]_{Predicate}]_{VP}]_{S}.$$

I have outlined some of the information that speakers have—usually not consciously—about sentences in their language. We have seen that this information can be given in a structural description which consists of two labeled bracketings, a surface structure and a deep structure. I have said that a grammar is an explicit representation of a speaker's knowledge of his language. Thus, a grammar, to be adequate, must represent the collection of structural descriptions of every well-formed sentence of the language. That is, we require of a grammar of a language that it finitely specify an enumeration of all and only the grammatical sentences of the language and that to each sentence the grammar assign all its structural descriptions.

Historically, there was a time not long ago when linguists ignored the deep structure of a sentence and considered its structural description to be just its surface structure. These linguists imposed laxer constraints on a grammar, since for structural descriptions, it could supply simply labeled bracketings of sentences. The field of algebraic linguistics started with the study of these grammars, and we will look only at them here. For studies of grammars that provide both deep and surface structures, see [**10**], [**11**].

Suppose then that one wishes to assign to each of the infinite number of sentences of a language a labeled bracketing of that sentence. What could a grammar capable of doing the job look like? It is only necessary to finitely specify a set of labeled bracketings as the set of structural descriptions of the sentences in the language, because each labeled bracketing uniquely determines the sentence to which it is assigned; merely delete the brackets and their labels. Thus a grammar can satisfy the requirements we place on it here, if it can specify the set of structural descriptions of the sentences of a language.

Just about the simplest system that one can think of for finitely specifying a set of labeled bracketings is to give a finite set of conditions of the type:

> a phrase of type A may have as its immediate constituents the sequence of elements $\alpha_1, \cdots, \alpha_n$

where each α_i is either a specified word of the language or any phrase of a specified grammatical category. These statements form the grammar. An arbitrary well-formed labeled bracketing is specified by the grammar as being a structural description of a sentence of the language described by the grammar if every phrase of the labeled bracketing satisfies some condition of the grammar and the outermost pair of brackets is labeled with the symbol for sentence.

It turns out that this system is exactly a theory of grammar that linguists proposed. Some linguists allowed in a grammar only the type of condition given above while others allowed conditions of the more general sort:

(14) a phrase of type A may have the sequence of elements $\alpha_1, \cdots, \alpha_n$ as its immediate constituents provided that to its left is the sequence $\beta_1, \cdots, \beta_k$ and to its right is the sequence $\gamma_1, \cdots, \gamma_l$.

Both of these versions were called the theory of *immediate constituent analysis*.

Knowing all this, we are now in a position to search for a mathematical model of these linguistic grammars. We can then study the properties of the model and extend these to the grammars we are modeling, as is usual in applied mathematics. Chomsky discovered an appropriate model in the semi-Thue system (cf. [**5**, pp. 81-101]). I present the definition of such systems here with slight modifications in content and terminology. Recall that in a grammar we must specify a set of words or letters out of which one can form sentences and a set of grammatical categories to which the phrases of the sentences belong. Thus the *total vocabulary* V must consist of

(a) a *terminal vocabulary* V_T of words or letters

and

(b) a *nonterminal vocabulary* V_N of grammatical categories. Then $V = V_T \cup V_N$ and $V_T \cap V_N = \emptyset$.

We will be interested in strings over V. Note that when we have two strings $\phi = \alpha_1 \cdots \alpha_m$ and $\psi = \beta_1 \cdots \beta_n$, where $\alpha_i, \beta_j \in V$ $(1 \leqq i \leqq m, 1 \leqq j \leqq n)$, ϕ *concatenated with* ψ (written "$\phi\psi$") $= \alpha_1 \cdots \alpha_m \beta_1 \cdots \beta_n$ is also a finite string over V. If we have two sets L, L' of strings over V, we can form their *product* $LL' = \{\omega \mid \omega = \phi\psi, \phi \in L, \psi \in L'\}$. If we let $L^0 = \{e\}$ (where "e" stands for the empty string of length zero), and $L^{i+1} = LL^i$ for $i \geqq 0$, then we can define the *Kleene star* of L, $L^* = \bigcup_{i=0}^{\infty} L^i$.

Note that V can be identified with the set of strings of length one over V, and thus V^* is the set of all finite strings over V.

A rewriting rule over V is a pair (ϕ, ψ) where $\phi, \psi \in V^*$. An *unrestricted rewriting system* (URS) is an ordered quadruple $G = (V_T, V_N, S, \rightarrow)$ where (a) V_T and V_N are finite, nonempty, disjoint sets, (b) $S \in V_N$ and (c) $\rightarrow$ is a finite set of rewriting rules over $V = V_T \cup V_N$ (i.e., $\rightarrow \subseteq (V_T \cup V_N)^* \times (V_T \cup V_N)^*$ is finite). Naturally V_T is called the terminal vocabulary of G, V_N its nonterminal vocabulary. S is its *initial symbol* and $\rightarrow$ is the set of *rules* of G. If $(\phi, \psi) \in \rightarrow$, we write $\phi \rightarrow \psi$.

Remember that we want to model linguistic grammars which in general comprise a finite set of statements of the type (14). If we simply write (14) as the rewriting rule

$$\beta_1 \cdots \beta_k A \gamma_1 \cdots \gamma_l \rightarrow \beta_1 \cdots \beta_k \alpha_1 \cdots \alpha_n \gamma_1 \cdots \gamma_l, \tag{15}$$

then we can model a linguistic grammar as a URS. We now wish to define the language which a URS generates so that a linguistic grammar and its model determine the same language. First, a sequence $\omega_1, \cdots, \omega_n$ of strings over V is a (*weak*) ϕ-*derivation* of ψ in G if $\phi = \omega_1$, $\psi = \omega_n$ and for $i = 1, \cdots, n-1$, there are $\chi_1, \chi_2, \chi_3, \chi_4 \in V^*$ such that $\omega_i = \chi_1 \chi_2 \chi_3$, $\omega_{i+1} = \chi_1 \chi_4 \chi_3$ and $\chi_2 \rightarrow \chi_4$ is a rule of G. If there is a ϕ-derivation of ψ, we write $\phi \Rightarrow \psi$. A string $x \in V_T^*$ is a *sentence generated* by G if $S \Rightarrow x$ (where S is the initial symbol of G). The *language generated* by G ($L(G)$) is $\{x | x$ is a sentence generated by $G\}$.

Next we need to define the manner in which a URS will associate labeled bracketings with the sentences it generates. Notice that not just any URS can model a linguistic grammar. In fact for a URS to be a model of a linguistic grammar it is necessary for each rule to have the property that exactly one symbol on its left-hand side is replaced by a sequence of one or more symbols on its right-hand side, all other symbols on the left-hand side occupying the corresponding positions on the right-hand side. In other words, each rule of the URS must meet the condition:

(16) Given the rule $\chi_1 \rightarrow \chi_2$, there are $\phi, \psi, \omega \in V^*$ and $A \in V_N$ such that $\omega \neq e$, $\chi_1 = \phi A \psi$ and $\chi_2 = \phi\omega\psi$.

In fact we can and will require further that ϕ, ψ, ω, and A in condition (16) be uniquely determined by χ_1 and χ_2. Such a URS is called a *context-sensitive* (CS) *grammar*. With a CS grammar then, applying a rule $\phi A \psi \rightarrow \phi\omega\psi$ amounts to building up a type A phrase in a legitimate context by giving its immediate constituents. If we were to have the rule surround ω with matched brackets labeled 'A' each time such a rule is applied, we would obtain a structural description assigned by G to a sentence of $L(G)$. So given $G = (V_T, V_N, S, \rightarrow)$, we let $L = \{ [_A | A \in V_N\}$ and $R = \{]_A | A \in V_N\}$, and let d: $(V_T \cup V_N \cup L \cup R)^* \rightarrow (V_T \cup V_N)^*$ be given by

(a) $d(\alpha) = \alpha$ if $\alpha \in V_T \cup V_N$
$\qquad = e$ if $\alpha \in L \cup R$

(b) $d(\phi\psi) = d(\phi)d(\psi)$.

d is called the *debracketing function*. Then a *strong ϕ-derivation of ψ* in G is a sequence $\omega_1, \cdots, \omega_n$ of strings in $(V_T \cup V_N \cup L \cup R)^*$ such that $\phi = \omega_1$, $\psi = \omega_n$ and for $i = 1, \cdots, n-1$, there are χ_1, χ_2, σ, τ, ω, and A such that

(a) $\omega_i = \chi_1\sigma A \tau\chi_2$, $\quad \omega_{i+1} = \chi_1\sigma[_A\omega]_A\tau\chi_2$,

(b) $d(\sigma A \tau) \rightarrow d(\sigma[_A\omega]_A\tau)$ is a rule of G,

and

(c) $d(\omega) = \omega$.

A string ϕ is a *structural description generated* by G if there is a strong S-derivation of ϕ in G and $\phi \in (V_T \cup L \cup R)^*$. Write $\mathscr{L}(G)$ for the set of structural descriptions generated by G. It is now easy to state how G "associates" structural descriptions with sentences. G simply assigns ϕ to $d(\phi)$ for each $\phi \in \mathscr{L}(G)$. Note that G can assign more than one structural description to a given sentence.

We will study the class of all CS grammars in the next section. Note that the heavier restriction on rules in linguistic grammars which I stated first can be extended to URS's. A *context-free* (CF) *grammar* G is a URS which has only rules of the sort $A \rightarrow \phi$ where $A \in V_N$ and $\phi \neq e$. This is a heavier restriction than (16), and thus every CF grammar is a CS grammar, but not conversely. We will say that a language L is a *context-sensitive* (respectively, *context-free*) *language* if there is a CS (CF) grammar G such that $L = L(G)$. Then CF grammars meet a restriction essentially stronger than (16) because there are CS languages which are not CF. Similarly, restriction (16) is essentially stronger than no restriction at all, since there are languages generated by URS's which are not CS languages. It is well known (cf. [**4**]) that the languages which are generated by URS's are exactly the recursively enumerable languages.

IV. **Properties of context-sensitive languages.** In this paper we will prove some theorems about CS grammars and CS languages. It will make our discussion easier if we explicitly adopt the following notational convention, which I have followed all along. In discussing grammars, symbols will have the following uses:

	Atomic Symbols	Strings	
Lower Case Greek	$\alpha, \beta, \gamma, \cdots$	$\cdots, \phi, \psi, \omega$	From $V = V_T \cup V_N$
Lower Case Latin	$a, b, c, \cdots$	$\cdots, x, y, z$	From V_T
Upper Case Latin	$A, B, C, \cdots$	$\cdots, X, Y, Z$	From V_N
	Early Letters	Late Letters	

'S' will usually be the initial (nonterminal) symbol.

Now we are prepared to prove our first result.

THEOREM 1 (CHOMSKY [**3**]). *Every CS language is a recursive set of strings, but not conversely.*

PROOF. Let L be a CS language and let $G = (V_T, V_N, S, \rightarrow)$ be a CS grammar which generates L. Given an arbitrary $x \in V_T^*$, we want a procedure for deciding whether $x \in L$. Notice that if $x \in L$, then for every S-derivation $\phi_1, \cdots, \phi_n$ of x, each ϕ_i is no longer than x. If for some j, ϕ_j were longer than x, then (since no rule of G decreases the length of a string to which it is applied) ϕ_{j+1} would be longer than x, and hence we see that ϕ_n would be longer than x. But this is impossible since $\phi_n = x$.

Now V_T and V_N are finite, so there are only a finite number of different strings which can appear in an S-derivation of x. Furthermore, if there is an S-derivation of x, there is one in which no string appears twice. Thus there is only a finite number (less than $[(n^l)^{n^l+1} - 1]/(n^l - 1)$, where l is the length of x and $n =$ cardinality of $V_T \cup V_N$) of sequences of strings over $V_T \cup V_N$ which we need to check to find out whether there is an S-derivation of x. Since we can enumerate the possible derivations and mechanically check each one to see whether it is an S-derivation of x, there is a mechanical procedure for deciding for any $x \in V_T^*$ whether or not $x \in L$. So L is a recursive set of strings.

For the other part of the proof, we will exhibit a recursive set of strings which is not a CS language. Let $x_1, x_2, x_3, \cdots$ be a mechanical enumeration of V_T^* without repetitions and let $G_1, G_2, \cdots$ be a mechanical enumeration of all CS grammars. Let $L = \{x_i \mid x_i \notin L(G_i)\}$. For any x, we can enumerate V_T^* until we get $x_i = x$ and enumerate CS grammars until we get to G_i, then check whether $x \in L(G_i)$. Since $x \in L$ if and only if $x \notin L(G_i)$, L is recursive. But L is not CS, since if it were then some G_i would generate L. But $L \neq L(G_i)$ for any i, since $x_i \in L$ if and only if $x_i \notin L(G_i)$.

Remark. In fact, every CS language has a primitive recursive characteristic function, but there are languages which have primitive recursive characteristic functions and yet are not CS.

Notice that the only property of CS grammars we used in proving Theorem 1 is that for each rule $\phi \to \psi$, ψ is no shorter than ϕ. It turns out that this is the essential property of CS grammars.

Theorem 2 (Kuroda [7]). *If G is a URS having only rules $\phi \to \psi$ in which ψ is no shorter than ϕ, then $L(G)$ is a CS language.*

Proof. Let G be such a URS. We will call n the *order* of G if n is the smallest number such that for each rule $\phi \to \psi$ of G the length of ψ is less than or equal to n. If the order of G is greater than 2 we can find a URS G' of order $n - 1$ such that $L(G) = L(G')$ and G' has the property of the hypothesis. Let G' have all rules of G with right-hand sides shorter than n and if $\alpha_1 \cdots \alpha_m \to \beta_1 \cdots \beta_n$ is a rule of G, let A be a new nonterminal symbol and if $m = n$ let $\alpha_1 \cdots \alpha_{m-1} \to \beta_1 \cdots \beta_{n-2}A$ and $A\alpha_m \to \beta_{n-1}\beta_n$ be rules of G', if $m < n$ let $\alpha_1 \cdots \alpha_m \to \beta_1 \cdots \beta_{n-2}A$ and $A \to \beta_{n-1}\beta_n$ be rules of G'. Clearly $L(G) = L(G')$ and for each rule $\phi \to \psi$ of G', ψ is no shorter than ϕ and the length of ψ is less than or equal to $n - 1$. Thus G' is the URS wanted.

From this result applied $n-2$ times, we can find a URS G^* of order 2 such that G^* has the hypothesized property and $L(G) = L(G^*)$. We can further require that each rule of G^* (or of G) is of the sort $X \to Y$ or $A \to a$ since each occurrence of a terminal symbol 'a' in a rule other than $A \to a$ can be replaced by a new nonterminal B if the rule $B \to a$ is added to the grammar. Thus G^* contains only rules of the types $A \to a$, $A \to B$, $A \to CD$, and $AB \to CD$. Of these only $AB \to CD$ cannot appear in a CS grammar. But if we have a new nonterminal E, then we can replace the rule with $AB \to EB$, $EB \to ED$, and $ED \to CD$. We thus obtain a CS grammar G'' such that $L(G) = L(G'')$, proving the theorem.

EXAMPLE. $L = \{a^n b^n a^n \mid n \geqq 1\}$ is a CS language. Let

$$G = (\{a, b\},\ \{S, A, B\},\ S, \to)$$

where

$$S \to aSBA, \qquad S \to abA, \qquad AB \to BA,$$

$$bB \to bb, \qquad bA \to ba, \qquad aA \to aa.$$

$L(G) = L$ and G is a URS such that for each rule $\phi \to \psi$, ϕ is no longer than ψ. Thus L is CS.

LEMMA 1 (SCHEINBERG [12]). *L is not CF.*

PROOF. Suppose L is CF. Let G be a CF grammar which generates L. We can assume without loss of generality that for all $A \in V_N - \{S\}$, there are ϕ and ψ such that $A \to \phi A \psi$ is a rule of G, and $\phi\psi \neq e$. For if for some $A \neq S$ there is no such rule, we can simply replace each rule $B \to \omega_1 A \omega_2 \cdots \omega_{m-1} A \omega_m$ by all rules $B \to \omega_1 \phi_{i_1} \omega_2 \cdots \omega_{m-1} \phi_{i_{m-1}} \omega_m$, where $A \to \phi_{i_1}, \cdots, A \to \phi_{i_{m-1}}$ are (not necessarily different) rules of G. Then the grammar does not use the nonterminal A but generates the same language as G, so that we can eliminate A from V_N. Now let i be the length of the longest ϕ such that $S \to \phi$ and let $\phi_1, \cdots, \phi_n$ be an S-derivation of $a^i b^i a^i$. Let $A \to x$ be the rule which rewrites ϕ_{n-1} to give ϕ_n. There are ψ and ω not both $= e$ such that $A \Rightarrow \psi A \omega$ (if $A = S$, note that $S \Rightarrow \phi_{n-1}$ and $\phi_{n-1} = \psi S \omega$ where $\psi\omega \neq e$ by the choice of i) and there are w and y such that $\psi \Rightarrow w$ and $\omega \Rightarrow y$. So $\phi_{n-1} = vAz$ and $\phi_n = vxz$. Now $\phi_{n-1} \Rightarrow v\psi^k A \omega^k z$ for $k \geqq 0$, so $\phi_{n-1} \Rightarrow vw^k x y^k z$ for $k \geqq 0$. Thus $S \Rightarrow vw^k x y^k z$, hence $vw^k x y^k z \in L$ for $k \geqq 0$. Thus $w \in \{a\}^*$ or $w \in \{b\}^*$ and also $y \in \{a\}^*$ or $y \in \{b\}^*$. Suppose

(a) $w \in \{a\}^*$ and $y \in \{a\}^*$.

Then $vwxyz = a^{i+j} b^i a^{i+l}$ where either j or $l > 0$. Contradiction.

(b) $w \in \{a\}^*$ and $y \in \{b\}^*$.

Then $vwxyz = a^{i+j} b^{i+l} a^i$ where either j or $l > 0$. Contradiction.

(c) $w \in \{b\}^*$ and $y \in \{b\}^*$.
Then $vwxyz = a^i b^{i+j} a^i$ where $j > 0$. Contradiction.

(d) $w \in \{b\}^*$ and $y \in \{a\}^*$.
Then $vwxyz = a^i b^{i+j} a^{i+l}$ where either j or $l > 0$. Contradiction.

Thus L is not CF.

We have now proved the remark in §3 that there are languages which can be generated by a URS but are not CS and that there are languages that are CS and not CF.

Let us next look at a closure property of CS languages.

LEMMA 2. *If L_1 and L_2 are CS languages, then so is $L_1 \cup L_2$.*

PROOF. Let $G_1 = (V_{T1}, V_{N1}, S_1, \rightarrow_1)$ and $G_2 = (V_{T2}, V_{N2}, S_2, \rightarrow_2)$ be CS grammars such that $L_1 = L(G_1)$, $L_2 = L(G_2)$ and $V_{N1} \cap V_{N2} = \emptyset$. Let '$S$' be a new nonterminal symbol and $G = (V_{T1} \cup V_{T2}, V_{N1} \cup V_{N2} \cup \{S\}, S, \{S \rightarrow S_1, S \rightarrow S_2\} \cup \rightarrow_1 \cup \rightarrow_2)$. Then $L(G) = L(G_1) \cup L(G_2) = L_1 \cup L_2$. This result is trivial to prove, but it is not obvious whether the intersection of two CS languages or the complement of a CS language is CS. We will come back to these matters later.

Next we will see that the CS languages are related to a certain type of automaton defined by Myhill [**9**]. A *linear bounded automaton* (LBA) is a quintuple $M = (A, \Sigma, S_0, \mu, F)$ where

(a) A is a finite set (the *alphabet*);
(b) Σ is a finite set (the *states*);
(c) $S_0 \in \Sigma$ (the *initial state*);
(d) $\mu \subseteq A \times \Sigma \times \Sigma \times A \times \{0, +1, -1\}$ is finite (the *move function*);
(e) $F \subseteq \Sigma$ (the *final states*).

An LBA can be thought of as having a linear tape which is scanned by a control unit which has a finite number of states. Each position on the tape can be occupied by a symbol of the alphabet. When the automaton is scanning a symbol a in state S, it can replace a by b, and go into state S' moving the tape l squares left if $(a, S, S', b, l) \in \mu$. A *configuration* of M is a triple (x, S, y) where $x, y \in A^*$, $S \in \Sigma$. A configuration k' *follows from* a configuration k in M if there are $x, y \in A^*$, $a, b \in A$, $S, S' \in \Sigma$ such that

(a) $k = (x, S, ay)$, $k' = (x, S', by)$ and $(a, S, S', b, 0) \in \mu$,
(b) $k = (x, S, ay)$, $k' = (xb, S', y)$ and $(a, S, S', b, 1) \in \mu$,

or

(c) there is a $c \in A$ such that $k = (xc, S, ay)$, $k' = (x, S', cby)$ and $(a, S, S', b, -1) \in \mu$.

A configuration $k = (x, S, y)$ is *accepting* if $S \in F$ and $y = e$. Given $x \in A^*$, the *initial configuration* with respect to x is (e, S_0, x). M

accepts a string $x \in A^*$ if there is a sequence $k_1, \cdots, k_m$ of configurations such that k_1 is the initial configuration with respect to x, k_m is accepting and k_{i+1} follows from k_i for $i = 1, \cdots, m-1$. That is, M *accepts* x if, when x is written on its tape and M is put in its initial state scanning the first symbol of x, M can compute until it reads off the tape at the right and in a final state. Notice that M uses only the amount of tape needed to write the input. The *language* (L(M)) *accepted* by M is the set of all strings M accepts. The main difference between this definition and Myhill's is that he required that μ be a FUNCTION from $A \times \Sigma$ into $\Sigma \times A \times \{0, +1, -1\}$. We will call an LBA *deterministic* if for each $a \in A$, $S \in \Sigma$ there is at most one $S' \in \Sigma$, $b \in A$, $l \in \{0, +1, -1\}$ such that $(a, S, S', b, l) \in \mu$. We will be considering the larger set of *nondeterministic* LBA's.

THEOREM 3 (LANDWEBER [8]). *Every LBA accepts a CS language.*

PROOF. Let $M = (A, \Sigma, S_0, \mu, F)$ be an LBA. Let $V_T = A$ and $V_N = (A \times A \times (\Sigma \cup \{T\})) \cup \{S'\}$ where S' and T are new symbols; that is, the nonterminal symbols are 'S'' and triples (a, b, α) where $a, b \in A$, $\alpha \in \Sigma \cup \{T\}$.

Let $G = (V_T, V_N, S', \rightarrow)$, where

(i) $S' \rightarrow S'(a, a, T)$,

(ii) $S' \rightarrow (a, a, S_0)$,

(iii) $(a, b, S) \rightarrow (a, c, \widetilde{S})$, if $(b, S, \widetilde{S}, c, 0) \in \mu$,

(iv) $(a, b, S)(d, f, T) \rightarrow (a, c, T)(d, f, \widetilde{S})$, if $(b, S, \widetilde{S}, c, 1) \in \mu$,

(v) $(a, f, T)(d, b, S) \rightarrow (a, f, \widetilde{S})(d, c, T)$, if $(b, S, \widetilde{S}, c, -1) \in \mu$,

(vi) $(a, b, S) \rightarrow a$, if there are $\widetilde{S} \in F$ and $c \in A$ such that $(b, S, \widetilde{S}, c, 1) \in \mu$,

(vii) $(a, d, T)f \rightarrow af$,

for all $a, d, f \in A$.

We see that $L(G) = L(M)$ as follows: the first two rules of G produce a string

$$(a_1, a_1, S_0)(a_2, a_2, T) \cdots (a_n, a_n, T).$$

The next rules perform a computation of M but preserve a copy of the 'input'. A CF rule of type (vi) will apply to yield the first terminal symbol; if this rule applies anywhere except at the right end of the string, the nonterminals to its right can not be converted to terminal symbols. Thus the derivation will yield a terminal string only if this rule is applied at the right end of the string and nonterminals to its left are rewritten by rules of type (vii). Since G is a URS in which all rules $\phi \rightarrow \psi$ have ψ no shorter than ϕ, $L(G)$ is CS.

THEOREM 4 (KURODA [7]). *If L is a CS language, there is an LBA M such that $L = L(M)$.*

PROOF. Let G be a CS grammar of order 2 such that $L = L(G)$. Let $A = V_T \cup ((V_T \cup V_N) \times \{0,1,2,3\}) \cup \{\$\}$,

$$\Sigma = \{S_0, S_1, S_2, S_3, S^*, S^{***}\} \cup \{S^*_\alpha, S^{**}_\alpha \mid \alpha \in V_T \cup V_N\}$$
$$\cup \{S_A, S_{A,C} \mid A, C \in V_N\},\ F = \{S_3\},$$

and

$$\begin{aligned}
\mu = {} & (a, S_0, S_1, (a,0), 1) && \text{for all } a \in V_T \\
& (a, S_1, S_1, (a,1), 1) && \text{for all } a \in V_T \\
& (a, S_1, S_2, (a,2), 0) && \text{for all } a \in V_T \\
& (a, S_0, S_2, (a,3), 0) && \text{for all } a \in V_T \\
& ((\alpha, i), S_2, S_2, (\alpha, i), \pm 1) && \\
& ((a,i), S_2, S_2, (A,i), 0) && \text{if } A \to a \\
& ((B,i), S_2, S_2, (A,i), 0) && \text{if } A \to B \\
& ((C,i), S_2, S_B, (A,i), -1) && \text{if } A \to BC \\
& ((B,1), S_B, S^*, \$, -1) && \text{if } A \to BC \\
& ((B,0), S_B, S^{***}, \$, 1) && \text{if } A \to BC \\
& ((\alpha,1), S^{***}, S_2, (\alpha,0), 0) && \\
& ((\alpha,2), S^{***}, S_2, (\alpha,3), 0) && \\
& ((\alpha,1), S^*, S^*, (\alpha,1), -1) && \\
& ((\alpha,0), S^*, S^*_\alpha, \$, 1) && \\
& ((\beta,1), S^*_\alpha, S^{**}_\beta, (\alpha,0), 1) && \\
& (\$, S^*_\alpha, S_2, (\alpha,0), 0) && \\
& ((\alpha,1), S^{**}_\beta, S^{**}_\alpha, (\beta,1), 1) && \\
& (\$, S^{**}_\beta, S_2, (\beta,1), 0) && \\
& ((D,i), S_2, S_{A,C}, (B,i), -1) && \text{if } AB \to CD \\
& ((C,i), S_{A,C}, S_2, (A,i), 0) && \text{if } AB \to CD \\
& ((S,3), S_2, S_3, \$, 1) &&
\end{aligned}$$

for all $\alpha, \beta \in V_T \cup V_N$, and $i = 0,1,2,3$.

$L(M) = L(G) = L$ since M first reads through the input, marking the left end, checking that only terminal symbols appear and guessing which is the rightmost symbol and marking it. M then undoes rules of G, trying to rewrite the tape until it has just S on it. If it succeeds, then in its final state it reads past the symbol it marked as rightmost; if the symbol was indeed the rightmost symbol, M has accepted the input. It is not clear whether one can make M deterministic.

Now with theorems 3 and 4, another closure property of CS languages becomes obvious.

LEMMA 3 (LANDWEBER [8]). *If L_1 and L_2 are CS languages, then so is $L_1 \cap L_2$.*

PROOF (DUE TO KURODA [7]). Let M_1 and M_2 be LBA's such that M_1 accepts L_1 and M_2 accepts L_2. There is an LBA M which makes two copies of its input, simulates M_1 operating on one copy and, if M_1 accepts the input, then simulates M_2 operating on the other copy. M accepts the input if and only if both M_1 and M_2 do.

This leaves us with two interesting open questions.

(1) Is the complement of a CS language CS?

(2) Do deterministic LBA's accept the same set of languages as nondeterministic LBA's?

These questions are not independent. It turns out that a negative answer to (1) implies a negative answer to (2). These questions are of considerable interest to mathematical linguists. An answer to either one will have considerable impact on the field.

Bibliography

1. Y. Bar-Hillel, M. Perles and E. Shamir, *On formal properties of simple phrase structure grammars,* Z. Phonetik Sprachwiss. Kommunikat. **14** (1961), 143-172; also appears as Chapter 9 in Y. Bar-Hillel, *Language and information,* Addison-Wesley, Reading, Mass., 1964.

2. N. Chomsky, *Three models for the description of language,* IRE Trans. Inform. Theory, IT-**2** (1956), 113-124.

3. ———, *On certain formal properties of grammars,* Information and Control **2** (1959), 137-167.

4. ———, "Formal properties of grammars" in *Handbook of mathematical psychology,* edited by D. Luce, R. Bush and E. Galanter, Wiley, New York, 1963.

5. M. Davis, *Computability and unsolvability,* McGraw-Hill, New York, 1958.

6. S. Ginsburg, *The mathematical theory of context-free languages,* McGraw-Hill, New York, 1966.

7. S.-Y. Kuroda, *Classes of languages and linear-bounded automata,* Information and Control **7** (1964), 207-223.

8. P. S. Landweber, *Three theorems on phrase structure grammars of type* 1, Information and Control **6** (1963), 131-136.

9. J. Myhill, *Linear bounded automata,* Wright Air Development Division. Tech. Note, 1960, 60-165.

10. P. S. Peters, Jr. and R. W. Ritchie, *On the generative power of transformational grammars,* (to appear in Information Sciences).

11. ———, *On restricting the base component of transformational grammars,* (forthcoming).

12. S. Scheinberg, *Note on the Boolean properties of context free languages,* Information and Control **3** (1960), 372-375.

UNIVERSITY OF TEXAS AT AUSTIN

XI
COMPUTER SCIENCE

A. H. Taub

Computer Science

The modern, stored program digital computer is about twenty years old. Its advent and the remarkable improvements in its components and circuitry have created a deep interest in the theory of automata and the methodology of problem formulation and solution by such devices. Out of the study of computers and the study of new formulations of various problems there has emerged the field of computer science, which includes (but is not limited to) such subfields as computer circuitry, machine organization and logical design, numerical analysis, theory of programming, theory of automata and switching theory.

Computer science is closely related to mathematics; indeed, numerical analysis is a branch of mathematics, while many of the problems arising from the design and use of computers are intimately associated with questions in combinatorial mathematics, abstract algebra and symbolic logic. But an even more fundamental relationship also exists. The inherent structure of a computer forces one who successfully studies or uses it to strive for the type of generality, abstraction and close attention to logical detail that is characteristic of mathematical arguments.

A brief review of the logically distinct units of a computer system is useful in understanding the nature and the growth of various subfields of computer science. Inasmuch as such a device is a general-purpose one and automatic (i.e. independent of the human operator after the solution of a problem starts), it should contain certain main organs: (1) a central processing unit (an arithmetic unit), (2) a memory, (3) a control unit and (4) input-output devices.

The function of the central processing unit (CPU) is to perform various operations on arrays of bits contained in it. These arrays of bits are usually called words and may represent numbers or non-numerical data drawn from many diverse fields. The crucial point is that with modern electronic techniques, the CPU can perform its operations with fantastic speeds, 10^6 to 10^8 operations per second for some CPU's and some operations. Thus if the computer is to be used in a manner commensurate with its abilities, it must have available to it at speeds comparable to its execution speeds a sequence of operators and operands. Such sequences are kept in the memory of the computer.

There must be an organ in the computer which will automatically call forth the various operators and operands and execute the former. This device is called the control of the computer. The list of distinct operations which the control can execute is called the code of the computer.

These three units are internal to the computer itself. There must also exist devices, the input-output organ, whereby the human operator and the machine can communicate with each other.

The design and construction of devices with the properties listed above has stimulated great interest in physical phenomena which may be exploited to construct components for computers. Thus many computer laboratories have studied and are continuing to study the possibility of using lasers and masers as computer memories. The possibility of using cryogenic techniques in the construction of computers has also been under intensive examination.

The design and construction of computers has also stimulated a great deal of theoretical work ranging from the "practical" aspects of switching theory to the theory of automata. I would characterize the first limit mentioned above as being concerned with problems such as the following: What is the minimum number of elements with the ability to perform prescribed elementary logical operations (for example, to accept as input two logical variables x and y and to have as an output the single logical variable "x and y") needed to construct a device capable of forming a given Boolean function of n Boolean variables?

Automata theory may be said to have originated with Turing's work [1] on the general definition of what is meant by a computing automaton. Present day problems in this logical-mathematical theory are concerned with abstract models of devices whose behavior at a particular instant of time depend not only on the present input to the machine, but generally on the entire past history of the device including past inputs. In the main it is essentially a chapter in formal logic and as such has the property characterized by von Neumann [2] of being "cut off from the best cultivated portions of mathematics, and forced onto the most difficult part of the mathematical terrain, into combinatorics".

In the paper from which the above quotation was taken, von Neumann predicted that automata theory would differ from formal logic in two respects: "(1) The actual length of 'chains of reasoning', that is, the chains of operations, will have to be considered and (2) the operations of logic... will have to be treated by procedures which allow exceptions with low but nonzero probabilities". He expected that the need to take account of point (2) would bring the theory of automata closer to analysis than to combinatorics. This has not happened as yet. Current research in automata theory is concerned with problems arising in connection with the use of computers, in contrast to the design of computers. Thus many workers in the field are studying, among other topics, the theory of programming languages, algebraic coding theory and pattern recognition.

When computers are used in problem "solving" they must be furnished with programs: lists of instructions which must be scanned by the control and executed. von Neumann and Goldstine [3] have pointed out that if this were just a linear scanning of the sequence of instructions which remain unchanged in form, then matters would be simple. Programming a problem for the machine would consist in translating a meaningful text from one language (for example, the language of mathematics in which the planner will have conceived the problem) into another language (that one of the code of the computer).

This is not the case. Because computers execute orders with great speeds, it is necessary to use iterative and inductive algorithms for problem solving. Then the relation between the program and the mathematically conceived procedure of solution is not a statical one, that of translation, but highly dynamical: A program stands not simply for its contents at a given time in reference to a given set of memory locations, but more fully for any succession of passages of the control through it with any succession of modified contents to be found by the control there; all of this being determined by all other orders of the sequence of instructions (in conjunction with that instruction now being executed by the control).

The theory of programming is in part concerned with techniques for providing a dynamic background to control the automatic evolution of a meaning. von Neumann and Goldstine [**3**], who so characterized programming, viewed that subject as a new branch of formal logics and indicated methods for mastering various parts of it. Their work and that of a large body of subsequent workers has been devoted to techniques involved in preparing problems for solution by a computer.

Problem oriented programming languages such as ALGOL, FORTRAN and a host of others have been devised. In addition special languages such as COBOL, SNOBOL and LISP have been created for dealing with special types of problems.

The existence of these languages and the compilers they require as well as the desire to have efficient use of computers has led to the introduction of operating programs, monitoring systems and executive systems; that is, programs that oversee the use of the computing equipment. This development has led to the creation of a special class of programmers who are called system programmers and who have created a branch of programming called systems programming.

Although the original impetus for developing modern computers arose from the need for better, bigger and faster "arithmetic engines", computers are now being increasingly used for such nonnumerical tasks as the simulation of various complex systems, the solution of problems involving only complicated logical operations and even the design of new computers and computer systems. The above list of uses is by no means exhaustive and many of the uses have grown into sophisticated bodies of knowledge with an associated theory that is given a label as a subfield of computer science.

Since other lectures will cover some of these fields in detail, I shall confine the remainder of my time to problems arising when one attempts to understand the errors involved when one uses a modern computer to obtain numerical solutions of mathematically formulated underlying physical or engineering problems. J. von

Neumann and H. H. Goldstine [4] have pointed out the following four sources of errors in obtaining numerical solutions of an underlying physical or engineering problem:

(I) *Errors of formulation*. These arise due to the fact that the mathematical formulation of an underlying physical or engineering problem represents such a problem only with certain idealizations, simplifications and neglections. In other words the problem that is actually formulated to be solved mathematically, the problem that is called the rigorous mathematical problem, is in itself an approximation to some other problem.

There are many examples of these approximations. Thus in the theory of fluid mechanics the notion of a perfect fluid, one without viscosity and without heat conductivity, is an approximate representation of a real compressible fluid. In certain circumstances it represents the behavior of a compressible fluid adequately and its importance arises from this fact. However in other cases it is completely inadequate to deal with the physical problem involved and heat conductivity and viscosity have to be admitted into the theory. Even the theory of a viscous heat conducting compressible fluid, which is much more involved than that of a perfect fluid, is inadequate for some problems. These problems require the replacement of the representation of a fluid as a continuum by the representation of a fluid as a collection of molecules, as is done in the kinetic theory of gases. Thus in this one field we find that successively finer representations of the physical problem have to be made and that the mathematical tools needed for making these refinements are quite different in character.

It is the relationship between the errors of formulation and those listed below that I want to stress today.

(II) *Errors in values of parameters*. The mathematical formulation of a problem, with the idealizations discussed under (I), may involve parameters whose values have to be inserted in a numerical computation. However, these parameters may not be known with sufficient accuracy and thus will introduce errors which may then "infect" the solution.

The discussion of the effects of this source of errors leads to the following mathematical problem: Are the solutions of the rigorous mathematical problem (the approximation to the underlying problem) continuous functions of the parameter? Depending on the answer to this question the importance of the errors of this category can be assessed.

(III) *Errors of truncation*. The mathematical problem described under (I) may involve transcendental functions and operations which will have to be evaluated by use of a finite sequence of elementary arithmetic operations. For example, the exponential function and the trigonometric functions are usually evaluated by the use of polynomial or rational function approximations and integrals and derivatives are evaluated by quadrature formulas and finite difference approximations respectively.

Thus the rigorous mathematical problem which is an approximation to an underlying problem is further approximated for computational reasons. The interaction between these two approximations is not sufficiently stressed by the people who prepare problems for and make use of computers to obtain numerical solutions of problems.

The approximation of one problem by another raises two mathematical questions of classical numerical analysis: (1) the question of convergence and (2) the question of stability. These questions can be illustrated by considering a problem involving ordinary or partial differential equations.

When a differential equation is replaced by a finite difference equation, a mesh is introduced over the independent variables. That is, only discrete sets of values of the independent variables are used in the problem. The question arises as to the behavior of the solution of the difference equations as the number of elements in this set increases. One must search for those discrete formulations of the problem which have the property that their solution converges to the solution of the transcendental problem as the number of elements in the discrete set becomes infinite.

The stability problem involves the question as to whether the solution of the differential equations and the approximating difference equations are continuous functions of the initial and boundary conditions. This question is of importance in view of the fact that because of round-off errors, which will be discussed below, it is impossible to satisfy these conditions exactly. Thus one needs some assurance that the errors introduced from this source will not amplify but will indeed decrease in importance, that is, will damp out.

(IV) *Round-off errors*. These errors arise because the "elementary arithmetic operations" of a computing machine are not rigorously and faultlessly performed. Thus in a digital machine real numbers x and y are replaced by digital approximations $\bar{x}$ and $\bar{y}$ respectively. The quantities $x \pm y$ may then have corresponding digital representations, that is, we may have

$$\overline{(x \pm y)} = \bar{x} \pm \bar{y}$$

However, as far as products and quotients are concerned, these are governed by the relations

$$\overline{(xy)} = \bar{x}\bar{y} + \eta'^{(p)} = xy + \eta^{(p)},$$
$$\overline{(x/y)} = \bar{x}/\bar{y} + \eta'^{(q)} = x/y + \eta^{(q)},$$

where $\eta'^{(p)}$ (and $\eta^{(p)}$) and $\eta'^{(q)}$ (and $\eta^{(q)}$) are the round-off errors of multiplication and division respectively.

Round-off errors are as old as the art of computation itself. If multiplications and divisions were not rounded off, even elementary computations would soon lead to the use of reams of paper. However, with the advent of high-speed computers great numbers of such arithmetic operations become possible and one must consider the cumulative effect these errors have on the validity of the results obtained.

There is as yet no general theorem governing the effect of round-off errors and every problem formulated for machine solution has to be analyzed separately to obtain an estimate of what the round-off error is and what importance it has for the problem at hand.

The problems with which the theory of round-off and truncation errors should be concerned are closely related to the problem posed by Kronecker when he insisted that mathematics be formulated in a manner which involved finite constructions. When we formulate problems for solution by computing machines, we are attempting to follow Kronecker's dictum with at least one important modification: namely, arithmetic is not being carried out faultlessly. The saving grace, if any, is that the errors committed in the arithmetic processes are known.

The formulation of such a theory presents a challenge to pure and applied mathematicians as deep and as important as any problems presently being dealt with. Unfortunately, many people are unaware of this challenge. Many mathematicians seem to feel that problems connected with computation are concerned solely with arithmetic, a subject whose avoidance was partly responsible for their becoming interested in mathematics, and therefore such problems are to be shunned.

I now wish to turn to the discussion of the interaction between the errors of formulation and the errors of truncation. My thesis is that such an interaction exists and that it may be profitably exploited to obtain easier and more correct computing algorithms for dealing with various problems. I shall illustrate this thesis by examples. Before doing this I should point out that the existence of present day (and future) computing machines makes (and will make) the exploitation of this interaction possible in some cases where without such machines it is not feasible to do anything very different from what has been done in the past.

I shall first discuss what may be considered a ridiculous example but what I hope will not long be a ridiculous one. This example is concerned with a method for dealing with problems in the theory of fluid dynamics. That theory deals with a continuum over which there are various fields defined: the velocity field, the pressure field and the density field. The physicist considers the fluid as a collection of molecules moving about in space in accordance with certain laws of motion and introduces these various fields in terms of averages of other quantities defined for the molecules themselves.

When a problem in fluid dynamics is formulated in terms of a continuum and fields defined over this continuum which satisfy certain differential equations and when these equations are replaced by finite difference equations, we are in effect replacing the fluid by a discrete collection of particles. One should then ask what relation this collection of particles has with the physicist's molecules. Indeed one should ask: Why go through this chain of approximations at all? Can we not deal with the physicist's molecules directly and won't this give more significant results than the former procedure?

The answer to the second question seems to be that with existing computers we cannot keep track of enough molecules for a long enough time to make it into a feasible method for handling fluid dynamics problems. However, this answer needs to be looked at closely. One reason for this is that no sharp estimates are at hand as to when the law of large numbers becomes operative. Will an assembly of thousands of molecules behave essentially the same as 10^{23} molecules as far as the central limit theorem of probability is concerned or are 10^{23} molecules really needed?

The work of Nordsieck and Hicks [**5**] on the application of

Monte Carlo techniques to the solution of the Boltzmann equation shows that computers may be effectively used in obtaining the molecular velocity distribution under conditions far from equilibrium. Hence computers enable one to deal with many fluid dynamical problems in a quite novel manner.

There are many instances in which discrete problems are approximated by differential equations which in turn are approximated by difference equations and then solved by use of computing instruments. Thus errors of formulation and truncation are needlessly introduced, for with the advent of modern high-speed computers the original discrete problem could be handled directly.

Many boundary valued differential equation problems originate from variational principles. This fact can be used to obtain simpler and more easily solved discrete approximations to the defining equations for the unknown functions. This is another illustration of the fact that the interaction between the formulation process and the truncation process may be exploited to obtain more meaningful and more accurate approximations to problems posed for numerical solution by computers. I shall not discuss this point further but refer you to a paper [**6**] where this approach has been used on Sturm-Liouville differential equations.

I shall devote the remainder of my remarks to the interaction of round-off errors and truncation errors. Consider the problem of solving the equation

$$x = G(x) \tag{1}$$

on a computer. We shall assume that x is a real scalar variable and $G(x)$ is a function such that there exists real numbers r and b for which

$$|G(x) - r| \leqq b|x - r|$$

where $0 \leqq b < 1$. This condition insures the convergence to r of the sequence of x_n's defined by

$$x_{n+1} = G(x_n), \qquad n = 0, 1, \cdots. \tag{2}$$

When one wishes to find the numbers x satisfying equation (1) one may attempt to do this by generating on a computer the sequence of x_n's satisfying equation (2). However, because of truncation errors one will attempt to solve an equation of the form

$$V_{n+1} = H(V_n), \qquad n = 0, 1, \cdots, \tag{3}$$

where

$$H(x) = G(x) + \xi(x) \tag{4}$$

and

$$|\xi(x)| \leqq a, \tag{5}$$

a being a constant. The function $H(x)$ is some algebraic approximation to the function $G(x)$ which in turn may be a transcendental function. The function is the truncation error and a is a bound for it.

Descloux [7] has shown that for any V_0 the sequence V_n defined by equation (3) is bounded and all its points of accumulation V satisfy the inequality

$$|V - r| \leqq |a|/(1 - b).$$

He has further shown that the scheme given by equation (3) is the best possible one for solving equation (2) in the following sense: for given a and b there exists a function $H(x)$ for which it is impossible to find an algorithm using only H, a and b providing closer points of accumulation to r than the algorithm (3).

Because of round-off errors a computer will not generate the sequence V_n. If the computer uses fixed point arithmetic it will generate a sequence of integers

$$y_{n+1} = [G(y_n) + \xi_n]_R$$

where $[\quad]_R$ is called a rounding procedure and $[x]_R$ is any integer-valued function of x satisfying the inequality

$$|[x]_R - x| < 1.$$

The normal rounding procedure is defined as

$$[x]_N = [x + 0.5].$$

This is the procedure that is usually used in hand computations and is incorporated in many computers. Descloux showed that there exists an N such that for any y_0 the sequence of integers defined by

$$y_{n+1} = [G(y_n) + \xi(y_n)]_R$$

satisfies

$$|y_n - r| \leqq |a|/(1 - b) + 1/2(1 - b) \tag{6}$$

for $n > N$; furthermore, for given a and b, there exists a function G and errors ξ for which the bound is attained.

Equation (6) then tells us what accuracy can be expected in solving equation (1) by the iterative scheme (2) when given truncation and round-off errors are introduced. Descloux has obtained the corresponding result when floating point arithmetic is used. I shall not give it here. The point I wish to stress is that the result may be in error by a significant amount.

Let us consider an example. Suppose we wish to solve the equation

$$x = (7/8)\,x.$$

The problem is a trivial one. However, let us suppose that we do not know that the solution is $x = 0$ but attempt to solve the equation by generating the sequence

$$V_{n+1} = (7/8)\,V_n.$$

If we use fixed point arithmetic with the decimal or binary point at the right of the registers we shall generate the sequence of integers

$$y_{n+1} = [(7/8)\,y_n + 1/2].$$

Thus starting with $y_0 = 8$, we obtain $y_1 = 7$, $y_2 = 6$, $y_3 = 5$, $y_4 = 4$, $y_5 = 4$, $y_n = 4$, $n \geqq 5$. Thus the binary machine solution will be 4×2^{-p} where p is the number of binary bits in the machine representation of numbers. Of course if p is large this error is tolerable. On the other hand the result

$$x = 4 \times 2^{-p}$$

is due to the fact that

$$4 = 1/2(1 - b) = 1/2(1 - 7/8).$$

Hence if b is close to one, the error may very well be intolerable.

Equation (6) furnishes one with a basis for evaluating different truncating schemes. Thus if $|a| \ll 1/2$ for each of two truncating schemes, the error in the result will be mainly determined by the round-off and hence there is no reason for using the more elaborate truncation procedure.

The question as to whether round-off procedures can be devised which lead to better results has also been considered by Descloux. He extended some results of A. Nordsieck and showed that there exists a round-off procedure such that the y_n satisfying

$$y_{n+1} = y_n + [G(y_n) + \xi_n - y_n]_A$$

are such that for any y_0, there exists an N such that

$$|y_{n+1} - r| < a/(1-b) + 1$$

for $n > N$. That is, the round-off error can be reduced to one bit in the last place. The round-off procedure is called "anomalous rounding" for it violates one's intuition. Its rules are

$$[x]_A : \text{for } |x| \leqq 1, \ |[x]_A| \geqq x,$$
$$\text{for } |x| \geqq 1, \ |[x]_A| \leqq x.$$

Thus when x is small enough one rounds up, that is, increases the error due to rounding. No one has yet been able to extend these results to the case of vector problems; that is, to the case where x is a vector and $G(x)$ is a vector valued function. Thus there remains an important open problem in understanding the interaction between round-off errors and truncation errors.

References

1. A. Turing, *On computable numbers with applications to the Entscheidungsproblem,* Proc. London Math. Soc. **42** (1936), 230-265.

2. J. von Neumann, "General and logical theory of automata" in *In cerebral mechanisms in behavior, The Hixon symposium (September 1948, Pasadena)* edited by L. A. Jeffress, Wiley, New York, pp. 1-31. Reprinted in *Collected Works of J. von Neumann,* Vol. V, Pergamon Press, Oxford, England, 1963.

3. H. H. Goldstine and J. von Neumann, "Planning and coding problems for an electronic computing instrument" in *Collected works of J. von Neumann,* Vol. V, 1963, pp. 80-151.

4. J. von Neumann and H. H. Goldstine, *Numerical inverting of matrices of high order,* Bull. Amer. Math. Soc. **53** (1947), 1021-1099. *Collected works of J. von Neumann,* Vol. V, Pergamon Press, Oxford, England, 1963, pp. 479-572.

5. A. Nordsieck and B. L. Hicks, *Monte Carlo evaluation of the Boltzmann collision integral,* 5th Internat. Sympos. on Rarefied Gas Dynamics, Oxford University, England, July 1966.

6. C. C. Farrington, R. T. Gregory and A. H. Taub, *On the numerical solution of Sturm-Liouville differential equations,* Math. Tables Aid Comput. **11** (1957), 131-150.

7. J. Descloux, *Note on the round-off errors in iterative processes,* Math. of Comp. **17** (1963), 18-27.

University of California, Berkeley

W. F. Miller[1]

Computer Science

Part I: Laboratory Information and Control Systems

Introduction. This paper is concerned with the information handling and control systems being developed in scientific research laboratories. In particular it deals with the functions and properties of current systems. Subsequent discussions will deal with particular computational problems and the methods available to us to handle these problems.

Large scientific experiments have become so complex that the information handling and the control of the experiment have been highly automated.[2] The motivations for automating are the same ones that are motivating automation in industry, viz. economy, reliability, and speed. In the laboratory, speed means rapid feedback to the experimenter of the partial results which permit him to decide on the future course of the experiment. The fast information handling and control systems put the experimenter back in intimate contact with his experiment.

Let us dwell a little longer on the scope of some large-scale experiments, in order to emphasize the mass of data and the type

[1] This work is supported in part by the U.S. Atomic Energy Commission and in part by the National Science Foundation.

[2] Whereas this paper discusses systems developed for the physical science laboratories, similar systems have been developed for biological and medical laboratories where experiments are also becoming very complex [1].

of considerations that are important to a laboratory. A single physics experiment on a large accelerator or a biology experiment on a nuclear reactor may involve a team of four or five scientists and an equal number of technicians not including the staff that operates the accelerator or reactor. The experiment may take weeks or even months to set up and may then run almost continuously for an equal length of time. The amount of data collected in this time will be enormous. For example, a six-month (1000 hours) experiment employing a spark chamber or an array of spark chambers can easily generate two million stereo pairs of spark chamber photographs. Usually one does not analyze all the photographs, but even analyzing twenty-five percent of them is a large task.

The point of all this is that (1) we generate too much data to be handled reliably by human hands, (2) the experiments last a long time on expensive equipment so that one must know as rapidly as possible whether he is getting useful data. The operating cost of a large accelerator is high enough that it makes it desirable to model the operation, that is, to generate cost functions. One can then optimize the schedule of experiments on the cost function.

Historical development. Let us take a look at the historical development of automatic data analysis and control systems. Here are the stages of development and the corresponding year of their accomplishment:

1. Digital Read out	1954
2. At-the-Site Data Reduction	1960
3. Direct Entry to Computer	1962
4. Data Analysis and Monitoring	1964
5. Computer Control of Equipment	1964
6. Decision Making and Control	1964
7. Experiment Analysis	Future

1. By digital read out we mean converting the experimental data to digital form on some medium such as punched paper tape or punched cards suitable for reading into a computer. Prior to this time, experimenters recorded readings of dials or counters in notebooks.

2. The at-the-site data reduction era came when we began to produce small computers which could be dedicated to a particular experiment or experimental group. The paper tapes or punched cards were read into the machines and at least preliminary results

were made available to the experimenter.

3. Direct entry came when the small computers were designed to that we could record data directly into their memories rather than having to pass through the intermediate stage of punched tape or punched cards.

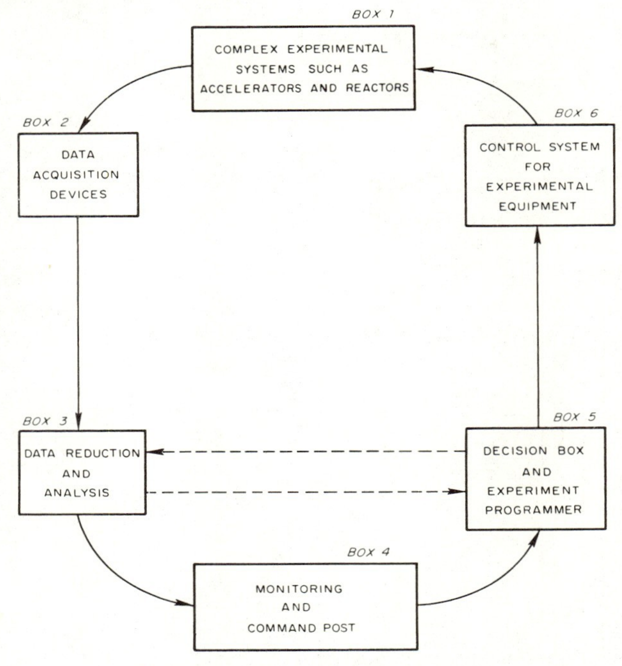

FIGURE 1. General Schematic for Closed Loop Data Analysis and Control System

4-5-6. Having surmounted the instrumentation and control programming problems in step 3, steps 4, 5, and 6 followed very quickly. Moreover, the computer industry entered the small computer marked in a competitive way about 1961-62. The input-output control and interface problems were now generally solved. A number of real-time data acquisition and control systems began to appear that embodied all or parts of the ideas in steps 4, 5, and 6. One system shown in Figure 2 embodied all three ideas. First, let us look at Figure 1. It shows a system that could be operated as a closed loop system depending on the judgment of the user at Box 4. Figure 2 is an example of a real system that worked in this mode. It carried out its first controlled experiment depicted in

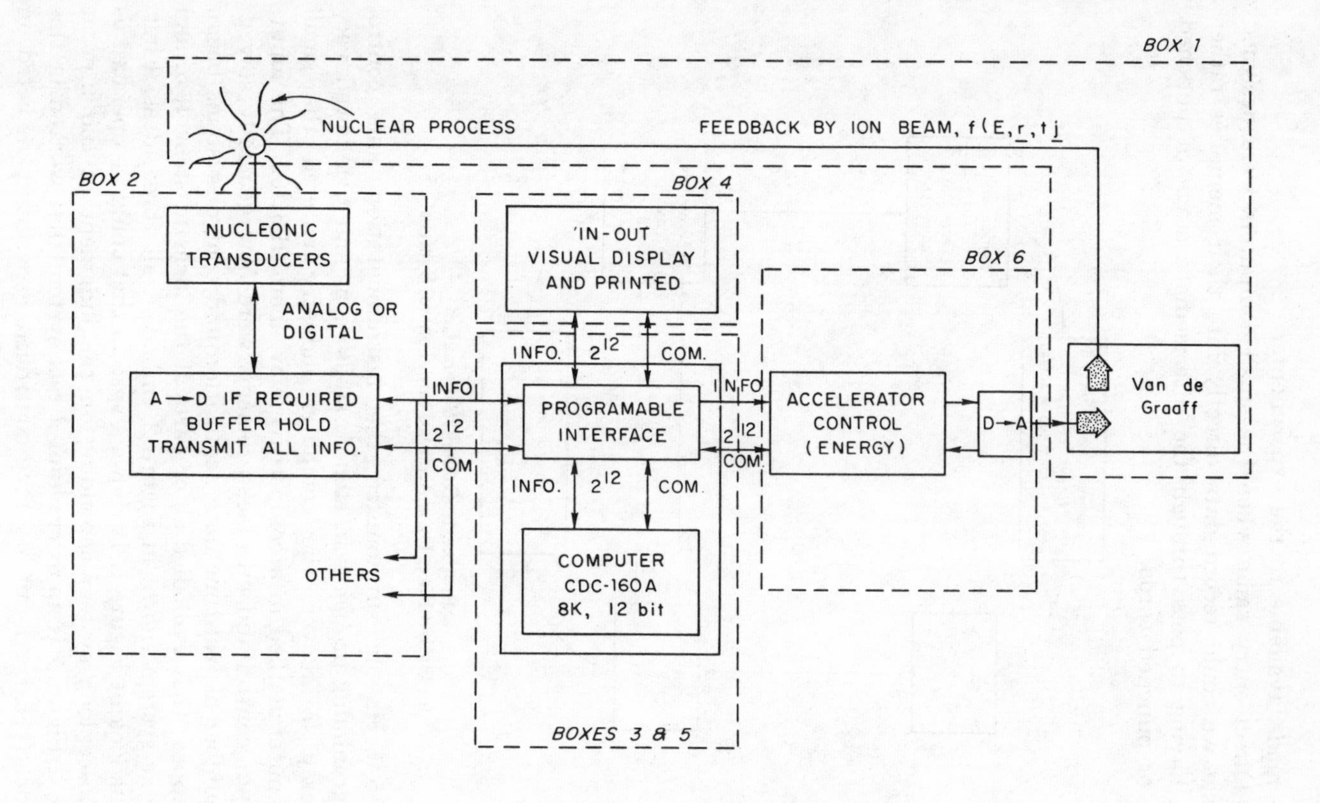

FIGURE 2. Closed Loop Data Analysis and Control System

Figure 3 in 1964 [2], [3]. Since that time the users have programmed about a dozen different experiments which work in a closed loop mode.

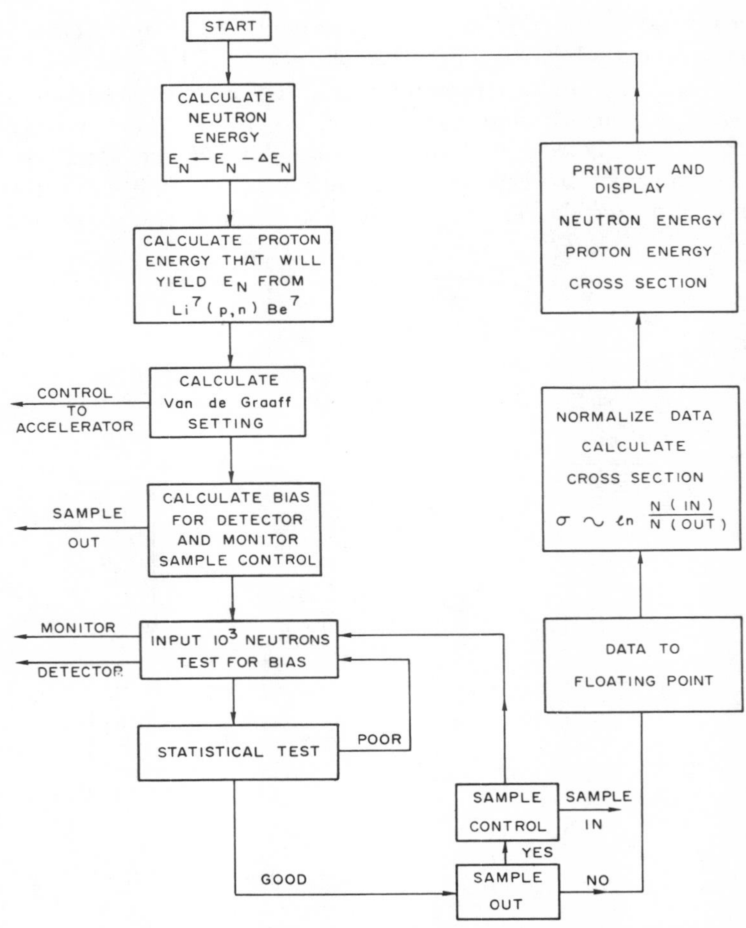

FIGURE 3. Schematic Flow Chart of Programmed Experiment

It is clear that such an elementary diagram as Figure 1 is a great oversimplification. The boxes could be somewhat rearranged and still properly represent a system, but the diagram does separate the various classes of mathematical and computer science research problems rather well. There are interesting mathematical and programming problems in all of these boxes. In succeeding discussions we shall be concerned principally with the problems in Boxes 3, 4, and 5.

Current systems. The systems described up to now have been aimed at control over short time intervals. The Van de Graaff example carried out experiments over a few days or weeks and did not employ massive amounts of data. We need to add a large file capability to Figure 1 to handle experiments of the kind we deal with at the Stanford Linear Accelerator Center. Figure 4 shows an appropriate modification of Figure 1 to reflect this requirement.

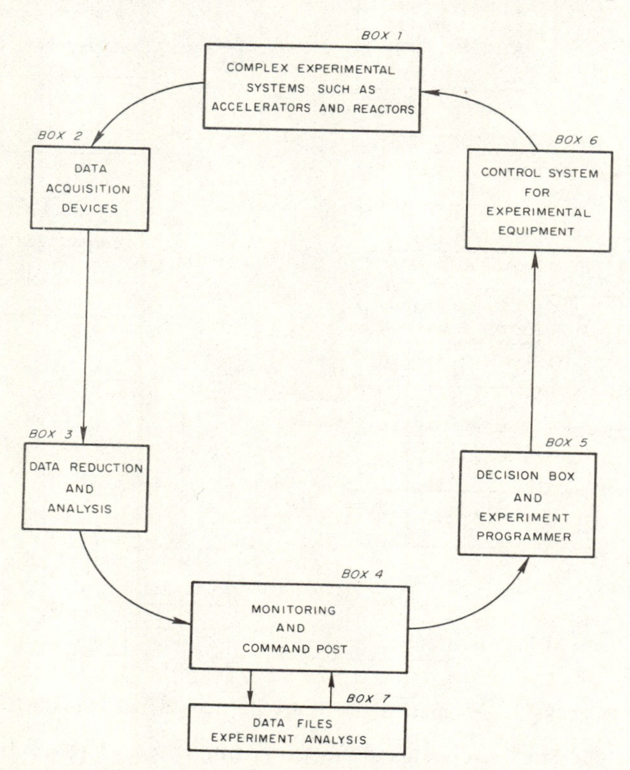

FIGURE 4. Control System Expanded to Include Large Data Files

For this we need files with a capability of 3×10^9 bits that can be examined interactively by an experimenter. The large file will contain the physics data extracted from the spark chamber or bubble chamber film or from particle counter systems. It consists of a record of each nuclear event found in the film data.[3] A typical experiment may consist of 10^6 stereo pairs of which 250,000 contain the events of interest. The data file would then consist of the 250,000 records where each record would contain the track coordinates of each branch of a track, the curvature of each track if the event took place in a magnetic field, the moments of all particles participating in the event, and perhaps a tentative hypothesis of the masses and energies of the particles. The physicist interacts with his file by some sort of statistical summarizing programs. There now exists a general purpose program called SUMX which permits the user to perform a variety of functions on the input files by introducing the appropriate parameters. The current "production" version of SUMX is an off-line version. The user submits to the computer a deck of control cards containing the parameters of his search and transformations, and the computer returns his output in the form of tables, single or two-dimensional frequency plots.

It has often been pointed out that doing data analysis is like doing an experiment. Often the physicist does not know what hypothesis to advance so he starts by guessing. On the basis of the results of his first guess he makes another guess until he is soon "zeroing in" on a reasonable hypothesis. This immediately suggests that we need an interactive file processor. At SLAC we have in limited use a simple version of On-Line SUMX. The On-Line SUMX permits a physicist to examine his data files from a cathode ray tube (CRT) display. He now specifies which data to work on and what transformations he wishes to perform, including the parameters, through keyboard entries at his display console. The histograms or scatter plots are returned to him on the display. From the console he can change the specification of the input file, change the "bin widths" of his histograms, or try different mass and energy hypotheses. The user can experiment with his data until he has arrived at the correct hypothesis. He may then have the summaries printed in tables and have the histograms plotted. Our On-Line SUMX is in a preliminary form, but we are clearly dedicated to refining it into one of the important tools of our Laboratory.

The ability to carry out interactively the analysis of the data

[3] Figure 5 shows a characteristic bubble chamber photograph. In this photograph one would be interested in the "two-prong" event seen on the center right.

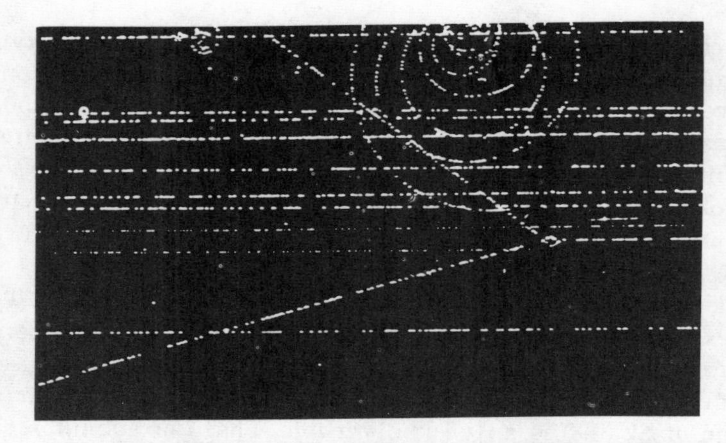

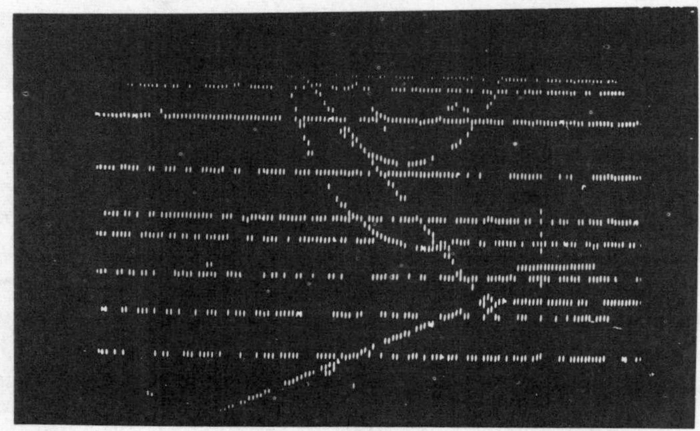

FIGURE 5. $5 \times 10\ mm^2$ Bubble Chamber Film and Playback

for whole experiments permits the physicist to guide the course of his experiment. In the past, the elapsed set-up time, the elapsed time to collect the data, and the elapsed time to analyze the data were all comparable. Rapid data analysis speeds up the setting up operation and permits the experimenter to alter his experiment if necessary while it is being conducted.

In any particular laboratory, one will certainly have all the elements of Figure 4; however, they may not be coupled as shown. In fact many of the boxes may not be coupled by data and/or control lines at all. Let us examine the systems we have at SLAC and how they are interconnected.

Figure 6 shows a schematic of the Stanford Linear Accelerator. The particles (electrons or positrons) are accelerated from the Gun to the Beam Switchyard (BSY) in pulses occurring 360/sec or fractions thereof. They may attain any energy up to 20×10^9 electron volts (20 BeV). Several beams may be fired down the accelerator at different energies and different repetition rates. The Beam Switchyard switches the appropriate beam into its intended experimental area where it may be further shared by more than one experiment. The accelerator is not under computer control, but the Beam Switchyard is, as well as the three magnetic spectrometers located in End Station A. There is direct control from the spectrometer computer to the BSY computer so that the spectrometer computer can, in principle, control the beam coming into its area. There is provision for, but not installed, a connecting line from the spectrometer computer to the principal computer facility. Also, there is a connection to a computer that collects data from and controls a filmless spark chamber. We are just now experimenting with this link. The bubble chambers and the film-generating spark chambers are linked to the central computer by the film they generate. Now information from the computer is printed out and carried to the experimenters, but eventually it will be sent by remote typewriters or CRT displays. Figure 7 shows schematically the SLAC computers and their interconnections.

At SLAC we do not have a complete closed loop system as shown in Figure 4 but we have several subsystems that look much like Figure 4. Control of the experiment is not a centralized thing since several experiments are going on concurrently. However, the magnetic spectrometer system and the wire spark chamber system have rather complete control of their own little subunits.

In the development of a physics center such as SLAC there are three phases of activity, each requiring somewhat different computational facilities and techniques. There is perhaps a fourth phase but its importance is not yet established.

The first phase is concerned with the design of the accelerator and the particle detectors, e.g. bubble chambers and spectrometers. Three important calculations are magnetic field calculations, particle trajectories, and radiation shower calculations. The mathematics of these problems is quite well in hand although there is a great deal of work involved in particularizing the methods for each problem. The particularizing can best be done by rapid

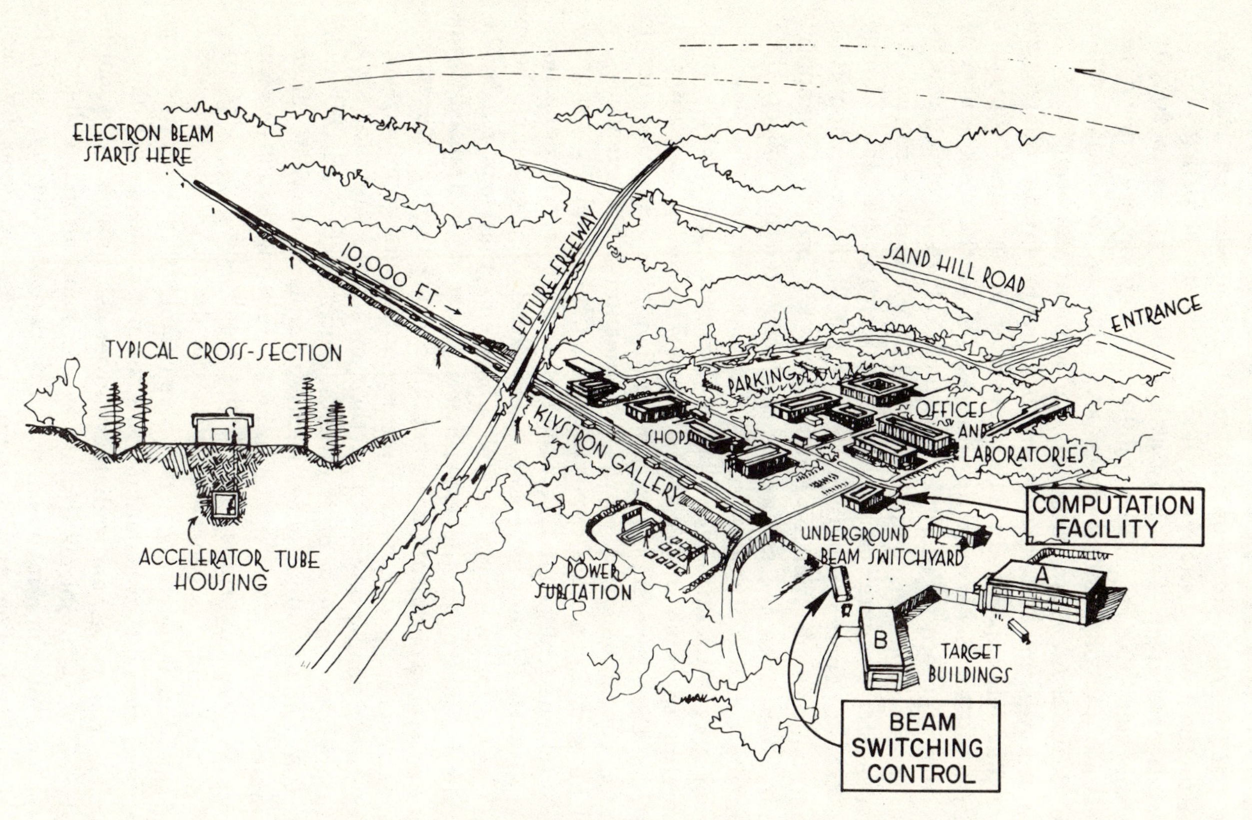

FIGURE 6. Schematic of Stanford Linear Accelerator

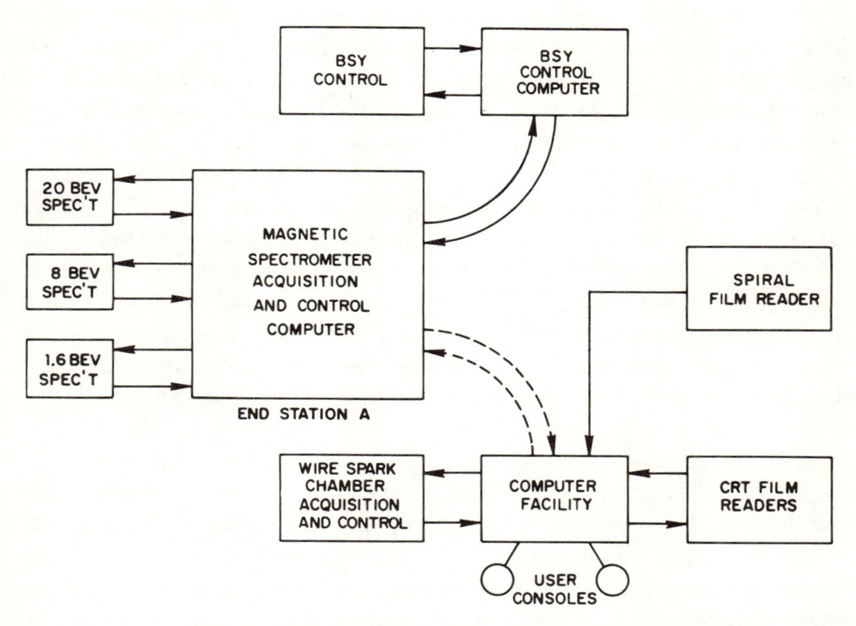

FIGURE 7. Schematic Showing Communication
Between Data Acquisition and Control Systems at SLAC

interaction between man and computer, especially by utilizing graphical displays.

The second phase is concerned with data acquisition and data analysis. During this phase the computer interest is concentrated on design of data handling and control systems, pattern recognition, curve fitting, hypothesis testing, and the like. The design of the systems and the mathematical and statistical approach to these problems deserves a great deal of attention because these decisions last a long time. The large data analysis codes that are employed have many man-years of work invested in them. The development of these codes is an engineering task much like the task of building the accelerator (on a smaller scale to be sure). Assumptions about the data formats, volume of data, physical parameters of the experiments, the pattern recognition schemes to be used, etc. become built into these codes and are difficult to change. Several of the data analysis codes for bubble chamber data analysis have taken twenty-five man-years to develop. Indeed, these codes may have been useful after the first six or eight man-years, but continuing development is required. In this second phase the control systems for the control of experiments are designed including the interaction languages for the man-machine part. From the computer viewpoint the second stage is very heavily a system era.

The third phase may be described as model building and testing. After the data has been analyzed and the physical parameters extracted, it must be incorporated into some physical model. It is not possible to describe all of the computational problems arising here but we shall describe one interesting calculation that involves a new area of computer science—algebraic calculations.

We shall briefly describe some work by Hearn [**4**], [**5**] on the calculation of the properties of elementary particle reactions. The calculations with which he is dealing involve the manipulation of rather complicated algebraic expressions which include both tensor and noncommutative matrix quantities. The expressions are used to calculate the cross section[4] for the reaction. The reactions are specified in terms of Feynman graphs [**6**] which display the evolution of the reaction with time. There are straightforward rules for the generation of the cross-section expressions from the Feynman graphs. Each edge of the graph represents a particle during the reaction and each vertex (of degree $\geqq 2$) represents an interaction.

The principal calculations are: (1) Follow the rules for going from Feynman graph to cross section. The cross section will be given in terms of the product of traces of sums and products of gamma matrices. Gamma matrices are 4×4 noncommuting matrices that express the transition rules permitted. (2) The computer must expand the original expressions, collect like terms. (3) Reduce the expression to something manageable. Calculations involving 1500 terms have been carried out and there is current interest in calculations that would involve 10,000 terms. There is nothing conceptually difficult in doing the calculation. For low-order calculations one can do them quickly by hand. However, with more powerful accelerators one can investigate more complex reactions calling for higher order calculations.

This type of algebraic calculation illustrates a kind of problem that is of great interest now—representation and manipulation of algebraic expressions.

The fourth phase that should be mentioned is the modeling of the operation. By a model we mean a cost function for a unit of operation, say, for one week. Given such a cost function one can then optimize the operation according to some desirable criterion. There are rather subtle costs involved in operating a multi-beam accelerator. The accelerating tubes, klystrons, are not all purchased

[4] The cross section is a measure of the probability that the specified particle will emerge with a given energy and direction.

from the same manufacturer so they have different lifetime characteristics and different maintenance contracts. The speed with which one replaces a klystron is dependent on the demands on the machine, but also has an impact on labor cost and klystron inventory. These may seem like trivial costs, but centers such as SLAC, and the future machine at Weston, Illinois, are sufficiently complex that an additional 10-percent efficiency means several million dollars per year.

PART II: GRAPHICAL INPUT-OUTPUT TECHNIQUES

Uses for graphical input and output. Any discussion of computation and its relationship to decision processes would be incomplete without a discussion of graphical displays. More of the cerebral cortex is devoted to visual processing than to any other of our senses [7]. It would seem most appropriate to utilize this facility of ours to the fullest in developing methods for communication between man and machines.

There are several ways in which graphical input-output devices can be used. For example:

(a) *To display messages and to edit text.* The airline reservation system is a simple example of such use. Inquiries may be made about the status of a particular flight, textual messages may be sent back to the inquirer, the inquirer may request a reservation, etc. Information retrieval systems may also be of this character. See Figure 1.

(b) *To display graphs of functions for monitoring a calculation and eventual recording.* A user may wish to view his output in a graphical form while his calculation is unfolding. The On-Line SUMX discussed in Part I is an example of such use. The user of On-Line SUMX supplies a hypothesis to be tested; the file searching programs select data on the basis of the parameters supplied. The histograms of the data are then displayed to the user. The user can ask for a hard copy of the histogram. See Figure 2.

(c) *To input initial guesses for mathematical calculations.* In minimization problems, for example, when one can often select the crude starting position more quickly by having a coarse plot of the function. In particular if there are a number of local minima the coarse plot will be very revealing and the eye will detect the lower of these and supply an initial value that will lead to the absolute minimum. See Figure 3.

(d) *To input geometrical data.* In engineering design of flow

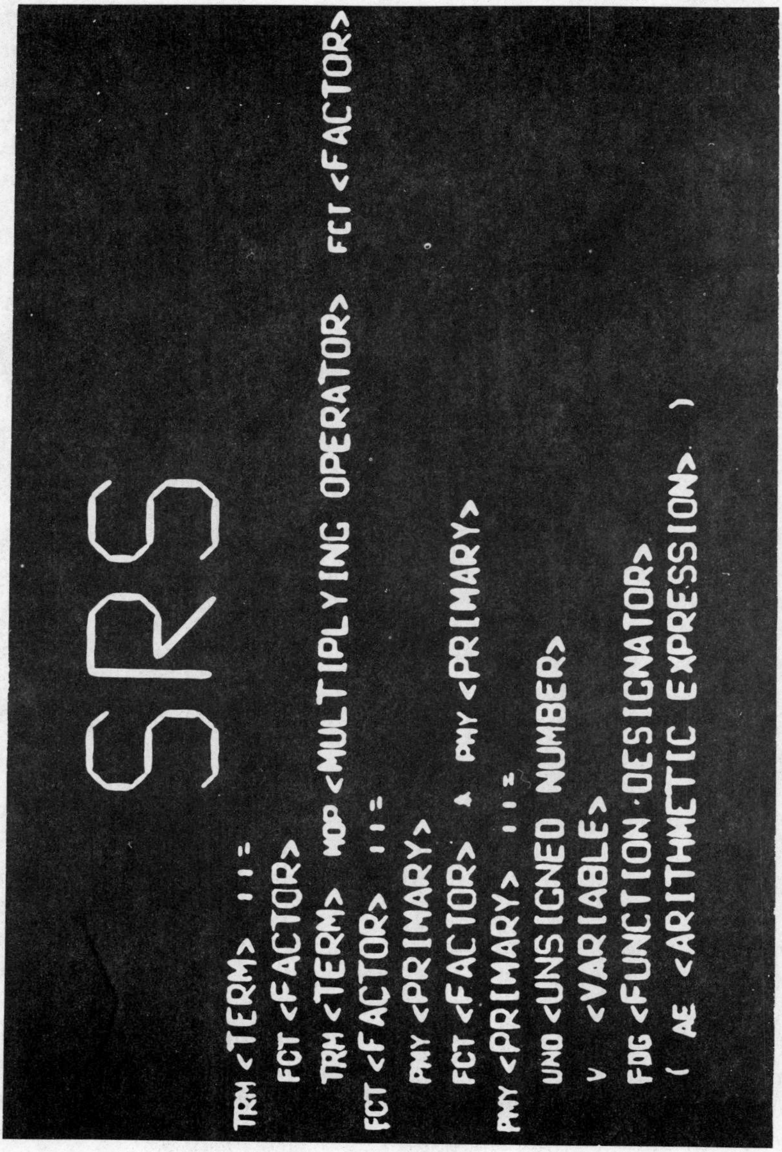

FIGURE 1. Information Retrieval on a Display

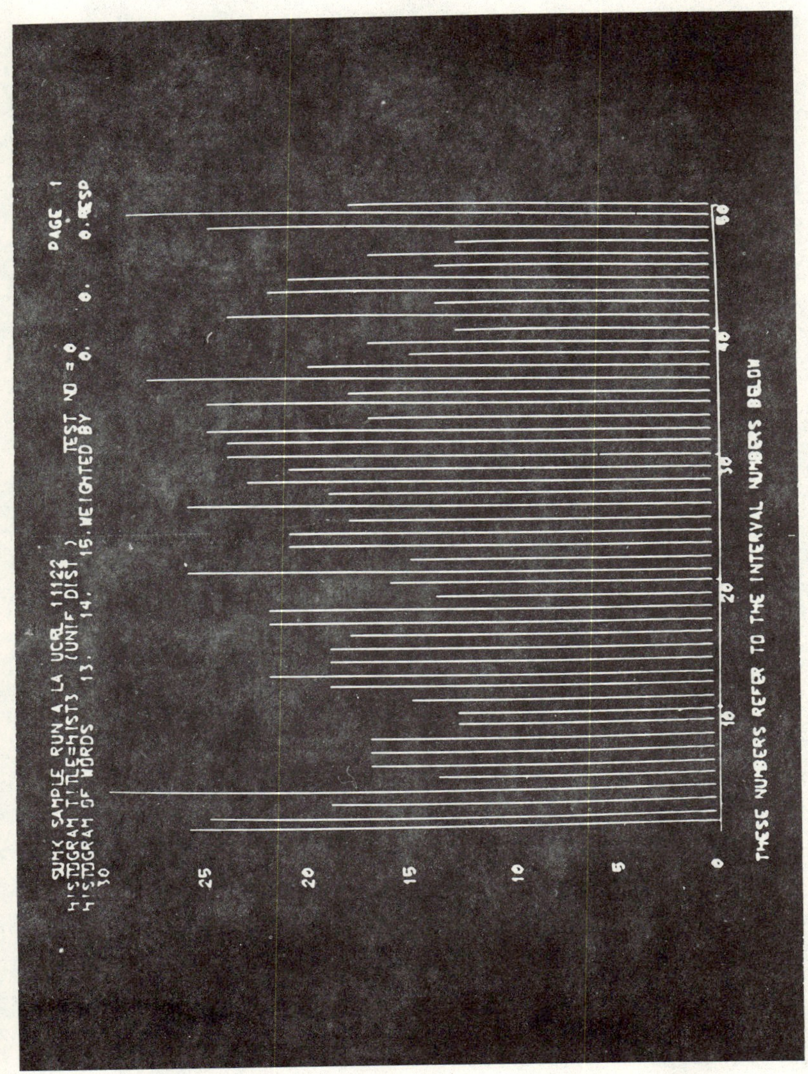

FIGURE 2. SUMX on a Display

FIGURE 3. Surface Plotted on a Display

```
. EXIT PROG                      MODE PARAMETER
. SCAN PICT      YES    . NPIC       000001
. CARDS PICT            . NUMAR      000015
                             SCANNER PARAMETER
  PROGRAM SWITCHES      . IMACH      000001
. IMOVF          YES    . ITHRES     000004
. IASCAN                . IDENS      000000
. IADD                           TAPE PARAMETER
. IDISP                 . ITAPNM     000000
. IWRTAP         YES    . KTAPE      000013
. MARK                  . IDATE      000000
. INDX                       DISPLAY PARAMETER
. IPRINT         YES    . IDTIM      000600
. ILOG           YES    . XM %       000025
. IDLOG          YES    . XA         000000
. ITIME                 . YM %       000100
. IWINDOW               . YA         000000
```

FIGURE 4. Control Panel

surfaces or magnet pole faces the experience and imagination of the designer can often be brought to his design programs more rapidly by a drawing than by functions or tables of data.

(e) *As a control panel.* A cathode ray tube (CRT) combined with some input medium such as a light pen may become a very flexible control panel. The control programs in the computer display several alternative courses of action to a user. The user indicates his choice by touching one of them with a light pen. By display of choices and touching "light buttons", the user can direct control of his system. See Figure 4.

An example of each of the above uses will be shown as part of the graphics demonstrations. (See Appendix I.)

Types of graphic display and input. The principal types of graphic input-output devices will be described here. For a more detailed discussion the reader is referred to [**8**] and [**9**]. [**9**] is highly recommended for those interested in equipment.

The cathode ray tube (CRT) is the principal display device used in conjunction with a computer. The CRT is comprised of an electron gun, a focusing and deflection system, and a phosphor coated screen or faceplate. The phosphor emits light under the impact of the electrons. Deflection systems are either electrostatic or magnetic. In the former case the electron beam is deflected by the presence of electrostatic charge collected on deflection plates. In the latter, the beam is deflected by magnetic field established by current in the coils of an electromagnet. Electrostatic deflection is faster than magnetic deflection but it is also less precise. Figure 5 shows schematically the construction of cathode ray tubes.

The CRT is activated by sending digital signals to its controller which are then converted to analog signals to set currents in the deflection coils, in the case of the magnetic deflection CRT. In simple CRT's each point on the faceplate is plotted by sending a digital signal for the Y-coordinate and a similar signal for the X-coordinate. To sustain the display, the computer must repeatedly send all the coordinates of the whole display.

A buffered CRT has a small memory associated with its controller which holds the coordinates of all points being displayed. The picture is automatically regenerated from the buffer memory so that the computer need only communicate changes to the controller.

With a simple CRT, if one wishes to display characters he must form the character by plotting points. More complex displays include a hardware character generator which performs the character formation. A user need only provide the identification of the character and the X, Y coordinates of its location.

An additional feature found on still more sophisticated displays is a vector generator. To generate a vector the user program sends to the controller the indication of vector mode and the X, Y coordinates of the end points of the vector. The electron gun is left "on" while the beam is deflected from the initial to final position thus generating a straight line or vector. Figure 6 shows a circle plotted in (a) point mode and (b) vector mode.

In summary then, a CRT display may be either buffered or unbuffered and it will plot points, characters, and vectors depending on the sophistication (and cost) of the device.

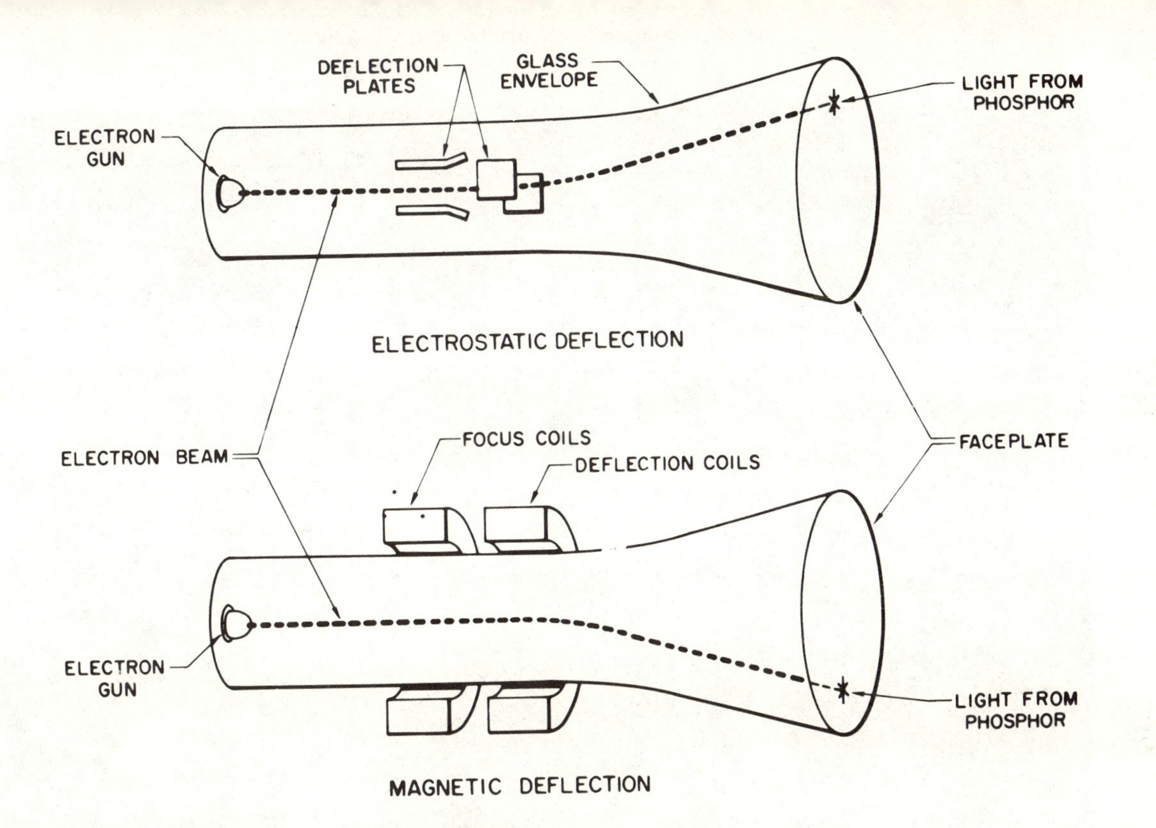

FIGURE 5. Schematic of CRT

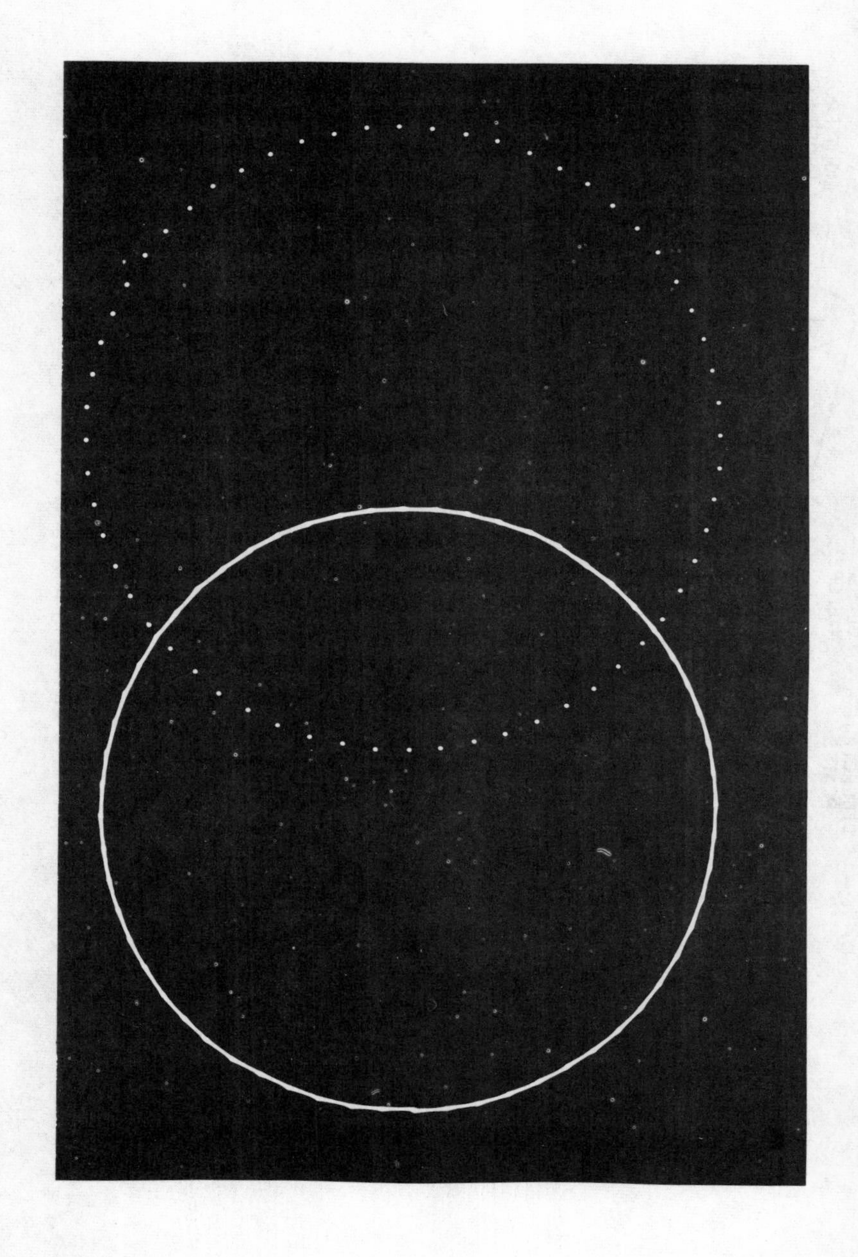

FIGURE 6. Circle in Point Mode and Vector Mode

There are several types of input devices that may accompany a display CRT. Common devices are:

(a) Typewriter-like keyboards,

(b) Toggle switches, push buttons or tuning knobs,

(c) Joysticks or trackballs,

(d) Light pens,

(e) Conductor tablets and styli.

In order to input from a keyboard, the current position is indicated by a cursor or marker. The position of the cursor is maintained either in hardware or in software. Pushing a keyboard button inserts the corresponding character at the position of the cursor. This character is inserted in the buffer for the display and may be read by the computer for subsequent use. The keyboard may also input control information. For example, by depressing the proper button one can signal the computer to read the buffer, or to change the scale on a picture.

Toggle switches may be used to insert control information or digital data whereas tuning knobs will supply analog data that must be converted to digital data. Joysticks and trackballs are used to control movements of a cursor. A joystick functions similarly to the stick control of an airplane. By pushing the stick forward analog control signals are supplied to the CRT control to move the cursor in the Y direction. Similarly, side movement of the stick moves the cursor in the X direction. If one wants to insert a point on the faceplate, a switch is closed when the cursor reaches the desired point. A trackball functions similarly to a joystick. A trackball is about the size of a bowling ball and fits in a socket of a table. Rolling the ball forward moves the cursor in the Y direction, and so on.

A light pen can be used to point to or sense light already displayed on the CRT faceplate. The light pen consists of a pencil size holder, a light sensitive device such as a photomultiplier in the tip of the pencil, and a timing circuit that correlates the light sensing and the positioning of the electron beam. This timing current permits the controller to determine where on the faceplate the light pen was pointing.

The RAND TABLET [**10**] and its commercial version the GRAFACON [**11**] are examples of conducting tablets. They are used in conjunction with a CRT display. A grid of lines is shown on the tablet and figures are sketched with a stylus. The stylus picks up signals from conductors embedded in the tablet and relays them to a decoding device. The controller then transfers the position information to a spot on the face of the accompanying CRT display as well as to the computer program using the input.

One additional note should be made concerning communication to and from the displays. If the display control has its own buffer memory, one can transmit data to and from the computer by telephone lines. If one wishes to change the display rapidly, telephone lines constitute a bottleneck. The data rate for telephone lines is 2400 bits per second. The code usually requires 11 or 12 bits per character. Thus one can transmit about 200 characters per second. One page on a display can hold about 1000 characters. If one is doing relatively slow editing the telephone rates are suitable. If one is editing a large amount of text by doing "page flipping", telephone lines require about five seconds per page. For These higher data rate requirements it is necessary to couple to the computer by coaxial cable and be located within about 2000 feet of the computer.

Control of a graphic system. The usefulness of a graphic input and/or output device depends greatly on the control system that couples it with the computer and the users' programs. This section describes one such system currently in use [**12**].

Figure 7 shows a schematic of the computer configuration used in the spectrometer data acquisition and control system. It shows where the cathode ray tube display, the light pen, the teletype, and some thumbwheel switches which are used to control the displays are connected. Figure 8 shows a schematic of the display organization for this system. The display consists of an unbuffered 20-inch CRT with point plotting, vector plotting and character generation capabilities, and in addition a light pen. Adjacent to the CRT are six sets of eight-digit thumbwheels with associated off-on switches and indicator lights. These thumbwheels are used to identify desired displays and to introduce parameters such as scale factors into a user's program.

With six thumbwheels one can select any six of 8^3 (eight-digit thumbwheels) displays maintained in the active display file. The identification for the desired displays can be set up on the thumbwheels. To call one into action the "on" button is pushed. One can have all six on at the same time, thereby causing a superposition of the pictures. There are times when it is quite convenient to

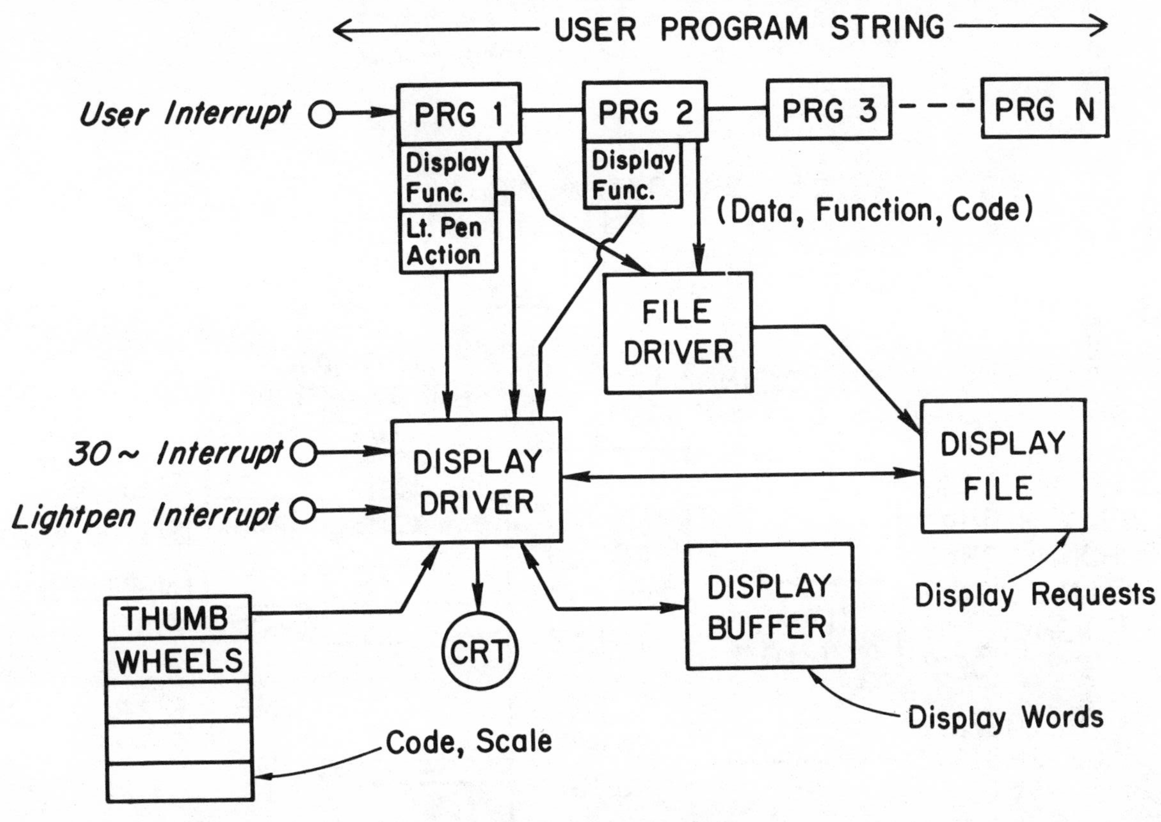

FIGURE 7. 9300 Configuration

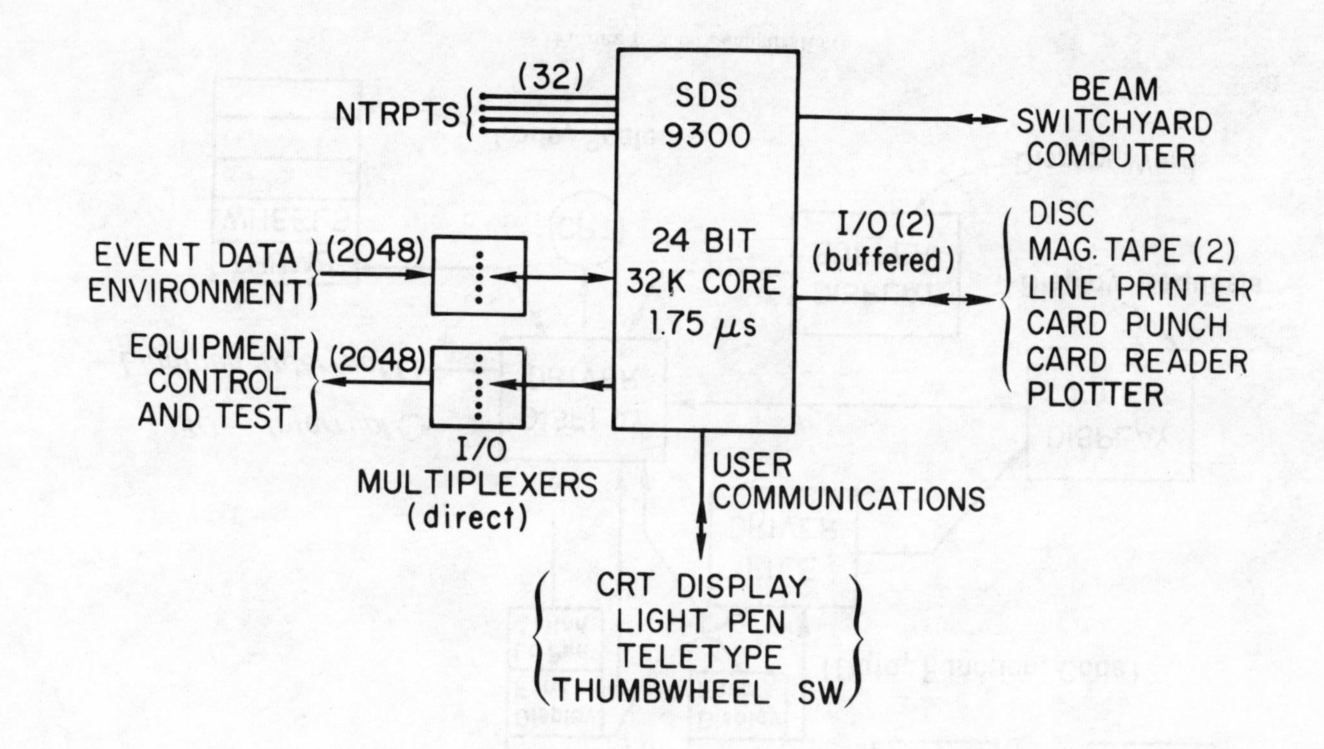

FIGURE 8. Schematic of Display Organization

superimpose different displays although this is not the usual mode of operation. It may be useful to have six programs displaying data in different regions of the faceplate. One can then use the display like a multiple console. The usual mode of operation, however, is to have the six programs most prominently used selected on the thumbwheel switches and have only one of them in an "on" status. The selected program would normally call for the CRT packing routine and transfer the data of that program to the CRT buffer. The complete picture is displayed from the computer over a data channel at a rate of about 30 frames per second. The display packing program need only be executed once unless new data from the program arrives and requires repacking of the picture data in the display buffer.

The displays are triggered by a 30-cycle clock interrupt. At the time of the interrupt the driver program interrogates the thumbwheel to see if there are any "on" switches. If there are none, the interrupt is terminated and the program continues. If there is a thumbwheel request "on" and this is the first encounter, a lower priority interrupt is set which calls the necessary packing routines. When the machine gets around to the lower priority interrupt, it transfers data from the display file through the user's transform routines to the display buffer. The next time the clock triggers the interrupt, the picture is shown on the face of the CRT.

Light pen requests can be made. If a light pen action is desired it must be specified with the data of the request (which is in the display file) as an additional routine. Upon light pen interrupt the associate display point in the buffer is traced to its corresponding file entry. If no light pen action has been requested the interrupt is ignored. Otherwise the user's second subroutine—that is, the light pen request subroutine—is called and it is supplied with the coordinates of the point detected by the light pen. The point that has been detected is intensified briefly before reserving display of the entire buffer.

Figure 9 is an example of a display made by this system. It permits a rapid monitoring of the status equipment, in this case the status of counters. For normal functioning of these scintillation counters there should be a buildup of the intensity to some plateau and decay as shown in the figure. Each counter is represented by a section across the surface of that plateau. If one of the counters is malfunctioning it will be readily visible in that its counter trace deviates markedly from those of the other counters.

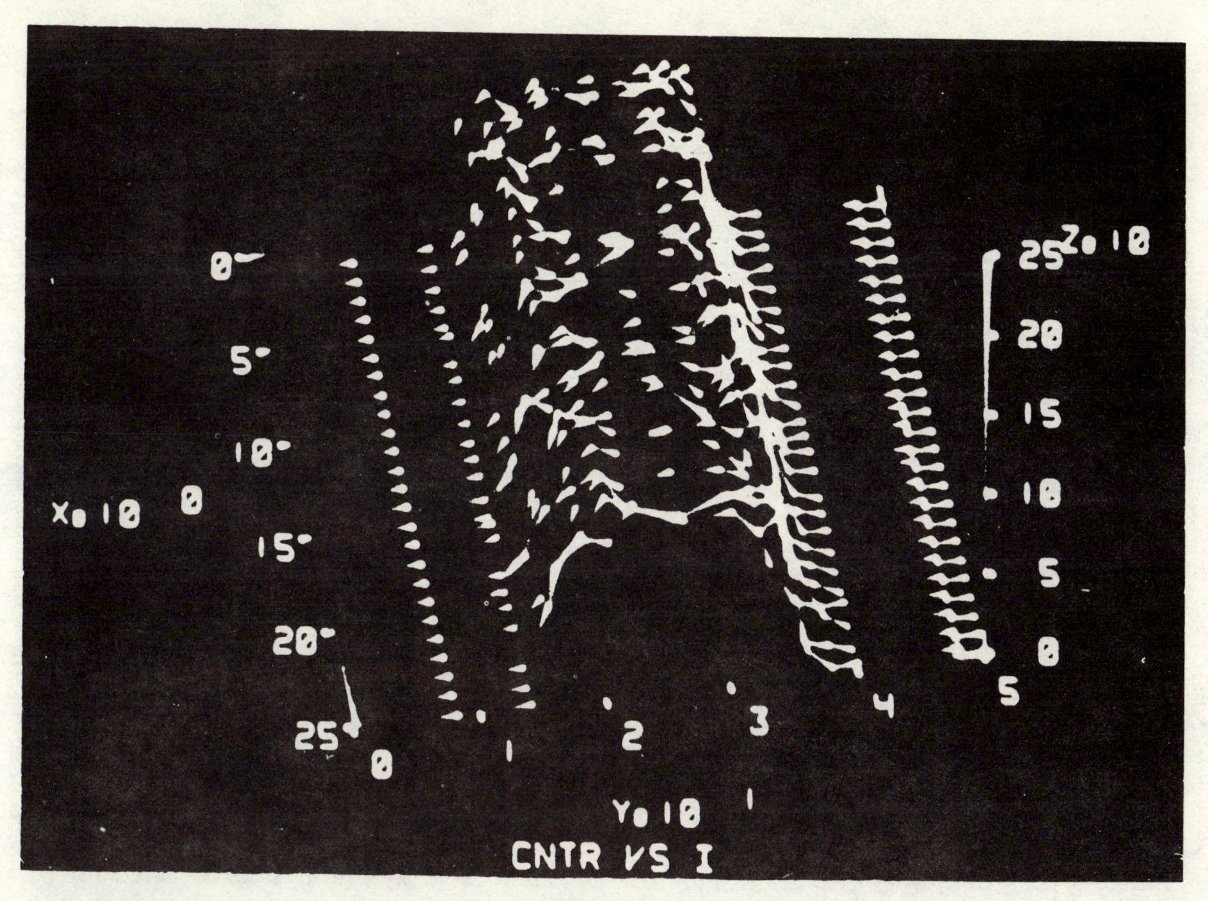

FIGURE 9. Counters

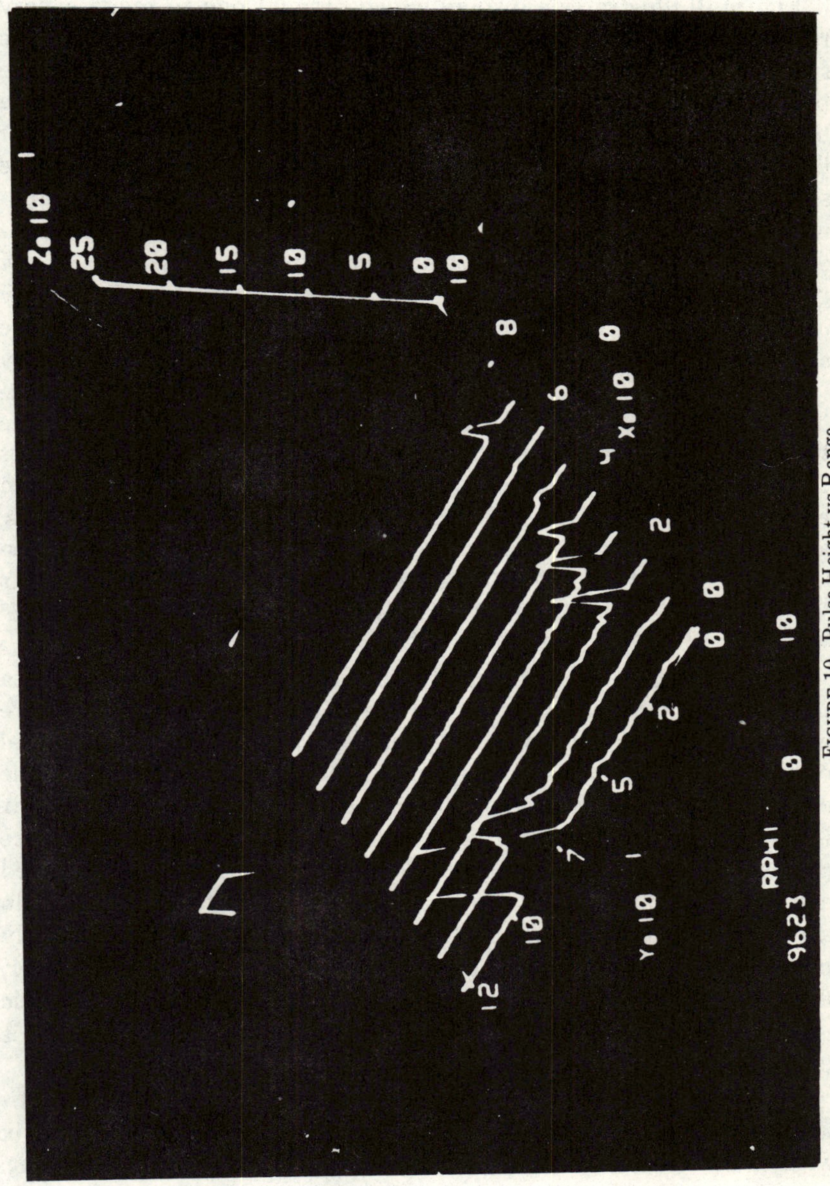

FIGURE 10. Pulse Height vs Range

Figure 10 illustrates a display which permits one to make a very rapid inspection of the quality of the particle beam entering the spectrometers. The display shows the frequency of counts for pulse height versus range. Intensity is on the verticle scale, the pulse height or energy is along the scale going from lower right to upper left, and range is the scale going from lower left to upper right. We see on this particular display the presence of three particle components. Electrons have very short range and generate large amounts of pulse height because of their showering characteristic. They generate showers very quickly and do not penetrate very deeply into the medium, so the large peak toward the left of short range and high intensity represents electrons. μ mesons have relatively long range, are weakly interacting, and generate very little pulse height. The small peak at the back of the photograph represents the presence of the μ meson. The peak near the center represents the presence of the π mesons. π mesons have longer range than electrons and are more strongly interacting than the μ mesons, and thereby generate more pulse height at the short range than the μ mesons. This type of display can be selected very quickly by the operator and he can deduce the quality of the beam that is available to him for his experiment.

Data structures. The choice of data structure for a graphic data processing application depends on the qualitative and quantitative aspects of the processing to be done. There are two fundamental choices with variations on each for explicit representation of the connectivity present in a picture. The two choices are a matrix (array) like structure or a list structure. There is a third choice discussed in Part III wherein the connectivity is not represented explicitly in the data structure but is represented implicitly in the syntactic description of the picture. The author is actively pursuing the third choice, the implicit representation; however, there are classes of problems not yet handled by the syntactic methods. For these pictures a direct explicit representation is (still) necessary.

The most general form of the matrix type structure is the multiword list structure [14]. The data are held in an $n \times m$ matrix with one row for each item in the picture, e.g. for each line, point, corner, etc. The first j columns are reserved for the attributes of the items, and the last $(m-j)$ columns are reserved for pointers that indicate the connectivity of the structure.

A multiword list structure works quite well if there are very few kinds of items in the picture. Figure 11 shows an example from

CONTENTS OF MULTI-WORD LIST ITEM PERTAINING TO A SPARK

Word	Data Mode	Information Contained in Word
1	Floating	X-coordinate of spark *
2	Floating	Y-coordinate of spark *
3	Floating	Z-coordinate of spark *
4	Fixed	Gap Number **
5	Fixed	Chamber Number **
6	Fixed	Local Degree of Vertex ***
7	Fixed	Pointer to 1st connecting spark ****
8	Fixed	Pointer to 2nd connecting spark ****
9	Fixed	Pointer to 3rd connecting spark ****
10	Fixed	Pointer to 4th connecting spark ****
11	Fixed	Pointer to 5th connecting spark ****
12	Fixed	Pointer to 6th connecting spark ****
13	Fixed	Pointer to 7th connecting spark ****
14	Floating	Distance to closest spark
15	Fixed	Pointer to closest spark *****
16	Floating	Distance to 2nd closest spark
17	Fixed	Pointer to 2nd closest spark *****

* Coordinate information is provided as input to LINK.

** This information is furnished by the subroutine SORT and depends on the configuration of the chamber(s).

*** This value will be N ($0 \leq N \leq 7$). It indicates the number of connected sparks. If non-zero, only the first N pointers will be used. All others will be set to zero.

**** If the contents of the word are zero, all remaining pointers will be set to zero. If non-zero, the value I lies in the range $1 \leq I \leq M$ and is the index of the multi-word list item pertaining to the connected spark.

***** The contents of these words will always contain a number I ($1 \leq I \leq M$) which is the index of the multi-word list item pertaining to the appropriate spark.

FIGURE 11. Multi-Word List

the processing of tracks in particle physics [13]. In most computers matrices are processed rapidly, an encouragement to use a matrix type structure whenever possible. If the connectivities become sparse and the number of types of items is large so that the length of the attribute list may vary greatly, the use of storage becomes rather inefficient. Many insertions and deletions of items, as in the case when one is constructing drawings, complicates the data structure and also leads to inefficient storage utilization. These factors lead one to consider other structures.

The CORAL [15] language and data structure has been designed to handle graphical problems. CORAL is an acronym for Class Oriented Ring Associative Language. The main components of the CORAL data structure are BLOCKS, RINGS, and RING LIST ELEMENTS. A block contains a block header, a group of ring list elements and data about the block. Figure 12 shows the structure of a block and of the ring list elements contained in a block.

The first cell in each block, the block header, contains the length of the block, the length of the list area, that is, the number of cells reserved for ring elements, and the block type as shown in Figure 12. The list area follows the block header and the data area follows the list area. The data cells may hold any data relevant to the block, e.g. coordinates of a point.

The ring list elements are also shown in Figure 12. Each ring list element has a forward pointer, pointing to the next list element in the ring with the last list element pointing to the first or ring start element to close the ring. Also each ring list element has a backward pointer. The backward pointers are of two types, distinguished by a bit in the pointer word. A backward pointer either points to the ring start element or to the ring list element two steps back. From any ring list element one can go back to the ring start in at most two steps. If the current backward pointer is a ring start pointer, then one can go to the ring start in one step. If not, one can take one step forward and then go to the ring start.

Figure 13 shows a CORAL structure for a triangle. There is a ring that associates the lines of the triangle. There is a ring for each point that associates that point with the lines that meet on the point. Deletion of a point will break the ring that includes that point causing deletion of the lines that meet at the point. Deletion of those lines will break the ring that forms the triangle, thereby deleting the triangle.

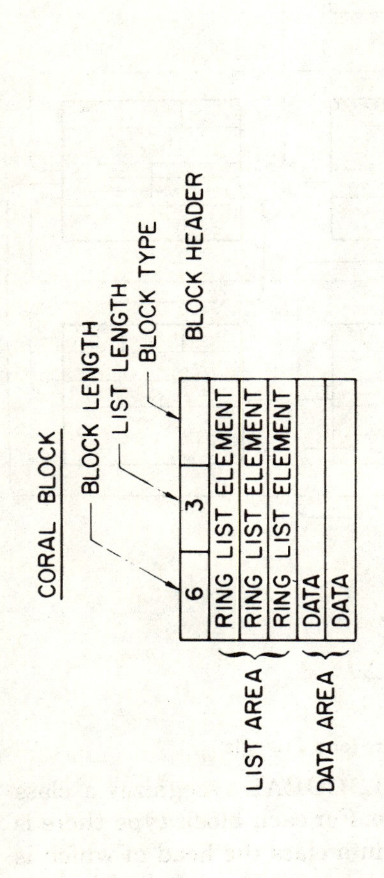

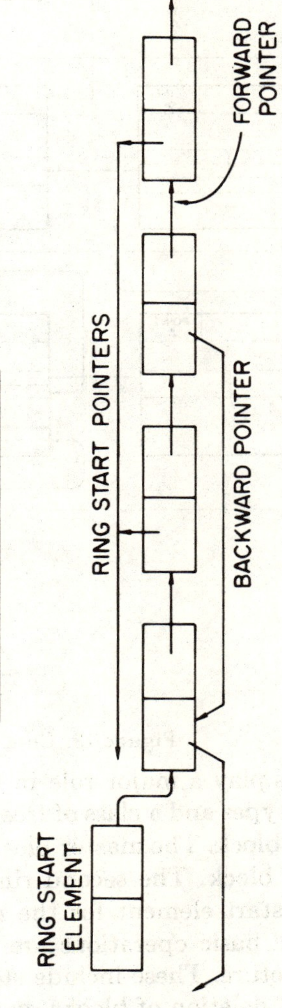

FIGURE 12. Coral Data Structure

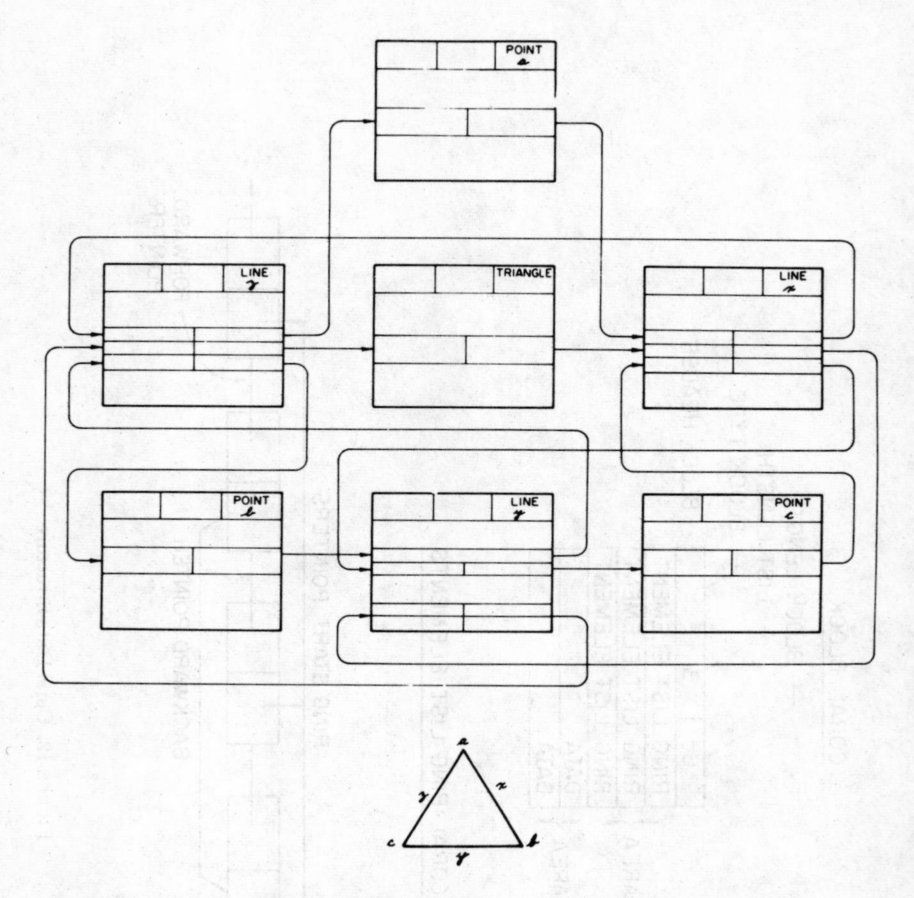

Figure 13. Coral Structure for a Triangle

Classes play a major role in CORAL. CORAL recognizes a class of block types and a class of free storage. For each block type there is a master block. The master blocks are in a class the head of which is a master block. The second ring list element of a master block is the ring start element for the ring of all blocks of the same type.

Certain basic operations are needed to accompany the CORAL data structure. These include storage allocation, ring tying, creation of blocks, deletion of blocks, movement of the ring, testing of block type and data, and so on.

PART III: PICTURE SYNTAX

Introduction. Work on the syntax of pictures has fallen into an early period and a late period, these periods separated by very few years. The early work includes that of Narasimhan in 1962 [**16**], [**17**], Ledley in 1963 [**18**], [**19**], and Kirsch in 1964 [**20**]. There has followed in the last few years another burst of activity, namely the work of Feder in 1966 [**21**], Shaw in 1967 [**22**], Miller and Shaw in 1967 [**23**], [**25**], and Clowes in 1967 [**28**]. Anderson [**26**] is also doing closely related work.

The motives for attacking pattern analysis by linguistic methods can be stated as follows:

(1) The structural aspects of recognition of patterns can be emphasized by a grammar. In doing pattern recognition it is often the structural relationship between one part of a pattern and another part of the same pattern which makes clear that there is a pattern present.

(2) With a suitable pattern description language, one can then bring to bear on the problem all the theory and techniques already developed in the processing of artificial languages.

(3) With a picture description language we can bring together the graphic display problems and the picture recognition problem under one formalism. In order to have a language which would describe patterns for generation and the patterns for recognition and permit the manipulations of such patterns one must have structures and rules for manipulation of the structures (a calculus).

(4) It would appear that the more fruitful way to develop a system is to find a grammar and let the grammar specify the data structure and the control necessary for such a system rather than the converse. As we will see the data structure problem becomes very simple once we can describe pictures syntactically.

Let us examine the earlier approaches and then discuss the approach being taken by Shaw and Miller and Shaw. Narasimhan has developed a hierarchic system of labels to specify points in patterns. This hierarchical labeling scheme has implicit in it a description language. The language is not written out explicitly in some metalanguage such as Backus Normal Form and thereby does not provide the opportunity to bring to bear the theory and techniques of translation which have been so successful in programming languages. The basic sets or primitives of the languages are specific topological properties like crossings, end points, bends, corners, and so forth. The description of the pattern is formed by listing statements which connect basic sets by line connectors and tables of primary connections. Grammar rules are applied to the tables of primary connections.

Ledley on the other hand did specify his grammar in terms of Backus Normal Form and thereby was able to apply the programming language syntactic analysis techniques to his pattern grammar. He does assume, however, that the source patterns are already expressed as strings in his pattern language. Going from the two-dimensional pictures to the strings is not specified and that is indeed a rather major part of any problem. Ledley made application of his grammar to classify chromosomes. He classified them in terms of boundaries of the chromosomes, which he supplied the program in string form.

Kirsch concentrated on trying to develop two-dimensional productions for the generations of simple geometrical figures. There are perhaps some good ideas contained therein. The idea of two-dimensional productions is an attractive one but seems not to have gone very far.

In this author's opinion these were important early attempts and have led the way to what is likely to be a very fruitful avenue of research.

Shaw and Miller and Shaw have emphasized a formal specification of the syntax of the pattern description language, a strong emphasis on preserving formal rules for the operators used in the language, and a standard form for sentences in the language. Early experience has shown that it was extremely important to be able to reconstruct from an explicit sentence in the language the pattern that has been recognized. It is also important in the generation problem to be able to generate directly from a sentence in the language the explicit pattern. It was also important to preserve all the connectivities (Miller and Clark [27]). The grammar that has been developed is suitable for describing any connected graph.

A picture description language. The formal description of the picture description language (PDL) developed by Shaw [23] and by Miller and Shaw [24] is given in this section. The next section discusses and illustrates its uses.

(A) *The primitives.* The language is applicable to n-dimensional pictures, $n > 1$, although we shall illustrate it for two-dimensional pictures.

A primitive class, p, may be defined as any two (n) dimensional object with a head and tail. The primitive class is given a name and a specification of its head and tail, and is defined by a Boolean function of a list of attributes. Each attribute may take on a set of values according to its definition. The most common Boolean function will be the conjunction of all the attributes on the attribute list.

EXAMPLE. We may define the primitive class of arcs of circles of any radius, negative curvature, and arc less than 180°. For example

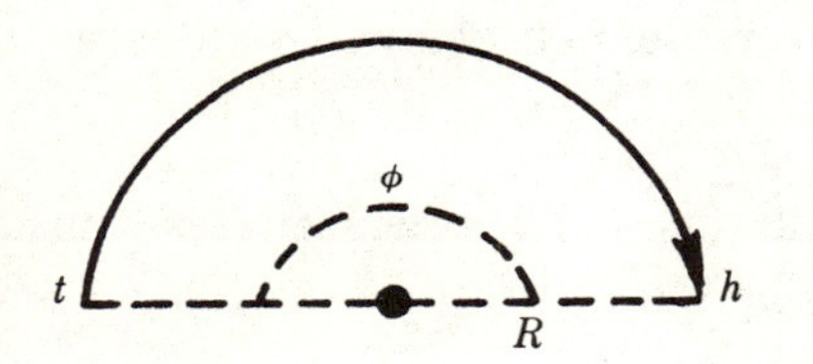

or

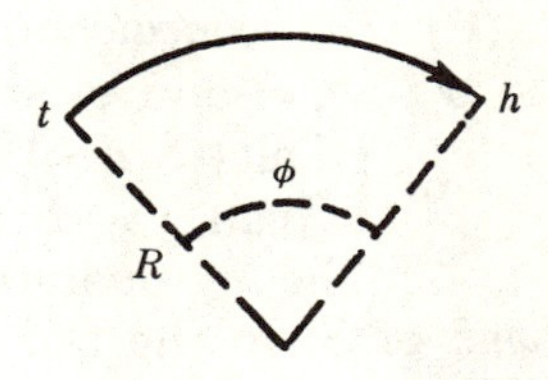

The attribute list for the primitive ARC has the form:

$$\text{ARC} \equiv (\text{ARC, Counterclockwise limit, Clockwise limit,}\quad \text{curvature} < 0, \phi < 180°).$$

This is a specific instance of the general form:

$$\text{Primitive} = (\langle\text{name}\rangle, \langle\text{tail specification}\rangle, \langle\text{head specification}\rangle, \langle\text{lst attribute}\rangle, \langle\text{2nd attribute}\rangle, \cdots, \langle n\text{th attribute}\rangle).$$

We use a superscript label to identify a particular member of a primitive class. An element $p^\dagger$ of class p has a value list containing specific values of the attributes in the attribute list of the class.

DEFINITION.

$$\text{Value}(p^\dagger) = ((\text{name} =)p, \vec{x}_{\text{Tail}}, \vec{x}_{\text{Head}}, (\text{Curvature} =) - 2, (\phi =)60°).$$

There is redundant information in the value list and, moreover, for some uses there may be irrelevant information on the value list. For generality of form and application this information is retained.

A primitive class may have a single element. It is convenient to define a special primitive, the null point primitive, λ.

A null point primitive has a head and tail at the same point. A blank picture may be described as a concatenation of null point primitives.

(B) *The syntax.* A sentence, S, in the language is defined by

1. $S \to p \mid (S\,\theta\,S) \mid (\sim S) \mid (\overline{S}) \mid T(\omega)S$.
2. $\theta \to + \mid \times \mid - \mid * \mid \sim$.
3. p is a primitive class described in B below.
4. $\{+, \times, -, *\}$ are concatenation operators described in C below.
5. $\{\sim, \overline{}, T(\omega)\}$ are unary operators described in D and E below.

(C) *The concatenation operators.* The concatenation operators $+, \times, -, *$, and $\sim$ are binary operators defined below. In all cases

$$6. \begin{cases} \text{Tail}\,((S_1\,\theta\,S_2)) = \text{Tail}\,(S_1) \\ \text{Head}\,((S_1\,\theta\,S_2)) = \text{Head}\,(S_2),\ \theta \in \{+, \times, -, *, \sim\}. \end{cases}$$

7. The $+$ operator: head to tail: $(S_1 + S_2)$ concatenates the head of S_1 to the tail of S_2.

EXAMPLE.

t —— h S_1

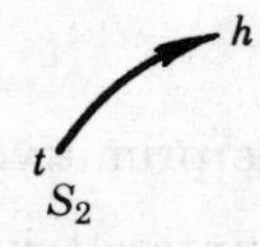

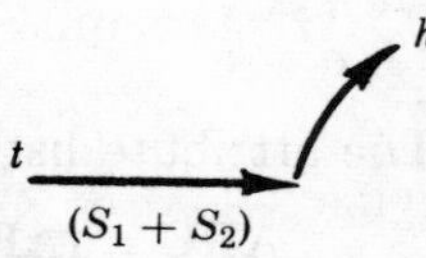

8. The $\times$ operator: tail to tail: $(S_1 \times S_2)$ concatenates the tail of S_1 to the tail of S_2.

EXAMPLE.

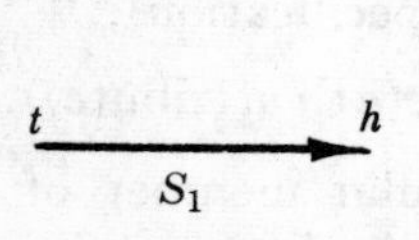

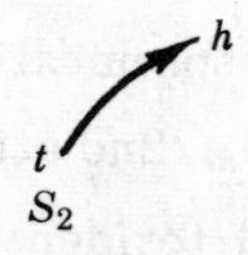

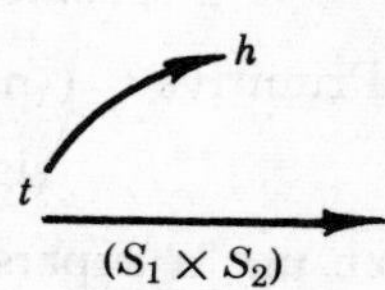

9. The $-$ operator: head to head: $(S_1 - S_2)$ concatenates the head of S_1 to the head of S_2.

EXAMPLE.

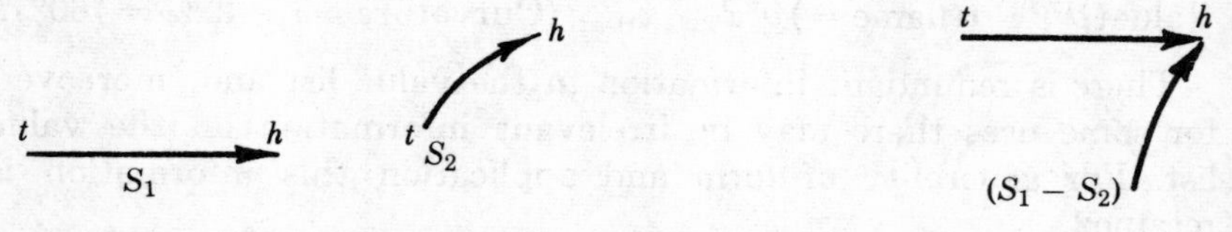

10. The $*$ operator: head to head and tail to tail: $(S_1 * S_2)$ concatenates the tail of S_1 to the tail of S_2 and the head of S_1 to the head of S_2.

EXAMPLE.

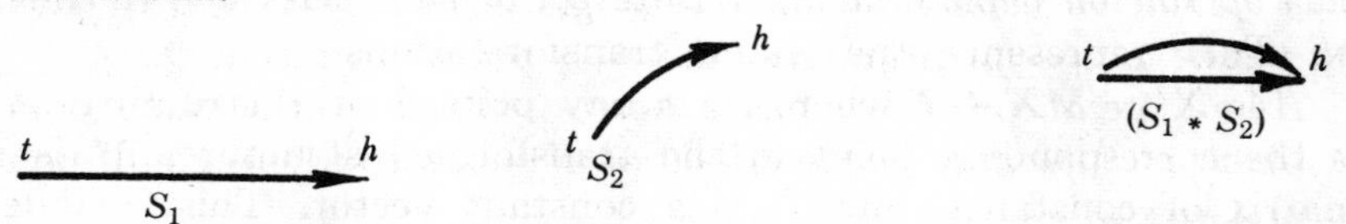

The $*$ operation may be undefined for some combinations of structures, S, just as the arithmetic operation a/b is undefined for certain values of a or b.

EXAMPLE.

VECT 1 $\equiv$ (VECT 1, tail at origin, head at upper right, unit vector in first quadrant)

VECT 2 $\equiv$ (VECT 2, tail at origin, head at lower left, unit vector in third quadrant)

(VECT 1 $*$ VECT 2) is undefined.

11. The binary $\sim$ operator: $(S_1 \sim S_2) \equiv (S_1 + (\sim S_2))$ where $(\sim S_2)$ is defined in D. That is, the binary $\sim$ means simply $+$ the unary $\sim$ of S_2, where the unary $\sim$ is as defined in D.

(D) *The unary operators.* The operators $\sim$ and $\vdash$ are unary operators defined as follows:

12. The unary $\sim$ operator: switches head and tail.

$$\text{Tail}((\sim S)) = \text{Head}(S)$$
$$\text{Head}((\sim S)) = \text{Tail}(S)$$

The structure remains the same.

EXAMPLE.

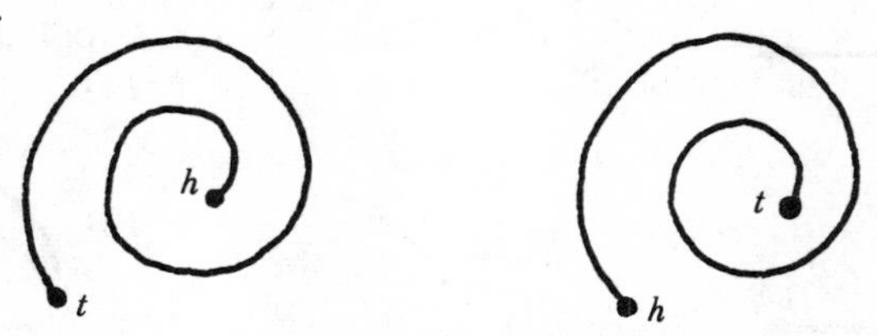

13. The $\vdash$ operator: blanks out all points.

$$\text{Head}(\overline{S}) = \text{Head}(S)$$
$$\text{Tail}(\overline{S}) = \text{Tail}(S)$$

All points in the structure are turned to null points.

EXAMPLE.

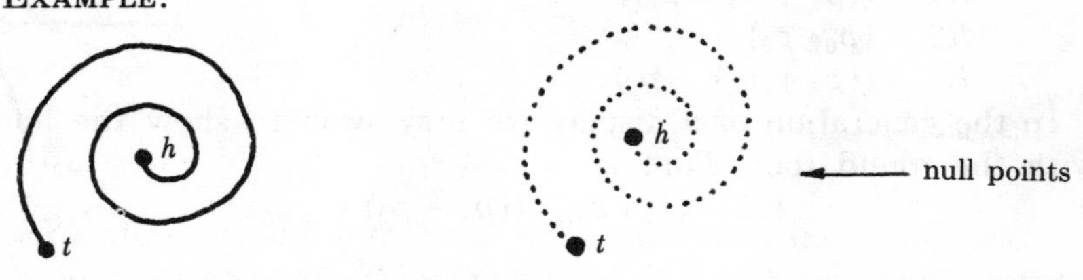

(E) *The transformation operator* $T(\omega)$ *is a unary operator which*

may operate on a particular primitive $p_j^\dagger$, *or on a class of primitives* p_j. $T(\omega)$ represents the affine transformations.

14. $\vec{X}' = M\vec{X} + T$ where $\vec{X}$ is any point in the structure, $\vec{X}'$ is the corresponding point in the transformed structure, M is a matrix of constants, and T is a constant vector. This includes stretching, rigid body rotations and translations, and shearing transformations. Although it may not be very useful, we define a transformation on a class.

$$T(\omega)(p) \equiv \{T(\omega)p^\dagger \mid p^\dagger \in p\}.$$

A common $T(\omega)$ will be isotropic stretching, i.e., scalar multiplication. We shall indicate that particular $T(\omega)$ by:

$$Cp_j;\ C \text{ is a positive constant.}$$

We observe that

15. $T(\omega)(S^\dagger(p_i^\dagger, p_j^\dagger \cdots)) = S^\dagger(T(\omega)p_i^\dagger, T(\omega)p_j^\dagger, \cdots)$.

(F) *Uses and illustrations*. Before proceeding with further description of the language, let us examine some uses and illustrations.

We can describe a class of houses with any of two types of roofs permitted. We need the following primitive classes which have a single member as shown.

A class of houses H is described by the following sentences in our language.

16. $H \rightarrow (R * B)$,
$R \rightarrow R1 \mid R2$,
$R1 \rightarrow ((p_1 + p_2) * p_3)$,
$R2 \rightarrow (p_6 * p_3)$,
$B \rightarrow ((p_4 + p_3) + p_5)$.

In the generation of a display we may wish to show the house with the round roof. That is

$$H2 \rightarrow (p_6 * p_3) * ((p_4 + p_3) + p_5).$$

We would show

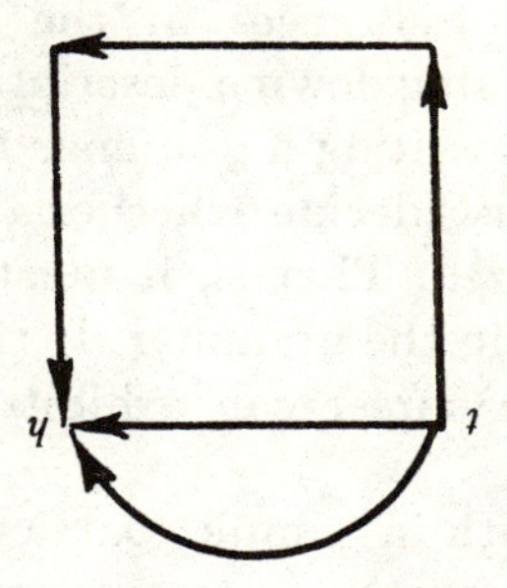

In a recognition problem we may ask the machine to recognize any house of class H. We call the process of carrying out the recognition according to the sentences 16 in the language *picture parsing*. A top-down parse of the sentences 16 is given by the following ALGOL-like procedure (without specifications and declarations).

BOOLEAN PROCEDURE HOUSE;
HOUSE ← **IF** R **THEN IF** NEXTOP(∗) **THEN** B **ELSE FALSE ELSE FALSE**;
BOOLEAN PROCEDURE R;
R ← **IF** R1 **THEN TRUE ELSE** R2;
BOOLEAN PROCEDURE R1;
R1 ← **IF** PRIMITIVE (p_1) **THEN IF** NEXTOP (+) **THEN IF** PRIMITIVE (p_2) **THEN IF** NEXTOP (∗) **THEN** PRIMITIVE (p_3) **ELSE FALSE ELSE FALSE ELSE FALSE**;
BOOLEAN PROCEDURE R2;
R2 ← **IF** PRIMITIVE (p_6) **THEN IF** NEXTOP (∗) **THEN** PRIMITIVE (p_3) **ELSE FALSE ELSE FALSE**;
BOOLEAN PROCEDURE B;
B ← **IF** PRIMITIVE (p_4) **THEN IF** NEXTOP (+) **THEN IF** PRIMITIVE (p_3) **THEN IF** NEXTOP (+) **THEN** PRIMITIVE (p_5) **ELSE FALSE ELSE FALSE ELSE FALSE**;
BOOLEAN PROCEDURE PRIMITIVE (X);

This is the basic primitive recognizer that searches the picture for the primitive labeled by the argument X.

BOOLEAN PROCEDURE NEXTOP (X);

This is a basic terminal symbol recognizer that searches the picture to see whether there is any possibility of concatenating another primitive as indicated by the operator. For example, NEXTOP (×) would try to find the start of a primitive which would be a candidate to × onto the primitive just found.

We may observe that the picture description language is in a sense a metalanguage. Sentences in the languages constitute a smaller language. In writing down a description of a class of pictures to be recognized, one is writing a grammar for this class of pictures. The picture parser must decide whether the object picture is in the class to be recognized. That is, it must recognize whether this picture has a structure in the grammar. If the answer to the picture parse is "TRUE", the parser can exhibit the particular sentence found.

Let us illustrate with a simple particle physics picture. We consider a class of interactions starting with a negative particle which may pass through the picture or may scatter from a positive particle or may decay into a neutral and another negative particle. Thus

17. $T_- \rightarrow t_- | t_- + (T_- \times T_+) | t_- + (T_- \times T_w)$ where T_- is a negative track *with* all subsequent events, t_- is a primitive negative track, T_+ is a positive track *with* all subsequent interactions, and T_w is a neutral track (not seen) with all subsequent interactions.

Let us consider only positive particles that continue through the chamber, that is,

18. $T_+ \rightarrow t_+$ where t_+ is a primitive positive track.

Let us consider neutral particles that may decay into pairs, that is,

19. $T_n \rightarrow t_n | t_n + (T_+ \times T_-)$.

A picture that must start with negative track would be represented by

20. $P \rightarrow T_-$ followed by (17), (18), and (19).

Sentences (17), (18), (19), and (20) in the picture description language constitute a grammar for the class of pictures we wish to recognize.

Figure 1 shows a picture that would be a sentence in the grammar (17)—(20) and the particular sentence.

The picture description language set forth *up to here* is suitable for describing pictures that can be described as trees as well as pictures with a simple connectivity. It can not yet describe all connected graphs. We introduce a labeling scheme which, with the unary operations $\sim$ and $\sqcap$, permit the description of any connected graph.

(G) *Labeling*. The labeling scheme presented here is edge-oriented and preserves the identity of the edge through the various transformations and manipulations which operators and operands may undergo. Consider the example in [**25**].

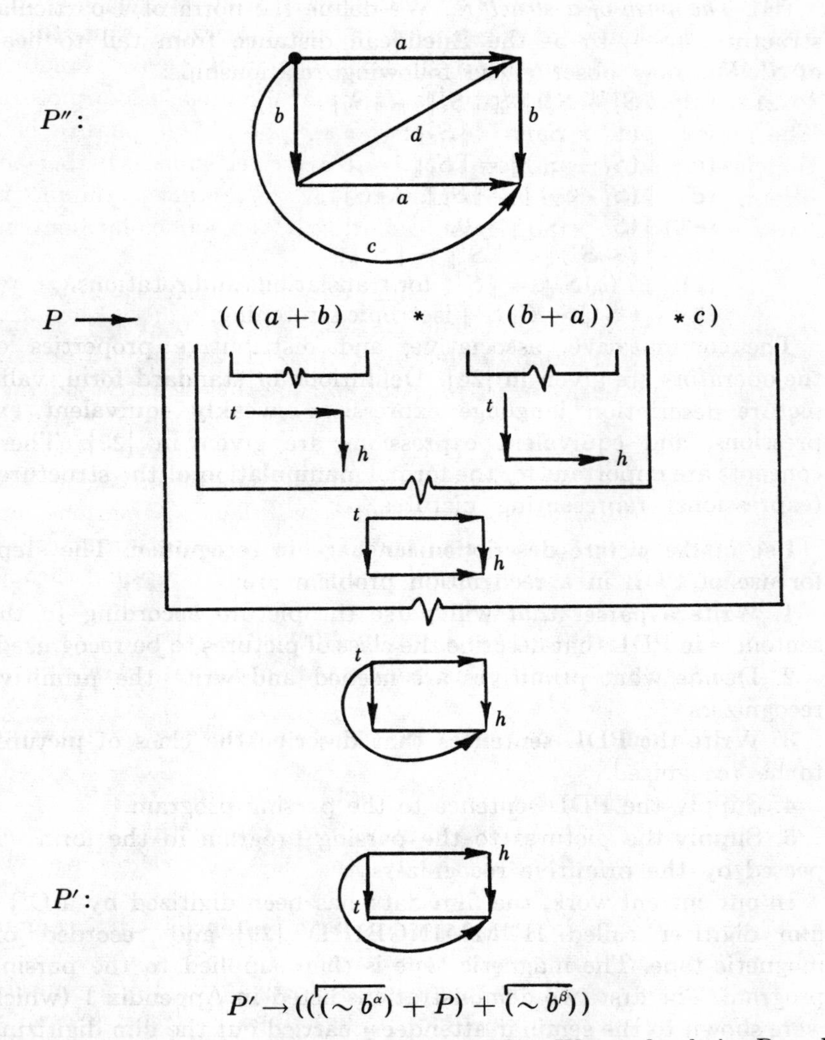

$$P' \rightarrow (((\sim b^{\alpha}) + P) + (\sim b^{\beta}))$$

where b^{α} and b^{β} must refer to the correct b's. We go back in P and label the b's in order that the head and tail of P' will be where we want them.

$$P \rightarrow (((a + b^{\beta}) * (b^{\alpha} + a)) * c),$$

$$P'' \rightarrow (P' * d).$$

The labeling preserves the correct identification even though we may perform various allowable operations such as commutation.

(H) *The norm of a structure.* We define the norm of a particular structure, $\|S^\dagger\|$, to be the Euclidean distance from tail to head of $S^\dagger$. We now observe the following relationships:

21. (a) $\|S_1^\dagger + S_2^\dagger)\| \leqq \|S_1^\dagger\| + \|S_2^\dagger\|$.
 (b) $\|(S_1^\dagger \times S_2^\dagger)\| = \|S_2^\dagger\|$.
 (c) $\|(S_1^\dagger - S_2^\dagger)\| = \|S_1^\dagger\|$.
 (d) $\|(S_1^\dagger * S_2^\dagger)\| = \|S_1^\dagger\| = \|S_2^\dagger\|$.
 (e) $\|(S^\dagger \sim S^\dagger)\| = 0$.
 (f) $\|(\sim S^\dagger)\| = \|\overline{S}^\dagger\| = \|S^\dagger\|$.
 (g) $\|T(\omega)S^\dagger\| = \|S^\dagger\|$ for translations and rotations.
 (h) $\|CS^\dagger\| = C\|S^\dagger\|$ isotropic stretching.

The commutative, associative, and distributive properties of the operators are given in [24]. Definitions of standard form, valid picture description language expressions, weakly equivalent expressions, and equivalent expressions are given in [25]. These concepts are important for the formal manipulation of the structures (expressions) representing pictures.

Use of the picture description language in recognition. The steps for use of PDL in a recognition problem are:

1. Write a parser that will parse the picture according to the sentences in PDL that describe the class of pictures to be recognized;
2. Decide what primitives are needed and write the primitive recognizers.
3. Write the PDL sentences that describe the class of pictures to be recognized.
4. Supply the PDL sentence to the parsing program.
5. Supply the pictures to the parsing program in the form expected by the primitive recognizers.

In our current work, the film data has been digitized by a CRT film digitizer called HUMMINGBIRD [29] and recorded on magnetic tape. The magnetic tape is then supplied to the parsing program. The first two demonstrations listed in Appendix I (which were shown to the seminar attendees) carried out the film digitizing and the picture parsing,. respectively.

[23] contains a picture description language that is suitable for describing characters of the English alphabet in block style. It also gives a picture description language suitable for describing a printed page.

In using the PDL for generation of displays on the CRT, the data structure for the picture is held implicitly in the string rather than explicitly in a data structure such as CORAL (Part II, this series). The effect of this is that one can replace searching of explicit data structures by parsing of the string representation of the picture. Insertion and deletion of structures in the picture are carried out by the substitutions in the string.

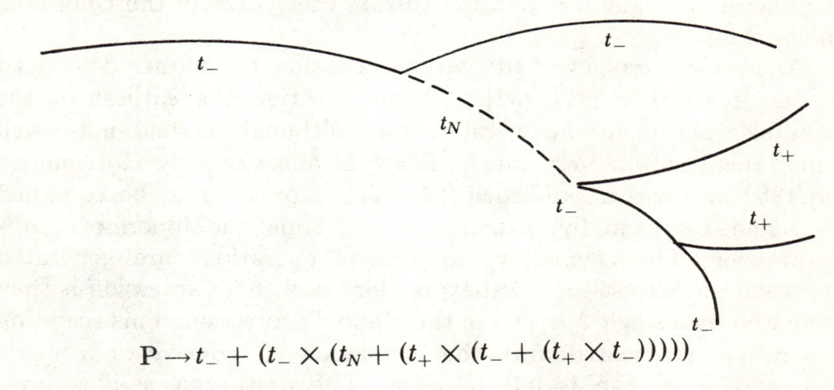

$$P \rightarrow t_- + (t_- \times (t_N + (t_+ \times (t_- + (t_+ \times t_-)))))$$

FIGURE 1. Particle Physics Picture

PART IV: TRANSLATORS FOR ARTIFICIAL LANGUAGES

Abstract. We shall discuss some of the string analyzing and compiling techniques that are currently in use. Let us quickly spell out some definitions, although it is not my intent to deal with the subject formally but rather to try to get across some understanding of how the techniques work.

In general, a language, L, will be a subset of a set of all strings of symbols from the alphabet with which we are dealing. The grammar (or the syntax) of L specifies which subset of strings of symbols from the alphabet are sentences in the language. The grammar of the language will often be described in some syntactic metalanguage. A commonly used syntactic metalanguage is the Backus Normal Form (BNF), which tells you how to either generate strings which are sentences of L or how to recognize whether a string from the alphabet A is a sentence in L.

In the early construction of compilers the syntax of the language was not explicitly displayed. The rules for interpreting a string or for recognizing a string in the alphabet were stated in English, and translators were constructed which would interpret these rules. For simple languages some very fast and efficient compiling can be done this way. As the number of operations and their interactive complexity within a language increases, these simple implicitly-defined languages are more difficult to cope with and, in particular, are much more difficult to change.

Since the classical paper of Irons [**30**] a much larger emphasis has been placed in explicit display of the syntax by means of a syntactic metalanguage and in utilizing the syntax in the compiling algorithms.

An excellent review of the various compiler techniques developed today is given in [**31**], [**32**]. Let me mention the earliest of the compiler efforts for historical reasons although I shall not dwell upon this example very much. The first effort was by Rutishauser in 1952 and was unpublished [**33**]. The expression to be compiled is scanned repeatedly, extracting each time the innermost subexpression. The elementary arithmetic operations are generated for each of these selected subexpressions and that expression is then replaced by a single operand in the original expression. This scanning operation is repeated until the expression is reduced to a single operand. The scan is left to right. The first scan assigns level numbers to each element of the expression, that is to each operand and to each operator. The innermost subexpressions are defined by the highest level number. These level numbers are updated as the subexpressions are replaced by the compiled operands.

Let us skip over to the table-driven compiler NELIAC, which is a dialect of ALGOL 58. This method is very fast but relatively restrictive. It still permits operation on pairs of operators only; a table can be constructed which shows the action to be taken for a match between current operator (COP) and next operator (NOP). The method also requires only a single temporary storage cell because of the simplicity of the language. No parenthetical nesting is permitted. Samuelson and Bauer in 1959 used a similar table technique in which they introduced the idea of a stack or, as they called it, a cellar. Two symbols at a time are compared just as in the NELIAC case but temporary results and operators are saved in a pushdown stack. For a complete discussion of table-driven compilers see [**34**].

One of the simplest syntax analysis techniques that is dependent upon having a phrase structure to the language and a syntactic linguistic description is the so-called top-down parsing method [**35**]. As a general method for compilers it has the following disadvantages. (1) Many false paths have to be tried before the correct one is found and any failure requires backtracking to the last successful recognition. (2) A corollary to this is that there is no systematic way to determine the success or failure of the method except by exhaustion. (3) It is difficult to insert the semantic rules such as code generators into the system. It essentially returns you a true or false statement about the parse. You can keep the information that is found, but it is not simple to insert the code generating rules as in the case of the table-driven compiler. Table-driven compilers—when used—get a match between two symbols. The entry in the table points to a subroutine that says what code to compile. (4) The writing of the particular top-down parsing rules for particular grammar must be done each time; that is, it is not simple to change the grammar as is the case for some of the later methods.

If, however, the grammar is relatively simple, as in the case of the Picture Calculus (Part III), and one has some feeling for the most probable paths with which the analyzer should search, one can state his syntax in such a way to help with the speed of the analysis.

Recently a great deal of attention has been given to the development of precedence grammars. The reader is referred to [**31**, Chapter II, p. 17-39].

APPENDIX I

Graphics demonstrations. The following demonstrations were shown in operation to the American Mathematical Society Summer Seminar Group on Thursday, August 10, 1967, 4:00 PM to 6:00 PM.

1. *Film digitizer control system.* This program addresses and by means of light pen controls the two Hummingbird film digitizers. The program then displays and outputs the unprocessed pictures. This system permits the user to try various parameters of the film digitizing system, such as the gray level at which to digitize, the density of the raster, the size of the scanning window in order to pick out certain size objects, etc.

2. *Picture parser.* This program illustrates the "parsing" or analysis of pictures. Given a syntax which formally describes a class of pictures of interest to the user the parser directs a search through the picture. The display alternates between the original picture and an abstracted version of what has been recognized and returned as legitimate sentences in the language of the grammar.

3. *Syntax retrieval system.* This program gives the user the ability to display the syntax of any programming language described in Backus Normal Form. ALGOL is the language being

displayed in the demonstration. The user may ask for the definition of a particular phrase class name, or may ask for all productions in which it appears. The light pen is used to delete information not currently of interest in the trace down a particular path.

4. *Visual, on-line transport.* This program uses the interactive capabilities of the display device in the dynamic design of beam optics systems.

5. *Conversational nuclear cross section evaluation.* Evaluation of new neutron cross section data requires major nontechnical efforts, often to the exclusion of detailed physics analysis. This system has been developed to accomplish such multistep operations as data compilation, formatting and display in a conversational mode.

6. *Draftsman's aid.* This program allows the user to define "graphic" pictures using the light pen as the principal input medium. Pictures may be constructed from basic entities; points, lines, arc, and circles.

7. *Debugging aid.* A programmer of assembly language routines may insert checkpoints in his program where he may then examine and/or change elements via the display device. He refers symbolically to these elements. In addition, any program interruption is intercepted at which point he may also exercise such options.

References

1. D. H. Fender, *Experiments in visual perception,* Proc. IBM Sci. Comput. Sympos. Man-Machine Comm., May 3-5 (1965), 135.

2. R. Clark and W. F. Miller, "Computer-based data analysis systems" in *Methods in computational physics,* Vol. 5, Academic Press, New York, 1966, p. 47-99.

3. W. F. Miller, *Computation and control in complex experiments,* Proc. IBM Sci. Comput. Sympos. Man-Machine Comm., May 3-5, (1965), 113.

4. A. C. Hearn, Comm. ACM **9** (1966), 573.

5. ———, *REDUCE user's manual,* ITP-247, Department of Physics, Stanford University, February 1967.

6. J. D. Bjorken and S. D. Drell, *Relativistic quantum mechanics,* McGraw-Hill, New York, 1965.

7. R. L. Gregory, *Eye and brain,* World University Library, McGraw-Hill, New York, 1966.

8. Fred Gruenberger (editor), *Computer graphics,* Thompson, 1967.

9. G. A. Michael, *Survey of graphic data processing equipment for computers,* AEC Sympos. Ser. 10, Use of Computers in Analysis of Experimental Data and the Control of Nuclear Facilities, U. S. Atomic Energy Commission Division of Technical Information, May 1967.

10. M. R. Davis and T. O. Ellis, *The RAND tablet: A man-machine graphical communication device,* Proc. 1964 FJCC, Vol. 26, Spartan, Baltimore, Md., p. 325.

11. H. J. Ridinges, *The GRAFACON man-machine interface,* Datamation **13** (1967), 44.

12. R. M. Brown, M. A. Fisherkeller, A. E. Gromme and J. V. Levy, *The SLAC high-energy spectrometer data acquisition and analysis system,* Proc. IEEE **54** (1966), 1730-1734.

13. Robert Clark and W. F. Miller, "Computer-based data analysis systems" in *Methods in computational physics,* Vol. 5, Academic Press, New York, 1966, p. 81.

14. W. T. Comfort, *Multiword list items,* Comm. ACM **7** (1964), 357.

15. W. R. Sutherland, *On-line graphical specification of computer procedures,* Tech. Rep. 405, Lincoln Laboratory, MIT, May 23, 1966.

16. R. Narasimhan, *A linguistic approach to pattern recognition,* Rep. No. 121, Digital Computer Laboratory, University of Illinois, Urbana, Ill., July 1962.

17. ———, *Syntax-directed interpretation of classes of pictures,* Comm. ACM **9** (1966), 166-173.

18. R. S. Ledley, *Programming and utilizing digital computers,* McGraw-Hill, New York, 1963, Chapter 8.

19. ———, *High-speed automatic analysis of biomedical pictures,* Science **146** (1964), 216-223.

20. R. A. Kirsch, *Computer interpretation of English text and picture patterns,* IEEE Trans. Elec. Comp. **EC-13** (1964), 363-376.

21. Jerome Feder, *Linguistic specification and analysis of classes of patterns,* Tech. Rep. 400-147, New York University School of Engineering and Science, New York, 1966.

22. K. J. Breeding, *Pattern grammar for a pattern description language,* Rep. No. 177, Department of Computer Science, University of Illinois, Urbana, Ill., May 1965.

23. A. C. Shaw, *A proposed language for the formal description of pictures,* GSG Memo 28, Computation Group, Stanford Linear Accelerator Center, Stanford, Calif., February 1967.

24. W. F. Miller and A. C. Shaw, *A picture calculus,* GSG Memo 40, Computation Group, Stanford Linear Accelerator Center, Stanford, Calif., June 1967.

25. A. C. Shaw, *A picture calculus—Further definitions and some basic theorems,* GSG Memo 46, Computation Group, Stanford Linear Accelerator Center, Stanford, Calif., June 1967.

26. R. H. Anderson, *Syntax-directed recognition of hand-printed two-dimensional mathematics,* Preprint presented at ACM Symp. on Interactive Systems for Experimental Applied Math., August 26-28, 1967.

27. Robert Clark and W. F. Miller, "Computer-based data analysis systems" in *Methods in Computational Physics,* Vol. 5, Academic Press, New York, 1966, p. 47.

28. M. B. Clowes, *A generative picture grammar,* Seminar Paper No. 6, Computing Research Section, Commonwealth Scientific and Industrial Research Organization, Canberra City, Australia, April 1966.

29. W. F. Miller and J. van der Lans, *System design for CRT film scanning and measuring,* Proc. 8th Nat. Sympos. Soc. Information Display, San Francisco, Calif., May 24-26, 1967.

30. E. T. Irons, *A syntax-directed compiler for ALGOL* 60, Comm. ACM **4** (1961), 51.

31. J. A. Feldman and David Gries, *Translator writing systems,* Tech. Rep. 69, Computer Science Department, Stanford University, June 9, 1967.

32. Saul Rosen (editor), *Programming systems and languages,* McGraw-Hill, New York, 1967.

33. A. C. Shaw, *Lecture notes on a course in systems programming,* Tech. Rep. 52, Computer Science Department, Stanford University, December 9, 1966, p. 103.

34. D. Gries, *The use of transition matrices in compiling,* Tech. Rep. 57, Computer Science Department, Stanford University, March 17, 1967.

35. T. E. Cheetham, Jr. and Kirk Sattley, *Syntax-directed compiling,* Proc. SJCC **25** (1964), 31.

STANFORD UNIVERSITY

Subject Index

Author Index for the Two Volumes

Roman numbers refer to pages on which a reference is made to a work of the author.
Italic numbers refer to pages on which a complete reference to a work by the author is given.
Boldface numbers indicate the first page of the articles in Part 1 or 2.
Roman numerals refer to Part 1 or 2.